Tel- 504505

... this vol... ...gs have
...no can make and use ...complex tools,
...cate in more complex ways, and engage in more complex
...ms of social life, than any other species in the animal kingdom.
Leading researchers from fields as diverse as biological and social
anthropology, archaeology, linguistics, psychology, neurology and
ethology, have come together to present a uniquely interdisciplinary
study of this central problem in human evolution.

TOOLS, LANGUAGE AND COGNITION IN HUMAN EVOLUTION

TOOLS, LANGUAGE AND COGNITION IN HUMAN EVOLUTION

Edited by

KATHLEEN R. GIBSON

Section of Anatomical Sciences,
University of Texas at Houston – Health Science Center,
and Department of Anthropology, Rice University

and

TIM INGOLD

Department of Social Anthropology,
University of Manchester

CAMBRIDGE
UNIVERSITY PRESS

PUBLISHED BY THE PRESS SYNDICATE OF THE UNIVERSITY OF CAMBRIDGE
The Pitt Building, Trumpington Street, Cambridge CB2 1RP, United Kingdom

CAMBRIDGE UNIVERSITY PRESS
The Edinburgh Building, Cambridge CB2 2RU, UK http://www.cup.cam.ac.uk
40 West 20th Street, New York, NY 10011–4211, USA http://www.cup.org
10 Stamford Road, Oakleigh, Melbourne 3166, Australia

First published 1993
First paperback edition 1994
Reprinted 1998

Printed in the United Kingdom at the University Press, Cambridge

Typeset in Times.

A catalogue record for this book is available from the British Library

Library of Congress Cataloguing in Publication data

Tools, language, and cognition in human evolution / edited by Kathleen
R. Gibson and Tim Ingold.
    p. cm.
    Revised versions of papers from "…an interdisciplinary
symposium, 'Tools, language, and intelligence: evolutionary
implications', held in Cascais, Portugal, March 15–24, 1990" – Pref.
    ISBN 0 521 41474 1
    1. Human evolution – Congresses. 2. Social evolution – Congresses.
3. Brain – Evolution – Congresses. 4. Language and culture –
Congresses. 5. Culture and cognition – Congresses. 6. Tools –
Social aspects – Congresses. 7. Tool use in animals – Congresses.
I. Gibson, Kathleen Rita, II. Ingold. Tim, 1948–
GN281.4.T66   1993
573.2 – dc20   92-1367   CIP

ISBN 0 521 41474 1 hardback
ISBN 0 521 48541 X paperback

Contents

Contributors

Christophe Boesch
Institute of Zoology, University of Basel, Rheinsprung 9, 4051 Basel, Switzerland

William Calvin
Department of Psychiatry and Behavioral Sciences, University of Washington, Seattle, WA 98195, USA

Iain Davidson
Department of Archaeology and Palaeoanthropology, University of New England, Armidale, NSW 2351, Australia

Dean Falk
Department of Anthropology, State University of New York, Albany, NY 12222, USA

Kathleen Gibson
Section of Anatomical Sciences (Dental Branch), University of Texas at Houston – Health Science Center, PO Box 20068, Houston, TX 77225, USA

Susan Goldin-Meadow
Department of Psychology, University of Chicago, 5730 South Woodlawn Avenue, Chicago, IL 60637, USA

Gordon Hewes
Department of Anthropology, University of Colorado, Hellems Building, Campus Box 233, Boulder, CO 80309, USA

Tim Ingold
Department of Social Anthropology, University of Manchester, Roscoe Building, Brunswick Street, Manchester M13 9PL, England

Daniel Kempler
6338 West 6th Street, Los Angeles, CA 90048, USA

Adam Kendon
43 West Walnut Lane, Philadelphia, PA 19144, USA

Jonas Langer
Department of Psychology, University of California, Berkeley, CA 94720, USA

Andrew Lock
Department of Psychology, Massey University, Palmerston North, New Zealand

William McGrew
Department of Psychology, University of Stirling, Stirling FK9 4LA, Scotland

Constance Milbrath
Centre for Social and Behavioral Sciences, University of California at San Francisco, 1350 Seventh Avenue, CSBS-317, San Francisco, CA 94143, USA

William Noble
Department of Psychology, University of New England, Armidale, NSW 2351, Australia

Sue Taylor Parker
Department of Anthropology, Sonoma State University, 1801 East Cotati Ave, Rohnert Pk, CA 94928, USA

Peter C. Reynolds
Peter C. Reynolds & Associates, P.O. Box 50217, Palo Alto, CA 94303, USA

Duane Rumbaugh
Department of Biology, Georgia State University, University Plaza, Atlanta, GA 30303, USA

E. Sue Savage-Rumbaugh
Department of Biology, Georgia State University, University Plaza, Atlanta, GA 30303, USA

Kathy Schick
Center for Research into the Anthropological Foundations of Technology, Indiana University, 419 N. Indiana, Bloomington, IN 47405, USA

Charles Snowdon
Department of Psychology, University of Wisconsin, 1262 West Johnson Street, Madison, WI 53706, USA

Nicholas Toth
Department of Anthropology, 108 Rawles Hall, Indiana University, Bloomington, IN 47405, USA

Elisabetta Visalberghi
Instituto di Psicologia Comparata, Via U. Aldrovandi 16b, 00197 Rome, Italy

Thomas Wynn
Department of Anthropology, University of Colorado, Austin Bluffs Parkway, PO Box 7150, Colorado Springs, CO 80933-7150, USA

Preface

In the summer of 1987, Edward Reed and Tim Ingold organized a symposium entitled 'Doing things with tools' at the Fourth International Symposium on Event Perception and Action in Trieste, Italy. Four of the participants, Tim Ingold, a social anthropologist, Bill McGrew, a primatologist, Kathleen Gibson, a biological anthropologist, and Tom Wynn, an archaeologist, spent much of the Trieste meeting sitting in sunlit cafés discussing a fundamental anthropological question, 'Were the evolution of tool-making and language interrelated phenomena?'

Subsequently, Ingold and Gibson submitted a conference proposal to the Wenner–Gren Foundation for Anthropological Research. The result was an interdisciplinary symposium, 'Tools, language and intelligence: evolutionary implications', held in Cascais, Portugal, March 15–24, 1990 (an overview of the conference by Gibson is published in *Man*[NS] vol. 26, pp. 255–264, 1991). Participants at the conference represented diverse fields, each of which must ultimately be integrated into our views of cognitive evolution – neurology, developmental psychology, linguistics, archaeology, primatology, biological anthropology and socio-cultural anthropology. Each participant was chosen for his or her breadth of interdisciplinary perspective and specific interests in interrelating questions of tool-use, language, social behaviour and cognition.

This volume is a result of that conference. It includes revised versions of all but two of the papers originally presented to the conference, those by Philip Lieberman and Patricia Greenfield. Also included is a contribution by Peter Reynolds, who was invited but unable to attend the conference itself.

The book does not pretend to answer all questions pertaining to the evolution of tool-use and language. What are needed, however, are conceptual models of human evolution – that is, models based on modern psychological, ecological and neurological theory that enable us to infer how ancient hominids may have behaved, given the specific conditions that obtained at various times in the past. This book is intended to pave the way towards the construction of such models, by providing a critical assessment of current perspectives and approaches in the relevant sciences. As the long and animated discussions at the conference revealed, there is much that has still to be resolved, and an exciting agenda lies before us for future research.

At a time when the different perspectives of biological anthropology and social-cultural anthropology seem further apart than ever, this book represents a rare editorial collaboration. There is, indeed much on which we, as editors, agree, for the project would not otherwise have been possible. Yet we also recognize that, coming as we do from different backgrounds and with different training – in biological and social anthropology respectively – we bring to the project perspectives and orientations that may be in some ways complementary, but in others divergent or even conflicting. We feel that these perspectives should be allowed to speak for themselves, and for this reason our editorial contributions – the general introduction, epilogue and section introductions – have been written by each of us independently rather than jointly authored. We make no claim to advocate a unified approach. We believe this diversity, however, is all to the good. For if there is one principle on which we are in complete and unqualified agreement, it is that progress in understanding the kind of beings that humans are, and how they came to be that way, can only come from a dialogue across the increasingly artificial boundary between natural and human sciences, both within and beyond the field of anthropology. This principle, above all, has provided the motive for the conference, and for the book.

We would like to take this opportunity to express our sincere appreciation to the Wenner–Gren Foundation for funding the conference (no. 110 in its International Symposium Series), and to the Foundation's President, Sydel Silverman, for making it all possible. We also thank Laurie Obbink for her expert organizational skills and aid throughout – from conference planning and organization to on-site arrangements and post-conference circulars. Fatima da Silva Lopez and Beth Giebus also provided invaluable on-site assistance in Portugal. We have benefitted from the detailed written record of the conference discussions kept by Paul Graves, who acted as monitor throughout the proceedings. Mary Ellen Thames of the University of Texas helped with the preparation of the manuscript for publication. We also thank the editorial staff at Cambridge University Press, and especially Sara Trevitt, for their patience and assistance. Above all, our thanks are due to all the participants at the conference whose labours and enthusiasm have borne fruit in the present volume.

Kathleen Gibson
Tim Ingold

Prologue

General introduction

Animal minds, human minds

KATHLEEN R. GIBSON

What is the nature of the human mind and how did it arise? Does the evolution of the human intellect represent an expansion of capacities already present in our animal predecessors as postulated by Charles Darwin? Or, are there major discontinuities between animal and human minds – gaps so vast as to demand spiritual or other dramatic explanations of the origin of human powers (Wallace, 1864, 1869)? These basic evolutionary questions remain unanswered more than a century after the publication of *The Descent of Man* (Darwin, 1871).

Although the twentieth century has witnessed major advances in our understanding of the evolution of human form, our knowledge of the evolution of language and cognition remains sketchy. We can trace the history of tool-making, but we do not know what tool-making demanded in terms of linguistic competence and sociality. We can chart evolutionary changes in the external morphology and size of the brain, but we lack understanding of the functional significance of these changes. We cannot even answer such basic questions as whether the evolution of the human mind required the addition of numerous discrete neural modules for faculties as diverse as language, self-awareness, and reciprocal altruism (Barkow, 1989; Tooby & Cosmides, 1989), or whether diverse human skills reflect the workings of a single, expanded mental system (Calvin, this volume).

One response to this dilemma is to give up in despair, as did the Linguistic Society of Paris in 1866 when it prohibited all speculations on the origin of language. Alternatively, we can begin to devise conceptual models of human behavioral evolution (Tooby & DeVore, 1987). In contrast to referential models which assume that early hominid behaviors resembled those of a single living species such as chimpanzees or baboons, conceptual models follow examples set by physics and other "hard" sciences. They derive behavioral, cognitive, genetic, ecological and other biological regularities from direct scientific observation and experimentation. These principles are then incorporated into evolutionary models in a predictive fashion.

Although some have attempted conceptual modeling of human evolution (e.g. Foster, 1990; Parker & Gibson, 1979; Tooby & DeVore, 1987), such endeavors remain tentative, because much of the scientific data have yet to be collected. We believe, however, that the development and application of conceptual models for interpreting the fossil and archaeological record are both desirable and feasible. Just as the deciphering of the evolution of human form required massive interdisciplinary efforts, so, too, the development of solid conceptual models of human cognitive evolution will require the integration of various disciplinary perspectives. Hence, the chapters in this volume provide data and theory from a range of disciplines pertinent to two critical evolutionary themes: the differences between animal and human minds and the degree of interrelatedness of behavioral domains. We hope that these chapters will motivate further fundamental research and provide stepping stones to future conceptual models.

Continuities versus discontinuities between animal and human minds

The midcentury dominance of discontinuity theories

Plato and Aristotle, like other Western thinkers of their time, considered the mind to be a manifestation of the soul – a spiritual entity which entered the body at birth, departed it at death, and provided it with powers of movement, thought and breath. Although Plato and Aristotle differed somewhat in their descriptions of the human soul, both agreed that it differed sharply from that of other living creatures. In the Aristotelian framework, three types of souls abounded in the living universe. A vegetable soul, found in living plants, animals and humans, gave the breath of life. An animal soul found in animals and humans provided sensory and motor capacities. The human soul added rational and creative thought (Aristotle, 1986). Eventually, Platonic and Aristotelian views of the differences between animal and human minds were embodied in the fabric of Christianity and, along with Judaic views of human dominion over the animals, they have been imparted to generations of Western scholars during their formative years.

Not all humans, however, have shared the dominant Western view of the qualitative uniqueness of our species. An undercurrent of Western thought has always stressed human fellowship with animals and unity with nature. Asian religious leaders postulated reincarnation, an evolutionary philosophy in which animal and human represent separate points on a continuum to spiritual perfection, and each can be reborn as the other. Ancient Hindus thought so highly of the mental powers of monkeys that they accorded them special status as helpers of the deities. Hunter-gatherers have often presumed an even closer kinship between animal and human, endowing animals with intellects and languages similar to their own. The Oubi of the Ivory Coast went so far as to consider chimpanzees and humans to be the descendents of two brothers, a handsome one, the founding father of humanity, and a smart one, the grandfather of the apes (Linden, 1976).

The Darwinian view of evolutionary continuity between animal and human minds profoundly countered dominant Western traditions and, for a time, scholars such as Darwin (1872) and Romanes (1882) eagerly catalogued similarities between animal and human behaviors. However, "they could not specify what was changing in evolution or the nature of the steps between different levels of behavior" (Lashley, 1949, cited by Hallowell, 1950). Perhaps for this reason, their views did not prevail, and until recently most Western scientists have followed their cultural heritage from the ancient Hebrews and Greeks. With the birth of physiology, the discovery of the circulation of the blood, and the understanding of the roles of cerebral oxygen and neural potentials, scientists abandoned discussions of the soul. They talked instead of brains, intellects, and language, but the terms of the discourse seemed to refer to souls in another guise. Sharp, qualitative mental boundaries were considered to separate human from animal. Animal behavior was described as instinctive; human behavior, as learned. Humans were said to be the only animals who could symbolize, construct sentences, and make tools. According to some, only humans had culture and possessed self-awareness.

The very rules on which some scientists based their work guaranteed a continuation of perceptions of wide gaps between animal and human behavior. Scientists, such as Darwin and Romanes, who professed to see continuities between animal and human behavior were accused of anthropomorphism, a major scientific taboo. Among behaviorists eager to avoid any taint of anthropomorphism, Morgan's Canon became the yardstick by which animal behavior was to be judged, and adherence to the Canon became the measure of the proper scientist (Morgan, 1894). Accordingly, animal behavior was presumed to be based on simple mental processes, unless it could be proven beyond question that higher cognitive skills were essential. In contrast, higher intelligence was considered to mediate nearly all human behaviors. Consequently, even when behaving in very similar ways, humans and animals were presumed to be using different mental mechanisms (see for example, Terrace, 1979).

Traditional disciplinary specializations also contributed to the perpetuation of views that gaps between animal and human behavior are unbridgeable. Linguistics, cultural anthropology, and developmental psychology defined themselves as studies of humanity, while zoology, ethology and comparative psychology focused almost entirely on animals. As a result, few scholars directly compared human and animal behavior using similar observational and analytical methods. Often, indeed, specialists in closely allied disciplines remained ignorant of each others' findings, sometimes deliberately so. As Montagu has remarked, "Physical and cultural anthropology developed as separate disciplines, a specialist in one of these disciplines, ... could not unfairly be described as one who agreed not to know what was going on in the other" (Montagu, 1962).

As a consequence of these varied forces, evidence for mental continuity actually declined after Darwin and Romanes (Cooper, 1935, cited by Hallowell, 1950), and, as noted by Hallowell (1950) discontinuity theories dominated scholarly discussions during the mid-twentieth century (e.g. Eiseley, 1956; Etkin, 1954; Huxley, 1941; Nissen, 1951; Simpson, 1949; Tappen, 1953; White, 1959). To be sure, some such as Hart (1938) and Marais (1939) spoke with a different voice, but they remained unheard because their views were counter to the wisdom of their times.

Modern evidence for behavioral continuities between animals and humans

Evolutionary processes, however, shape and mold existing structures and behaviors. Although they can undergo profound and sudden evolutionary changes, new structures and behaviors do not appear without evolutionary antecedents. Consequently, as Hallowell (1950) noted, theories which postulate unbridgeable gaps between animal and human hinder evolutionary reconstructions.

Thanks to generations of field and laboratory studies of animal behavior, the tide has turned. Today, few, if any, behavioral discontinuities appear to separate ape and human. One study more than any other precipitated this change and initiated the challenge to human discontinuity theories – Jane Goodall's work with the chimpanzees of the Gombe Stream Reserve (Goodall, 1986). In 1963, Goodall's photographs of chimpanzees making tools overturned one of the most lasting definitions of humankind, "man the toolmaker" (usually attributed to Benjamin Franklin, see Hewes, this volume). Goodall and her students eventually delineated many other previously unsuspected continuities between humans and chimpanzees including lethal intergroup conflicts and long-term behavioral bonds among genetic relatives.

Shortly after Goodall's paradigm-changing observations, a second major challenge to human uniqueness came from a quite different source. Beatrix and Allen Gardner, trained respectively in European ethology and behaviorist psychology (Parker, 1990), began teaching the chimpanzee, Washoe, to use elements of the American Sign Language for the Deaf. Since then, several chimpanzees, an orangutan, a bonobo, and a gorilla have demonstrated symbolic capacities (Gardner, Gardner & Van Cantfort, 1989; Miles, 1983; Patterson & Linden, 1981; Savage-Rumbaugh, 1986; Terrace, 1979).

Numerous other hallmarks of humanity have also fallen in recent decades. We now know that other animals can use syntax (Greenfield & Savage-Rumbaugh, 1990), emit sounds of environmental reference (Cheney & Seyfarth, 1990), recognize themselves in mirrors (Gallup, 1979), engage in deliberate deception (Byrne & Whiten, 1988), exhibit cross-modal perceptions (Ettingler & Blakemore, 1967), and use a tool to make a tool (Toth & Schick, 1991).

Unfortunately, the demolition of one behavioral discontinuity after another brings us no closer to charting the evolution of human cognition than did the discontinuity theories of early decades. If apes possess all behaviors that humans can think to define, then what, if anything, evolved? Do any significant cognitive differences exist?

Hints of emerging new paradigms

At first glance, chapters within this volume fit squarely within the camp of continuity theorists. Many contributors pose new challenges to concepts of human qualitative uniqueness. According to Boesch, mother chimpanzees in the Tai Forest of the Ivory Coast attempt to "teach" tool-using techniques to their young, and they provision their young long after weaning. Savage-Rumbaugh provides evidence that great apes readily comprehend English syntax. McGrew reports that wild chimpanzees can use series of tools in succession to meet single goals. Moreover, he challenges anthropocentric notions that the stone tools found at Plio-Pleistocene East African sites could not have been made by apes.

Despite, however, the continuing accumulation of evidence for the capacities of the apes, many contributors remain convinced that the differences between non-human species and human beings are vast (Calvin, Davidson & Noble, Falk, Gibson, Ingold, Langer, Parker, Reynolds, Snowdon, Visalberghi). Other animals possess elements that are common to human behaviors, but none reaches the human level of accomplishment in any domain – vocal, gestural, imitative, technical or social. Nor do other species combine social, technical, and linguistic behaviors into a rich, interactive and self-propelling cognitive complex.

Paradoxically, then, these chapters imply closer behavioral affinities between apes and humans than previously suspected while, at the same time, revealing great species divergences that cannot be neatly summarized by classic stereotypes, such as that of "man the tool-maker". The scientific challenge is to articulate these species similarities and differences in terms amenable to direct research and evolutionary reconstruction.

One side-effect of traditional disciplinary specializations has been a tendency to define tool-making, symbolism, syntax, culture and other capacities as all-or-nothing phenomena which an animal either does or does not possess. Perhaps, as several contributors recognize, it is time to abandon all-or-none definitions and begin to think of complex behaviors as existing in varied levels or degrees. In other words, perhaps the appropriate questions are not whether animals can make tools or use syntax, but how tool-making, tool-use, syntax and other capacities exhibit varying degrees of development. Thus, Parker delineates distinct levels of planning capacity, while Langer, Gibson, and Calvin focus on the behavioral implications of species differences in amounts of information-processing capacity and neural tissue.

Quantitative analyses and other ways of recognizing different levels of behavioral capacity represent relatively new approaches to cognitive evolution. Possibly, they will prove to be dead ends. For now, however, they hold promise for resolving the conceptual dilemmas posed by apes who seemingly can do anything that humans can do, yet fall far short of human achievements. Such analyses are also attractive because of their compatibility with the demonstrated differences in size between ape and human brains. Finally, if, as modern research suggests, apes possess the rudiments of human behavior, but humans are better able to organize conceptual schemes together to form new hierarchical levels of behavior (Gibson, 1990, this volume; Greenfield, 1991; Langer, this volume), then Lashley's (1949) lament that we cannot specify mechanisms for the transformation of animal skills into human abilities may no longer apply. We may well be entering a new era of potential conceptual reconstructions of hominid evolution.

Tools, language and cognition – one event or many?

The founding fathers of modern evolutionary studies were generalists whose interests spanned the realms of psychology, anthropology, and biology (e.g. Freud, Darwin, Broca). The twentieth century, however, has witnessed pronounced scientific specialization. Whole disciplines have focused entirely on individual human capacities such as language, hominid tool remains, or cognition. Even within a single field, such as primatology, scholars often study specific behaviors such as dominance, foraging, vocalization, or deception rather than examining the entire behavioral repertoire of a species. These disciplinary and subdisciplinary specializations foster the idea that each behavior is unique unto itself. Occasionally, however, a voice surfaces arguing that the entire human behavioral repertoire evolved together as one complex adaptive or cognitive whole: for example, Washburn's claim that "Tools, hunting, fire, complex social life, speech, the human way and the brain evolved together to produce ancient man of the genus *Homo*" (Washburn, 1960).

To understand the evolution of the human mind, we must first confront this fundamental question of whether human behavioral and cognitive domains are separately compartmentalized, as self-contained "modules", or whether they exist only as interrelated aspects of a total system. If individual human behaviors such as language, tool-use, imitation, and self-awareness are entirely unrelated to each other, then each may have evolved at a different time, and each must be represented by differing neural areas. Knowledge of the evolution of one behavior can tell us nothing about the evolution of the others, and the charting of mental evolution will require the separate reconstruction of the selective mechanisms and evolutionary time-tables affecting dozens of behavioral domains. If, on the other hand, behavioral domains are interrelated, then each may be represented by similar or overlapping neural areas, and from the knowledge of levels of evolutionary development in one domain, we may extrapolate to those of others.

Although scholars have delineated many domains in which human behavioral capacities exceed those of the apes, language and tool-making have usually been considered to be the defining characteristics of our species (Hallowell, 1956; LeGros Clark, cited Oakley, 1959; Huxley, 1941). Many have noted that language and symbolism profoundly alter our thought processes, and hence, shape and propel the human mind to fundamentally new levels (Eiseley, 1956; Etkin, 1954; Hallowell, 1950; Lieberman, 1991; MacPhail, 1987; Tappen, 1953; White, 1932, 1942, 1959). Tool-making has also facilitated major changes in human knowledge and thought, for example, by providing the means for keeping permanent records, for allowing scientific investigation and for long-distance transportation.

Unfortunately, speech and gesture do not fossilize. Hence, the evolution of language cannot be directly charted from early hominid remains. Brain size and external form do fossilize, however, and tool-use and manufacture leave traces in the archaeological record. Consequently, the delineation of clear links between language and tool behavior would provide a strong conceptual framework for modeling the evolution of language and cognition. The idea that such links may exist is not new (Hewes, this volume). However, their precise nature has been explored only cursorily and is subject to diverse interpretations.

Many psychologists and psycholinguists consider that tools, language and intelligence rest on common cognitive and developmental substrates. Piaget, for instance (1952, 1954, 1955), suggested that tool-use, logic, mathematics, and language rest upon similar mental constructional capacities. Case's (1985) recent expansion and quantification of Piaget's framework strengthens this view. Greenfield (1978) has also shown profound similarities between the maturation of language and of block constructional techniques. Others have noted developmental synchronies between language, object manipulation and social skills (Bates *et al.* 1979; Gopnick and Meltzoff, 1986, 1987). Turning from a developmental to an evolutionary context, cognitively oriented theorists have hypothesized that human tool-use and linguistic accomplishments differ from those of apes in hierarchical constructional domains similar to those which distinguish older from younger children (Gibson 1983, 1988; Reynolds, 1983, and Greenfield, in press).

Linguistically oriented scholars have pointed to similarities in design features of language and tools (Bronowski & Bellugi, 1970; Kitihara-Frisch, 1978) and suggested the presence of a grammar of tool-making analogous to the syntactical rules of language (Falk, 1980; Holloway, 1969; Lieberman, 1975; Montagu, 1976; see review by Ingold, 1986).

Neuroscientists have long noted commonalities in the neural control of speech and manual manipulations. Both speech and skilled movements of the right hand reflect left hemispheric control of skilled movements (Kimura, 1979). Consequently, brain lesions which cause language deficits may also cause difficulties with object manipulation and gesture. On the basis of these and other neurological considerations, several scholars have suggested that selection for skilled movements of the hand may have also led to fine control of oral movements (Bradshaw & Nettleton, 1982; Calvin, 1982, this volume; Lieberman, 1984; MacNeilage, Studdert-Kennedy and Lindblom, 1984).

Quite apart from sharing similar neurological, cognitive, and structural substrates, language and tool-use may have mutual feedback and facilitative relationships. Thus, as long ago as 1927, de Laguna doubted whether complex tool-making, necessitating conceptual thought and advanced planning, could have existed among hominids who had not yet learned to speak. Many anthropologists have reiterated and expanded upon the potential interrelationships between the planning involved in tool-using and making endeavors and speech (Gowlett, 1984; Hallowell, 1950; Isaac, 1978; Oakley, 1954; Revesz, 1956).

Other anthropologists suggest that species-wide obligatory tool-use may have served as a selective agent for the origin of language. Linguistic communication in tool-using species would have permitted more precise communication between mothers and young in feeding and other care-taking endeavors (Borchert & Zihlman, 1990; Parker & Gibson, 1979) and facilitated discussions among adults of rendezvous sites and hunting and gathering plans (Gibson, 1988; Peters, 1974; Washburn, 1960). Alternatively, co-operative tool-use, rather than individual tool-using and tool-making endeavors, may have selected for gestural or vocal linguistic capacity (Leroi-Gourhan, 1943; Yau, 1989).

Building on these ideas, some scholars have speculated on the levels of intelligence and language possessed by the makers of stone tools (Foster, 1990; Gibson, 1985, 1988; Gowlett, 1984; Isaac, 1976; Montagu, 1976; Parker, 1985, Toth, 1985; Wynn, 1979, 1981). These works must all be seen as preliminary, however, because our understanding of the relationships between behavioral domains remains weak. Nor is it universally accepted that stone tools reflect linguistic or cognitive processes.

Contributors to this volume provide fresh perspectives on these recurring themes. In particular, attention is focused on interrelationships between speech and gesture, on the nature of the interactions between tools, imitation, and social behavior; and on developmental and neurological linkages among behavioral domains. In the following paragraphs, I review briefly some of the principal issues raised in the five sections into which the chapters in this volume have been divided. Needless to say, there are many cross-currents that link these sections, and the boundaries between them are inevitably somewhat artificial.

Part I – Word, sign and gesture

Human language capacities manifest themselves in vocal, gestural and written domains, and much speculation focuses on which came first, gestural or vocal language. Gestural theories have intuitively appealed to many scholars (Oakley, 1951, Hewes, 1973, Yau, 1989) on varied grounds. For one, all of the great apes have well-developed gestural capacities, but their vocal skills appear to fall far short of those of *Homo sapiens*. This implies that the common ancestor of great apes and humans also possessed greater gestural than vocal capacities. Gestural theories also hold appeal on neurological grounds. Both tool-use and gesture reflect neurological control of the arm and hand. Hence, the advanced motor and cognitive controls needed for tool-use and tool-making might automatically provide increased gestural capacity.

Others, however, have suggested that the vocal capacities of the monkeys and apes have been grossly underestimated (Steklis, 1985), that vocal and gestural languages rest on similar neurological substrates (Calvin), or that vocalizations can also manifest themselves in gestural, as opposed to phonemic forms (Foster, 1983, 1990). If these views are correct, gestural theories may rest on shaky premises.

In this section, a number of scholars address these issues. Adam Kendon, a linguistic anthropologist, describes the range of gesture usage in modern human populations. Susan Goldin-Meadow, a psycholinguist, examines the development of gestural communication in congenitally deaf children deprived of language training. Sue Savage-Rumbaugh and Duane Rumbaugh, both of them psychologists, report on the auditory syntactic capacities of a bonobo, Kanzi. Charles Snowdon, also a psychologist, reviews current knowledge of avian and primate vocal capacities. Chapters in other parts of the book complement these contributions. In particular, Lock reviews the development of syntax as well as of gestural and vocal communication in human children. Kempler discusses the overlapping neurological control of gestural and vocal symbolism, and Calvin notes the potential of individual neurons to mediate movements of both the hand and the oral cavity.

Taken together, these chapters nudge us from our preconceptions of *Homo sapiens* as primarily vocal communicators and of the uniqueness of human vocal syntax. They also provide tantalizing evidence of potential neurological and cognitive overlap of vocal and gestural control. For the present, however, the view that apes exhibit more advanced gestural than vocal control holds firm.

Part II – Technological skills and associated social behaviors of the non-human primates

Which served as the prime mover in the evolution of the human brain – technology or social behavior? Alternatively, are the two so closely intertwined as to be inseparable evolutionary forces? In this section, four animal behaviorists provide pertinent laboratory and field data. Elizabetta Visalberghi describes tool-use without imitation in cebus monkeys, which are celebrated for their tool-using accomplishments. William McGrew asserts that chimpanzees provide the best referential models for human evolution and substantiates his claims by comparing the subsistence tools of chimpanzees and humans. Christophe Boesch describes the most accomplished population of ape tool-users yet observed by modern science, the chimpanzees of the Tai Forest.

These chapters are also complemented by those in other sections. Wynn notes the importance of imitation for the transmission of human tool-making skills. Reynolds, Toth and Schick, and Davidson and Noble elaborate on the differences between ape and human tool-using and tool-making endeavors. Reynolds, Ingold, Gibson, and Parker detail the profound interactions between human social behavior and technology. Langer discusses the importance of object manipulation for the genesis of the capacities for language and pictorial representation.

Collectively, these contributions suggest that, in both humans and apes, tool-using and tool-making traditions are inseparable from social behaviors such as sharing, teaching, and imitation. Consequently, neither tool-use nor social cognition can be considered in isolation as prime evolutionary movers. Rather, the two are parts of an interdependent behavioral complex. Nor do discrete behaviors such as tool-making, the use of a tool to make a tool, imitation or sharing distinguish apes and humans. Rather humans reach higher levels of performance in each domain and exhibit greater interdependence of social and technological skills.

Part III – Connecting up the brain

In this section, two biological anthropologists (Falk and Gibson) and two neurobiologists (Calvin and Kempler) reflect on the neurological control of human linguistic and technical skills. Kempler focuses on gestural and verbal deficits in Alzheimer's patients. Falk elaborates on the evolution of neural and manual lateralization. Calvin notes the potential contributions of brain expansion to motor control. Gibson suggests that human tool-use, social behavior, and language primarily reflect neural information processing capacities.

Overall, these chapters provide evidence that gesture, vocal language, tool-use and social behavior reflect complementary neural processes and some overlapping neural circuitry. They thus greatly strengthen views that the evolution of one or more of these behaviors could have enhanced the development of the others. Current neurological evidence, however, does not clearly demand the coevolution of behavioral domains.

Part IV – Perspectives on development

Species differences in adult behaviors reflect alterations in developmental processes. As a consequence, conceptual models must incorporate the findings of comparative developmental studies. It has, for instance, been hypothesized that changes in developmental rates (Gould, 1977) or in specific infantile "channeling" behaviors (Gibson, 1990) impinge on ultimate adult linguistic and cognitive capacities. Moreover, in modern humans certain tasks have to be mastered prior to the development of normal symbolic and technical capacities.

If such mastery is a prerequisite for language and tool behavior in modern humans, then ancestral forms could not have attained modern linguistic and technical competence prior to the evolution of the ability to master these precursor tasks. Developmental studies can provide evidence on what these precursors might be.

In this section, three psychologists (Lock, Langer, and Milbrath) and an anthropologist (Parker) apply their knowledge of development to phylogenetic questions. Lock describes human gestural, vocal, syntactic, and object manipulation behaviors. Langer compares ontogenetic processes and developmental rates in macaque, cebus, and human infants. Parker and Milbrath apply a developmental model of social planning skills to the phylogenetic record.

All agree that object manipulation, symbolism and language are strongly interwoven and interdependent in the development of the human child. Parker and Milbrath further note the critical interactions of social behavior, cognition and object manipulation skills in the planning of complex tool-using and social endeavors, while Lock discusses the difficulty of separating gestural and vocal behaviors in human infants. The authors disagree, however, on the extent to which these developmental frameworks can be applied to phylogenetic analysis. Lock takes a cautionary view, but Parker and Milbrath are more optimistic.

In sum, psychological and psycholinguistic data substantiate the idea that language, tool-use and social behavior are interrelated, and provide data of potential use in modeling human evolution.

Part V – Archaeological and anthropological perspectives

Modern archaeologists hotly debate the degree of technological, intellectual, and social sophistication of the Neandertals, the Plio-Pleistocene hominids and other prehistoric populations. For the most part, these debates ignore data on neurology, cognitive development, and ape behavior. Some also neglect to examine tool-use and tool-making in modern human groups. Chapters in this section attempt to reinterpret the archaeological and fossil records in light both of interdisciplinary data and the behavior of people in contemporary human societies.

Wynn, Reynolds, and Ingold focus on tool-use, tool-making, and sociality in modern human populations. Toth and Schick, as well as Davidson and Noble, dwell on the potential significance of the stone tools of ancient hominids for understanding their cognitive capacities. All agree that modern technological achievements require strong interactions between social and technological domains of activity. The contributors express divergent opinions, however, on the cognitive parallels between tool-making and language and on the necessity of language for the transmission of tool-making skills.

All concede that Upper Paleolithic peoples possessed modern linguistic and cognitive capacities. They differ, however in the levels of cognitive capacity which they attribute to earlier hominids. Toth and Schick suggest that even the Plio-Pleistocene hominid makers of Olduwan tools were cognitively advanced compared to the great apes. Wynn would place the first expansion of human cognition somewhat later, with the advent of *Homo erectus*. In contrast, Davidson and Noble consider all pre-Upper Paleolithic peoples to have had little more intelligence than the apes.

These divergent interpretations of the archaeological record could be considered to provide a disappointing end for those who hoped this volume might provide the definitive "just-so" story of hominid evolution (including at least one of the editors). Perhaps, however, the divergence should be looked at in a more positive light. The various opinions expressed by the archaeologists reveal areas in need of future research and indicate a field that is moving from "just-so" stories to a fundamental scientific examination of the diverse neurological, psychological and developmental data that can throw light on the origins of the human mind.

References

Aristotle (1986). *De Anima (On the Soul)*. London: Penguin Books.

Barkow, J. (1989). *Darwin, Sex and Status: Biological Approaches to Mind and Culture*. Toronto: University of Toronto Press.

Bates, E., Benigni, L., Bretherton, I., Camaioni, L. & Volterra, V. (1979). *The Emergence of Symbols: Cognition and Communication in Infancy*. New York: Academic Press.

Borchert, C. M. & Zihlman, A. L. (1990). The ontogeny and phylogeny of symbolizing. In *The Life of Symbols*, ed. M. LeC. Foster & L. J. Botscharow, pp. 16–44. Boulder: Westview Press.

Bradshaw, J. L. & Nettleton, N. C. (1982). Language lateralisation to the dominant hemisphere: Tool use, gesture, and language in hominid evolution. *Current Psychological Reviews*, 2, 171–92.

Bronowski, J. S. & Bellugi, U. (1970). Language, name and concept. *Science*, 168, 699.

Byrne, R. & Whiten, A. (1988). *Machiavellian Intelligence: Social Expertise and the Evolution of Intellect in Monkeys, Apes, and Humans*. Oxford: Oxford University Press.

Calvin, W. H. (1982). Did throwing stones shape hominid brain evolution? *Ethology and Sociobiology*, 3, 115–24.

Case, R. (1985). *Intellectual Development: Birth to Adulthood*. New York: Academic Press.

Cheney, D. L. & Seyfarth, R. M. (1990). *How Monkeys See the World*. Chicago: University of Chicago Press.

Cooper, J. M. (1935). The scientific evidence bearing upon human evolution. *Primitive Man*, 8, 1–56.

Darwin, C. (1871). *The Descent of Man and Selection in Relation to Sex*, 2nd edn. New York: Appleton and Company (1930).

Darwin, C. (1872). *The Expression of the Emotions in Man and Animals*. London: John.

De Laguna, G. A. (1927). *Speech: Its Function and Development*. New Haven: Yale University Press.

Eiseley, L. C. (1956). Fossil man and human evolution. In *Current Anthropology: A Supplement to Anthropology Today*, ed. W. L. Thomas, pp. 61–78. Chicago: University of Chicago Press.

Etkin, W. (1954). Social behavior and the evolution of man's mental faculties. *The American Naturalist*, 88, 129–42.

Ettlinger, G. & Blakemore, C. B. (1967). Cross-modal matching in the monkey. *Neuropsychologia*, 5, 147–54.

Falk, D. (1980). Language, handedness and primate brains: did the Australopithecines sign? *American Anthropologist*, 82, 72–8.

Foster, M. LeC. (1983). Solving the insoluble: Language genetics today. In *Glossogenetics: The Origin and Evolution of Language*, ed. E. de Grolier, pp. 455–80. New York & Paris: Harwood Academic Publishers.

Foster, M. LeC. (1990). Analogy, language and the symbolic process. In *The Life of Symbols*, ed. M. LeC. Foster & L. J. Botscharow, pp. 81–94. Boulder, Colorado: Westview Press.

Gallup, G. (1979). Self-awareness in primates. *American Scientist*, 4, 307–16.

Gardner, R. A., Gardner, B. T. & Van Cantfort, T. E. (1989). *Teaching Sign Language to Chimpanzees*. Albany: State University of New York Press.

Gibson, K. R. (1983). Comparative neurobehavioral ontogeny and the constructionist approach to the evolution of the brain, object manipulation and language. In *Glossogenetics: The Origin and Evolution of Language*, ed. E. de Grolier, pp. 37–61. New York & Paris: Harwood Academic Publishers.

Gibson, K. R. (1985). Has the evolution of intelligence stagnated since Neanderthal man? In *Evolution and Developmental Psychology*, ed. G. Butterworth, J. Rutkowska & M. Scaife, pp. 102–14. Brighton, England: Harvester Press.

Gibson, K. R. (1988). Brain size and the evolution of language. In *The Genesis of Language: A Different Judgement of Evidence*, ed. M. Landsberg, pp. 149–72. Berlin: Mouton de Gruyter.

Gibson, K. R. (1990). New perspectives on instincts and intelligence: Brain size and the emergence of hierarchical mental constructional skills. In *"Language" and Intelligence in Monkeys and Apes: Comparative Developmental Perspectives*, ed. S. T. Parker & K. R. Gibson, pp. 97–128. Cambridge: Cambridge University Press.

Goodall, J. (1986). *The Chimpanzees of Gombe: Patterns of Behavior*. Cambridge, MA: Harvard University Press.

Gopnick, A. & Meltzoff, A. N. (1986). Relations between semantic and cognitive development in the one-word stage; the specificity hypothesis. *Child Development*, 57, 1040–53.

Gopnick, A. & Meltzoff, A. N (1987). The development of categorization in the second year and its relation to other cognitive and linguistic developments. *Child Development*, 58, 1523–31.

Gould, S. J. (1977). *Ontogeny and Phylogeny*. Cambridge, MA: Harvard University Press.

Gowlett, A. J. (1984). Mental abilities of early man: a look at some hard evidence. In *Hominid Evolution and Community Ecology*, ed. R. Foley, pp. 167–92. London: Academic Press.

Greenfield, P. M. (1978). Structural parallels between language and action in development. In *Action, Gesture and Symbol: The Emergence of Language*, ed. A. Lock, pp. 415–45. London: Academic Press.

Greenfield, P. M. (1991). Language, tools and the brain: The ontogeny and phylogeny of hierarchically organized sequential behavior. *Behavioral and Brain Sciences*, 14, 531–95.

Greenfield, P. M. & Savage-Rumbaugh, E. S. (1990). Grammatical combination in *Pan paniscus*: processes of learning and invention in the evolution and development of language. In *"Language" and Intelligence in Monkeys and Apes: Comparative Developmental Perspectives*, ed. S. T. Parker & K. R. Gibson, pp. 540–78. Cambridge: Cambridge University Press.

Hallowell, A. I. (1950). Personality structure and the evolution of man. *American Anthropologist*, 52, 159–73.

Hallowell, A. I. (1956). The structural and functional dimensions of a human existence. *Quarterly Review of Biology*, 31, 88–101.

Hart, C. W. M. (1938). Social evolution and modern anthropology. In *Essays in Political Economy in Honour of E. J. Urwick*, ed. H. A. Innes, pp. 99–116. Toronto: University of Toronto Press.

Hewes, G. (1973). Primate communication and the gestural origin of language. *Current Anthropology*, 14, 5–24.

Holloway, R. L. (1969). Culture, a *human* domain. *Current Anthropology*, 10, 395–412.

Huxley, J. (1941). *Man Stands Alone*. New York: Harper and Brothers.

Ingold, T. (1986). Tools and *Homo faber*: construction and the authorship of design. In T. Ingold, *The Appropriation of Nature: Essays On Human Ecology and Social Relations*, pp. 40–78. Manchester: Manchester University Press.

Isaac, G. L. (1976). Stages of cultural elaboration in the Pleistocene: possible archaeological indicators of the development of language capabilities. In *Origins and Evolution of Language and Speech*, ed. S. R. Harnard, H. D. Steklis & J. B. Lancaster, pp. 275–88. New York: New York Academy of Sciences.

Isaac, G. L. (1978). The first geologists – the archaeology of the original rock breakers. In *Geological Background to Fossil Man*, ed. W. W. Bishop, pp. 139–47. Edinburgh: Scottish Academic Press.

Kimura, D. (1979). Neuromotor mechanisms in the evolution of human communication. In *Neurobiology of Social Communication in Primates: An Evolutionary Perspective*, ed. H. D. Steklis & M. J. Raleigh, pp. 197–219. New York: Academic Press.

Kitihara-Frisch, J. (1978). Stone tools as indicators of linguistic abilities in early man. *Annals of the Japan Association for Philosophy of Science*, 5, 101–9.

Lashley, K. E. (1949). Persistent problems in the evolution of mind. *Quarterly Review of Biology*, 24, 29–30.

Leroi-Gourhan, A. (1943). *L'homme et la matiere*. Paris: Albin Michel.

Lieberman, P. (1975). *On the Origins of Language*. New York: Macmillan.

Lieberman, P. (1984). *The Biology and Evolution of Language*. Cambridge, MA: Harvard University Press.

Lieberman, P. (1991). *Uniquely Human: The Evolution of Speech, Thought, and Selfless Behavior*. Cambridge, MA: Harvard University Press.

Linden, E. (1976). *Apes, Men and Language*. New York: Penguin Books.

MacNeilage, P. F., Studdert-Kennedy, M. G., & Lindblom, B. (1984). Functional precursors to language and its lateralization. *American Journal of Physiology 246 (Regulatory Integrative Comparative Physiology)*, 15, R912–14.

MacPhail, E. (1987). The comparative psychology of intelligence. *Behavioral and Brain Sciences*, 10, 645–95.

Marais, E. (1939). *My Friends the Baboons*. London: Blond and Briggs.

Miles, H. L. (1983). Two way communication with apes and the origin of language. In *Glossogenetics: The Origin and Evolution of Language*, ed. E. de Grolier, pp. 201–10. New York & Paris: Harwood Academic Publishers.

Montagu, M. F. A. (1962). *Culture and the Evolution of Man*. New York: Oxford University Press.

Montagu, M. F. A. (1976). Toolmaking, hunting, and the origin of language. In *Origins and Evolution of Language and Speech*, ed. S. R. Harnard, H. D. Steklis, & J. B. Lancaster, pp. 267–74. New York: New York Academy of Sciences.

Morgan, C. L. (1894). *An Introduction to Comparative Psychology*. London: Walter Scott.

Nissen, H. W. (1951). Social behavior in primates. In *Comparative Psychology*, ed. C. P. Stone, pp. 423–57. New York: Prentice-Hall.

Oakley, K. P. (1951). A definition of man. *Science News*, 20, 69–81.

Oakley, K. P. (1954). Skill as a human possession. In *History of Technology*, vol. 1, ed. C. J. Singer *et al.*, pp. 1–37. Oxford: Oxford University Press.

Oakley, K. P. (1959). *Man the Tool-Maker*. Chicago: The University of Chicago Press.

Parker, S. T. (1985). Higher intelligence as an adaptation for social and technological strategies in early *Homo sapiens*. In *Evolution and Developmental Psychology*, ed. G. Butterworth, J. Rutkowska, & M. Scaife, pp. 83–100. Brighton, England: Harvester Press.

Parker, S. T. (1990). Origins of comparative developmental evolutionary studies of primate mental abilities. In *"Language" and Intelligence in Monkeys and Apes: Comparative Developmental Perspectives*, ed. S. T. Parker & K. R. Gibson, pp. 3–74. Cambridge: Cambridge University Press.

Parker, S. T. & Gibson, K. R. (1979). A model of the evolution of language and intelligence in early hominids. *Behavioral and Brain Sciences*, 2, 367–407.

Patterson, F. G. & Linden, E. (1981). *The Education of Koko*. New York: Holt, Rinehart, and Winston.

Peters, C. (1974). On the possible contribution of ambiguity of expression to the development of protolinguistic performance. In *Language Origins*, ed. R. W. Westcott. Silver Spring, MD: Linstock Press.

Piaget, J. (1952). *The Origins of Intelligence in Children*. New York: Norton.

Piaget, J. (1954). *The Construction of Reality in the Child*. New York: Basic Books.

Piaget, J. (1955). *The Language and Thought of the Child*. New York: World Publishing.

Revesz, G. (1956). *The Origins and Prehistory of Language*. London: Longmans, Green.

Reynolds, P. C. (1983). Ape constructional ability and the origin of linguistic structure. In *Glossogenetics: The Origin and Evolution of Language*, ed. E. de Grolier, pp. 185–200. New York & Paris: Harwood Academic Publishers.

Romanes, G. J. (1882). *Animal Intelligence*. London: Kegan, Paul, and French.

Savage-Rumbaugh, E. S. (1986). *Ape Language: From Conditioned Response to Symbol.* New York: Columbia University Press.

Simpson, G. G. (1949). *The Meaning of Evolution.* New Haven: Yale University Press.

Steklis, H. D. (1985). Primate communication, comparative neurology, and the origin of language re-examined. *Journal of Human Evolution*, 14, 157–73.

Tappen, N. C. (1953). A mechanistic theory of human evolution. *American Anthropologist*, 55, 605–7.

Terrace, H. S. (1979). *Nim, a Chimpanzee Who Learned Sign Language.* New York: Knopf.

Tooby, J. & Cosmides, L. (1989). Evolutionary psychology and the generation of culture. I. theoretical considerations. *Ethology and Sociobiology*, 10, 29–49.

Tooby, J. & DeVore, I. (1987). The reconstruction of hominid behavioral evolution through strategic modeling. In *The Evolution of Human Behavior: Primate Models*, ed. W. G. Kinsey, pp. 183–237. Albany: State University of New York Press.

Toth, N. (1985). The Oldowan reassessed: a close look at early stone artifacts. *Journal of Archaeological Science*, 12, 101–20.

Toth, N. & Schick, K. (1991). Early stone technologies and linguistic/cognitive inferences, paper presented at the American Association for the Advancement of Science meetings, Washington DC, February 17, 1991.

Wallace, A. R. (1864). The origin of human races and the antiquity of man deduced from the theory of 'natural selection'. *Journal of the Anthropological Society of London*, 2, clviii–xxxvii.

Wallace, A. R. (1869). Geological climates and the origin of species. *The Quarterly Review*, 126, 359–94.

Washburn, S. L. (1960). Tools and human evolution. *Scientific American*, 203, 63–75.

White, L. A. (1932). The mentality of primates. *Scientific Monthly*, 34, 69–72.

White, L. A. (1942). On the use of tools by primates. *Journal of Comparative Psychology*, 34, 369–74.

White, L. A. (1959). The concept of culture. *American Anthropologist*, 61, 227–51.

Wynn, T. G. (1979). The intelligence of later Acheulean hominids. *Man (N.S.)*, 14, 379–91.

Wynn, T. G. (1981). The intelligence of Olduwan hominids. *Journal of Human Evolution*, 10, 529–41.

Yau, S. (1989). Moulded gestures and guided syntax: a scenario of a linguistic breakthrough. In *Studies in Language Origins*, vol. 1, ed. J. Wind, E. G. Pulleybank, E. de Grolier & B. Bichakjian, pp. 33–42. Philadelphia: John Benjamins Publishing Company.

A history of speculation on the relation between tools and language

GORDON W. HEWES

Are tools and language related in a sense more profound than that both are attributes of human intelligence or culture? The connection has been seen as either that early language contributed materially to the emergence of human tool-using and tool-making, or that language was one of the outcomes of more advanced tool-making and tool-using in early hominids. That both language and unprecedented tool-related skills were the products of hominid cognitive and manipulatory superiority does not seem to be a very impressive idea, since it offers no explanation for the origin of such superiority. The idea of the miraculous or accidental birth of language and of human technology is scientifically unsatisfying, even if it happened in a creature with an unusually well developed brain and excellent manual co-ordination.

Graeco-Roman and other ancient speculation about the origins of language and tools was very limited because there was little acceptance of the notion that human beings had descended from animal-like ancestors which lacked language and at best exhibited minimal intelligent use of tools. To be sure, a great many people, ancient and recent, believed that other animals have language, or at least did have once upon a time. Gods and spirits likewise were supposed to have used language, and it was also reasonable to assume that language was bestowed on or taught to mankind by divine beings. When the question arose, particular inventions were also regarded as supernatural gifts, or as the work of culture-heroes. That language might have been triggered by tool-using, or the reverse, was an implausible notion. The tool-using powers of the gods and other supernatural beings were simply taken for granted. Typically the gods came to possess spears, bows or clubs, or, depending on the technology of the time, had swords, armour, helmets, chariots and so on. In some mythologies, there were regular workshops presided over by a smith-god, which supplied the other deities with tools and weapons. The deities also managed to equip themselves with clothing and ornament, and often sumptuous dwellings.

Rarely if ever was there any attempt to explain the origins of the artifacts utilized by the gods, any more than there were myths to account for their competence in language. That human children learned to speak from older individuals, and often needed special training in the use of certain tools and weapons, was well understood, but this was not regarded as a topic worthy of serious philosophical inquiry. Given such a background, common to all or nearly all cultural traditions, it is surprising that there was ever any reason to seek explanations either for language or for tool-making and tool-using.

Intellectual curiosity about some aspects of language appeared in a few ancient civilizations, notably in India and Greece. Plato's *Cratylus* (1926) reports a dialogue in which Socrates sets forth his ideas about language to Cratylus and Hermogenes, two philosopher friends. Plato, writing in the fourth century B.C., has Socrates alluding to gestural signs as alternatives to speech. Much of the dialogue has to do with the supposed origins of Greek words, often explained by onomatopoeia. Socrates uses the metaphor of the 'loom of language', thus likening language to a tool (*organon*). Hermogenes argued that the forms of words were merely conventional or arbitrary, against Socrates' view that names or words had been correctly assigned, since the gods would not have used incorrectly assigned ones. He cites the *Iliad* (xx.74), in which Homer notes that some of the words used by the gods are not the same as those now used by mortals. In a further expansion of the tool metaphor for language, the primordial 'namegivers' are described as 'artisans' (*demiurgoi*). The *Cratylus* does not assert that tool-using influenced or promoted the use of language, nor that language generated tool-using. The main point of the dialogue is simply that words have intrinsically correct meanings, a philosophical concern that was shared by the Confucians around the same period. Syntax is hardly mentioned, and is thus not regarded as analogous to the rule-sets which might be operative in tool-making.

Later Classical writers, such as Cicero, praised the capabilities of the human hand (in *De natura deorum*), and observed that words were 'notes of things' (*vocabula sunt notae rerum*), not a profoundly impressive point. Cicero's contemporary, Titus Lucretius Carus (who died around 55 B.C.) presented the first plausible account of the life of early mankind in his cosmological *De rerum natura*, with primitive humans in a beast-like condition without tools, weapons, fire, clothing, artificial shelters or language (Lucretius 1951). Of language (Book 5, line 1928 ff.) he wrote *at varios linguae sonitus natura subegit | mittere et utilitas expressit nomina rerum* – 'as for the various sounds of spoken language, it was nature that drove men to utter these, and practical convenience that gave the form to the names of objects'. Thus he rejected the notion that speech was a gift of the gods. He went on to note that speechless infants are 'led by their plight to employ gestures, such as pointing with the finger at objects in view', a commonplace observation which has only recently been re-examined by developmental psychologists. For Lucretius, the use of fire was discovered, and not simply stolen from the gods by a culture-hero. Modern as many of Lucretius' ideas seem to us, he did not perceive a deep relationship between tools and language, although both were supposed to have arisen from natural causes and for practical ends.

Most of Lucretius' account was rejected as atheistic by both later Classical and early Christian writers, although Origen (d.254) in *Contra Celsum* accepted the idea that the first humans were created naked and 'indigent', differing from the beasts only in their greater intelligence, and surviving by their wits. Such ideas were developed further by the Patristic writer, Caecilius Firmianus Lactantius (d. ca. 326), in *De opificio dei sive de formatione hominis*, a remarkably detailed survey of human anatomy, derived partly from the Stoic writers. Starting with human bipedalism (1974 edn.: Chap.5, 6:1) Lactantius observed that it freed the hands from locomotion, which can then perfect the powers of speech, while the hands are enabled to make and manipulate tools and weapons, replacing the natural body parts which perform analogous functions in other creatures. Read today, Lactantius sounds like an evolutionist. He hails the hands, as well as the voice, as 'the servants of reason and wisdom' and, contrary to many other ancient writers, he locates the seat of reason and the soul in the brain rather than in the heart (*ibid.*, 16:1).

Gregory of Nyssa (d. ca. 396) was yet another Church Father who dealt at length with the capabilities of the human body. In *Peri kataskeus anthropou* (1943, *De hominis opificio*), he described the hands and the speech organs in what can be called a comparative zoological fashion. Like Lactantius he contrasted the natural powers of animals' claws, hooves, horns, and tusks with the feebler human hands and feet, etc. More to the point here, he wrote that 'the hands are a special aid to him [man] for the needs of language', though this might be understood to mean simply that thanks to our hands we can express our language in writing. But somewhat cryptically, he added that 'for my part I have in mind something else when I speak of the usefulness of the hands for the formation of speech', namely, that 'if the body had no hands, how would the articulate voice be formed in it?' Like Lucretius, Gregory supported the idea that language was a natural emergent, rather than a specifically divine gift to mankind. To be sure, *Genesis* 2:19–20 can be interpreted to mean that Adam was already fully equipped to use language when he was invited to name the animals in Eden.

Augustine of Hippo (d.430) contributed to linguistic theory when, in his *Confessions* (1952), he described the process whereby an infant (he was writing about himself) first acquires speech, in part by pointing at things and being given their names by its mother. Although infantile deixis links the hand (and fingers) to the onset of speech, it does not involve tool-using.

The Biblical stories of Adam's acquisition of language and of the Confusion of Tongues at Babel fully sufficed for Jewish, Christian and eventually Islamic theologians and scholars for the next millenium and more, while the origins of human tool-making were ignored. No-one seems to have expressed any interest in why or how Tubal-Cain, for example, came to be an artificer of tools in bronze and iron (*Genesis* 4:22), or for that matter, how Noah acquired ship-building skills. Indic and Chinese thinkers generally lacked curiosity about the origins of language and tool-using, although the Chinese were interested in how their writing system had come about. For the Indians, Sanskrit was the language of the Cosmos, except for the Nyanya school, which regarded language as conventional in origin.

The maritime ventures of the fifteenth and sixteenth centuries led scholars to reconsider the possible primitive or savage state of the earliest humans, although no surviving humans were discovered that totally lacked speech, and even the most 'degraded' possessed some rudimentary tools and weapons. For certain writers, primitivity was explained as a deplorable regression, probably due to diabolic enthralment; others solved the problem by identifying existing primitive peoples as 'pre-Adamites'. Thomas Hobbes (1588–1679), in *Leviathan* (ch.13), was positively Lucretian, or ahead of his time, in envisaging early man in a state of savagery without arts, letters and society. Against this view, Descartes and his followers saw an unbridgeable gulf between mankind, endowed with soul, reason, and therefore language, and all other creatures, seemingly ruling out any continuity with speechless and automaton-like beasts. Human tool-using capacities, as usual, received no specific comment. William Petty (1623–1687) considered man's place in the scale of nature in some detail, placing the elephant ahead of the ape (or drill), though acknowledging the special vocal aptitude of the parrot. Although Mercati (d.1593) had identified prehistoric chipped flints as man-made, it was not until the seventeenth and eighteenth centuries that the notion of an age of stone tools and weapons, as originally suggested by Lucretius, was found to be reflected in the lifestyles of many peoples still living in parts of the Americas.

In the eighteenth century, Condillac, Maupertuis and de la Mettrie re-opened the question of glottogenesis, and contrary to the Cartesians, de la Mettrie proposed that an ape might be taught language, if necessary by the signs used to teach the deaf. Rousseau constructed a scenario dealing with cultural evolution, including the invention of tools, but without linking tools to language. His projected educational reform included instruction in tool-using, and 'manual arts' became important in various schemes before the close of the century. The numerous Enlightenment writers on human nature were impressed with the glorious progress of technology, but took its cognitive roots for granted, unlike their approach to the topic of language. The philosophes certainly did not undervalue the arts and crafts, as the lavishly illustrated volume appended to the *Encyclopèdie* testifies, but the question of how all man's tool-using skills had originated was still not considered to be a subject of profound intellectual importance.

It was an ex-artisan and one-time printer, Benjamin Franklin, who first wrote that 'man is a tool-using animal', oddly enough in a review of Boswell's *Life of Johnson* (1778) (see Boswell, 1887, p. 245). His respect for manual workmanship appeared in his self-composed epitaph, 'Benjamin Franklin, printer'. By this time description and museum collections of 'savage' tools and weapons were becoming familiar, but the Lucretian view of a stone age was not really validated until the Danish shell-midden evidence became available in the early nineteenth century. Not long after, Boucher de Perthes and others demonstrated the presence of crude stone tools in sites also containing the bones of now extinct mammals.

Just at this time, philologists were engaged in reconstructing ancient languages with rigorous new methods, and were apparently defusing some of the previous speculations about ultimate language origins so prevalent during the Enlightenment. The discovery of Neanderthal Man in 1856 revived the question of whether such folk might have had speech, perhaps long before the dawn of Indo-European languages. All this was going on when geology was overturning the absurdly short geochronology proposed by Biblical scholars. Did the stone tools which were now being found in abundance testify to the ability of their makers to master language? In 1863 Schleicher perceived the possible bearing of Darwinism on language, in an open letter to Haeckel. Thomas Carlyle echoed Franklin's aphorism, adding (in *Sartor Resartus*, 1841 Bk I, Ch 5) to 'Man is a tool-using animal' the sentence 'without tools he is nothing; with them he is all'.

The American missionaries Savage and Wyman, working in West Africa, reported in 1843 the existence of a new and very large anthropoid ape and, at the same time, that chimpanzees had been seen using stone tools to crack hard nuts (Savage & Wyman 1843–4). Though this was known to Darwin, along with his observations of the tool-using behaviour of certain Galapagos finches and of some of the chimpanzees in London Zoo, the impact of this information was largely delayed until Jane Goodall reported on the termite-fishing of the chimpanzees at Gombe, a century later (Goodall 1968).

The ensuing serious study of Stone Age artifacts in the first half of the nineteenth century, and then of the first finds of Stone Age skeletal remains – which, as in the case of the Neanderthal find of 1856, differed radically from those of modern humans – challenged some of the scholars who were then investigating the roots of ancient languages. Another ingredient of the new scientific view of mankind came from geologists, critical of scriptural geochronology. Did the earliest humans speak, or perhaps communicate gesturally as occasional eighteenth century writers had suggested? The prospect of applying Darwinism to language opened immense possibilities for speculation, enhanced by the publication of Darwin's *Descent of Man* (1871) and not restrained, except among the more orthodox linguists, by the ban on language papers proclaimed in Paris in 1867 (see below).

Romanes, an enthusiastic Darwinist, tried to train a chimpanzee in the London Zoo to count and to identify colours, as part of his larger programme of research on 'mental evolution in animals', which he began in 1883 and continued throughout the following years (Romanes 1892). At the same time, he doubted Emin Pasha's report from Africa, dated 1890, of wild chimpanzees carrying torches in the forest. Attempts to study 'language' in wild monkeys and apes were meanwhile made by Garner, who set up an observation cage in the Gabon forest, where in the 1890s he recorded vocalizations on a wax-cylinder phonograph. His results were pathetically meagre, but he must be credited with the first systematic effort to study primate behaviour in the field. It is significant that much of the subsequent study of wild primates focused on their vocalizations, when it did not concentrate on their social behaviour; object manipulation was seldom investigated until the time of the Gombe Stream chimpanzee project, although it occurred frequently in psychological laboratory situations.

The linguistic establishment firmly closed the door on serious speculation about language origins, or at least attempted to do so, in the notorious formal ban on language-origin papers issued in 1867 by the Linguistic Society of Paris. Most of the late nineteenth century and early twentieth century work on the problem of glottogenesis fell by default to anthropologists, psychologists and philosophers. Fortunately, the domain of linguistics at the time did not extend to neurology, and very important work relating to language was being done by neurologists, a medical specialty not yet so named when in 1861, Broca announced the localization of speech functions in the left cerebral hemisphere. His findings were quickly confirmed and expanded by Wernicke and Jackson. The constricted horizons of the Neogrammarians and subsequent Saussurians were bypassed by the investigators of language in the brain. It was now possible to conceive of a deep relationship in the striking lateralization of both language and skilled manual functions. For early man, there was now evidence that the makers of stone tools had been right-handed. The young Freud investigated aspects of language pathology in an 1891 paper on aphasia, a topic which is of central importance in neurolinguistics. Tool apraxias were known but seemed to be of less interest to brain researchers.

In 1868, before the Paris ban had diverted orthodox philologists from mischievous glottogonic speculation, Geiger had discovered, studying Indo-European roots, that language had preceded tool-making, because most roots for motor actions seemed to reflect hand, arm and limb movements, rather than tool-manipulation. Noire found fault with this, noting that roots for such concepts as 'to cut' were more likely to have arisen out of tool-using (1917: 139). In any case, Noiré favoured a gestural origin for language, obviating a concern with later vocal roots.

Darwin said little about language in *The Descent of Man* (1871) or in its later revisions. He was content to go along with Tylor's cautious speculations, first enunciated in his *Researches into the Early History of Mankind* (1865), which included a gestural base, and even with Horne Tooke's notion, advanced a century earlier, to the effect that language was an art, like brewing or baking. Darwin was aware that birdsong may involve considerable learning and was not innately governed. In *The Descent of Man*, Darwin did liken the perfection of the human hand to that of the human vocal organs, both of which he saw as being adapted to functions absent from the lower animals.

In 1876 Engels wrote a short piece on the role of labour in the transition from ape to man, though it was not published until 1896 (Engels 1934). In this he made much of the hand and its association with productive work. Speech, for Engels, arose simply when human beings had 'something to say to one another'. In a footnote to *The Origin of the Family, Private Property and the State* (1972 [1884]), Engels seems to agree with Lewis Henry Morgan's notion that language arose from gestural signs. As for apes, he wrote that 'no simian hand ever fashioned even the crudest of stone knives'.

Progress in brain research and in the localization of language-related controls led to the re-examination of some prehistoric crania, and of mandibles in which the genial tubercles were thought by some to indicate a tongue fully adapted to articulate speech.

Ludwig Noiré (1829–1889) was one of the few nineteenth century scholars to address the issue of the relation between tools and language in the full context of the then available prehistoric as well as linguistic evidence. Almost alone for the time, he wrote a whole book on the significance of tools in human evolution: *Das Werkzeug und seine Bedeutung fuer die Entwickelungsgeschichte der Menschheit* (1880). In this work, he argued that inarticulate vocalizations which had accompanied primordial tool-making and tool-using were gradually transformed through onomatopoeia into spoken syllables and words.

As early as 1868 Haeckel had postulated a speechless ape-man ancestor for mankind, which he accordingly named *Pithecanthropus alallus*. The generic portion of this term was applied, a quarter of a century later, by Dubois to the famous skull cap found by him in eastern Java, but with the specific cognomen 'erectus' because of its surprisingly modern-looking femur. Learned opinion was divided over whether such a low-browed and relatively small-brained creature could have had language, or for that matter, whether it could have been skilled enough to fashion tools (none were discovered at the site). The linguistic ability of the Neanderthalers, for which evidence was now augmented by finds in Belgium and soon thereafter in France, was also debated. Their brains were unusually large, but their jaws looked rather primitive. Neanderthal lithic workmanship (of the Mousterian type) was of good quality, complicating the task of determining Neanderthal intelligence. The next puzzling specimen to be discovered was the Heidelberg mandible, found in 1908. The absence of the genial tubercles, supposed to be indicators of a fully human attachment of the tongue, suggested that the possessor of this mandible lacked the capacity for speech.

Further important fossil hominid discoveries were made during the first three decades of the twentieth century (besides the Heidelberg mandible, these included the Taung Specimen, Rhodesian Man and Peking Man, along with the fraudulent Piltdown Man), but the connection between tools and language received limited attention during this period. Interest in cognitive capabilities was focused on the braincase rather than on artifactual evidence, which in the cases of these fossil finds was present only at Choukoutien. The ingeniously faked Piltdown fossil possessed a capacious braincase, and was 'found' with seemingly solid evidence of stone and bone tool-making, misleading an entire generation of palaeoanthropologists.

The re-evaluation of the early hominid fossil record began in earnest after further australopithecine finds. Dart's original juvenile specimen had been downgraded by the established authorities as a fossil anthropoid ape. The newer specimens showed the group to have been already bipedal, and possibly even makers of pebble tools, though this was disputed. What the suite of australopithecine discoveries did do was to stimulate reconstruction of the early hominid biogram, drawing in part on evidence from systematic field studies of wild higher primates, starting with baboons. So far as tool-using was concerned, efforts at reconstruction were based on the 'man the hunter' model, but some attention was given to the question of the possible possession by early hominids of a proto-language. The discussion also took note of the marked reduction in the hominid line of the canine teeth, one possible further indication of a switch from dental armament to extra-somatic tools.

In the 1950s several popular books and articles appeared on language origins, and on the archaeology of the early Palaeolithic now enlarged by fieldwork in Africa. By this time, conditions had become more favourable for studies of primate behaviour in the wild. A number of works appeared on the theme of 'man the tool-maker' (in particular the many writings of Kenneth Oakley, e.g. 1961), a theme which fitted well with the focus on 'man the hunter'. And in the late 1950s and 60s, Holloway and others began to seek signs of a language capacity in early hominid endocasts, also considering the evolutionary interrelation between brain function and tool-using. The Wenner-Gren conference, 'Anthropology Today', convened in New York in 1952 (see Kroeber 1953), included a section on the origins of language, a topic which had for many years been considered a little eccentric in serious scientific circles.

Space does not permit a detailed account of the rapidly growing body of work on early language and early tools, and of the relevance to this work of non-human primate behavioural evidence. Among the many scholars associated with this revitalized topic we can mention Golovin, Washburn, Leroi-Gourhan, Bunak and Jane Lancaster. After some abortive efforts by previous investigators to inculcate speech in chimpanzees, the Gardners and the Premacks embarked on experiments utilizing visual symbols (gestural signs or plastic tokens) rather than speech, and began to obtain striking results, stimulating further chimpanzee studies, and also work with orang-utans and gorillas. These projects generated some acerbic criticism from certain Cartesian-minded linguists who were unable to tolerate the idea of a rational language-using ape.

The revival of respectability for glottogonic research had been heralded in 1967 by Wescott (see Wescott 1967). Almost by coincidence, the equally neglected subject of gestural sign-languages was also re-opened at this time, notably by Stokoe and Bellugi, in time to reinforce the ape-language experiments which employed the sign language of the deaf, ASL. These developments led to my own proposal, in 1973, of an *explicit* formulation of the relation between tool-using and the emergence of language (Hewes 1973). The subject of glottogenesis had been taken up at the 1972 meeting of the American Anthropological Association, resulting in a volume of papers (Wescott 1974). By 1975 the topic had become very much more complex, and a large conference, sponsored by the New York Academy of Sciences, was held in that year, with a strong participation of neurologists, psychologists, prehistorians, palaeoanthropologists, and even a few linguists, covering the major issues surrounding the evolution of language and speech (see Harnad, Steklis & Lancaster 1976). Two papers, by Montagu and Isaac, dealt specifically with the bearing of tool-using on the emergence of language (Montagu 1976, Isaac 1976), but the question surfaced in several other papers as well, especially in connection with cerebral lateralization.

In 1981, Eric de Grolier organized a conference, 'Glottogenesis', which was held in Paris and sponsored by UNESCO. Two papers in the de Grolier volume theorized that language and tool-use were linked by common mental constructive capacities (Gibson, 1983; Reynolds, 1983). A few years later, in 1983, an international Language Origins Society was formally established, and scientific papers and symposia on glottogonic matters have been accumulating at a remarkable pace ever since. These have included a focus on the question of the relation between language and tools. Analyses of the forms of Palaeolithic stone artifacts have been used to determine whether formal variations reflect semantic classifications on the part of their makers (Wynn 1988, Dibble 1989). Intensive studies have been made of monkey and chimpanzee tool-using (e.g., Tomasello 1990). Beck has written a valuable survey of tool-using in animals in general (1986), and relevant material is to be found in recent volumes edited by Mellars & Stringer (1989), Parker & Gibson (1990) and Foster & Botscharow (1990). The peer commentary following the paper by Pinker & Bloom (1990), and the work by MacNeilage (1987) on the evolution of hemispheric specialization for manual functions and language, are further indications of the rapid progress now characteristic of this field.

In the past, the major impetus for work on this general question has presented the human capacity for language as the explicandum, whereas the tool-making and tool-using faculty has been taken for granted as an expectable attribute of a highly intelligent and manually dextrous 'tool-using animal', and therefore as requiring less scientific examination. This may reflect a deep logocentric philosophical bias on the part of western scholars, and one that has had a very long history. Now that this bias has been exposed for what it is, a whole new field of questions has been opened up, as the chapters in this volume reveal.

References

Augustine (1952). *Confessions*, trans. E. B. Pusey, Vol.18 (Great Books of the Western World). Chicago: Encyclopaedia Britannica.

Beck, B. B. (1986). Tools and intelligence. In *Animal Intelligence: Insights into the Animal Mind*, eds. R. J. Hoage & L. Goldman, pp. 135–47. Washington, DC: Smithsonian Institution Press.

Boswell, J. (1887). *Life of Johnson*, vol. III, ed. George Birckbeck Hil. Oxford: Clarendon Press.

Carlyle, T. (1841). *Sartor Resartus: The Life and Opinions of Herr Teufelsdröckh*, 2nd edn. London: Fraser.

Darwin, C. (1871). *The Descent of Man*. London: John Murray.

Dibble, H. L. (1989). The implications of stone tool types for the presence of language during the Lower and Middle Pleistocene. In *The Human Revolution: Behavioural and Biological Perspectives on the Origins of Modern Humans*, ed. P. Mellars & C. Stringer, pp. 415–31. Edinburgh: Edinburgh University Press.

Engels, F. (1934). *Dialectics of Nature*, trans. from German by C. Dutt. Moscow: Progress.

Engels, F. (1972 [1884]). *The Origin of the Family, Private Property and the State.* New York: Pathfinder Press.

Foster, M. L. & Botscharow, L. J. eds. (1990). *The Life of Symbols.* Boulder, Colorado: Westview Press.

Gibson, K. R. (1983). Comparative neurobehavioral ontogeny and the constructionist approach to the evolution of the brain, object manipulation and language. In *Glossogenetics: The Origin and Evolution of Language*, ed. E. de Grolier, pp. 37–61. New York & Paris: Harwood Academic Publishers.

Goodall, J. van Lawick (1968). The behaviour of free-living chimpanzees in the Gombe Stream Reserve. *Animal Behaviour Monographs* 1, 161–311.

Gregory of Nyssa (1943). *Peri kataskeus anthropou (De hominis opificio).* Trans. (*La création de l'homme*) and notes by J. Laplace & J. Danielou. Paris: Éditions du CERF.

Harnad, S., Steklis, H. & Lancaster, J. eds. (1976). *Origins and Evolution of Language and Speech.* Annals of the New York Academy of Sciences, vol. 280.

Hewes, G. W. (1973). An explicit formulation of the relation between tool-using and early human language emergence. *Visible Language,* 7(2), 102–27.

Isaac, G. L. (1976). Stages of cultural elaboration in the Pleistocene: possible archaeological indicators of the development of language capabilities. In *Origins and Evolution of Language and Speech*, ed. S. R. Harnad, H. D. Steklis & J. B. Lancaster, pp. 275–88. Annals of the New York Academy of Sciences, vol. 280.

Kroeber, A. L. ed. (1953). *Anthropology Today.* Chicago: Chicago University Press.

Lactantius, Caecilius Firmianus (1974). *De opificio.* Latin text and translation (*L'Ouvrage du dieu créateur*) by M. Perrin. Paris: Éditions du CERF.

Lucretius, Titus Carus (1951). *De rerum natura.* Cambridge: Cambridge University Press.

MacNeilage, P. (1987). The evolution of hemispheric specialization for manual function and language. In *Higher Brain Function: Recent Explorations of the Brain's Emergent Properties*, ed. S. Wise, pp. 285–309. New York: John Wiley.

Mellars, P & Stringer, C. eds. (1989). *The Human Revolution: Behavioural and Biological Perspectives in the Origins of Modern Humans.* Edinburgh: Edinburgh University Press.

Montagu, A. (1976). Toolmaking, hunting and the origin of language. In *Origins and Evolution of Language and Speech*, ed. S. R. Harnad, H. D. Steklis & J. B. Lancaster. pp. 267–74. Annals of the New York Academy of Sciences, vol. 280.

Noiré, L. (1880). *Das Werkzeug und seine Bedeutung fuer die Entwickelungsgeschichte der Menschheit.* Mainz.

Noiré, L. (1917). *The Origin and Philosophy of Language.* Chicago: Open Court.

Oakley, K. P. (1961). *Man the Tool-maker.* London: British Museum (Science).

Parker, S. T. & Gibson, K. R. eds. (1990). *"Language" and Intelligence in Monkeys and Apes: Comparative Developmental Perspectives.* Cambridge: Cambridge University Press.

Pinker, S. & Bloom, P. (1990). Natural language and natural selection. *Behavioral and Brain Sciences*, 13, 707–84.

Plato (1926). *Plato: Cratylus, Parmenides, Greater Hippias, Lesser Hippias.* Greek text and trans. by H. N. Fowler. (Loeb Classical Library). London: Heinemann.

Reynolds, P. C. (1983). Ape constructional ability and the origin of linguistic structure. In *Glossogenetics: The Origin and Evolution of Language*, ed. E. de Grolier, pp. 185–200. New York & Paris: Harwood Academic Publishers.

Romanes, G. J. (1892). *Darwin, and After Darwin. An Exposition of the Darwinian Theory and a Discussion of Post-Darwinian Questions*. London: Longmans, Green.

Savage, T. S. & Wyman, J. (1843–4). Observations of the external characteristics, habits and organization of the *Troglodytes niger*. *Boston Journal of Natural History*, 4, 362–86.

Tomasello, M. (1990). Cultural transmission in the tool use and communicatory signalling of chimpanzees. In *"Language" and Intelligence in Monkeys and Apes: Comparative Developmental Perspectives*, ed. S. T. Parker & K. R. Gibson, pp. 274–311. Cambridge: Cambridge University Press.

Tylor, E. B. (1865). *Researches into the Early History of Mankind and the Development of Civilization*. London: John Murray.

Wescott, R. W. (1967). The evolution of language: re-opening a closed subject. *Studies in Linguistics*, 19, 67–82.

Wescott, R. W. ed. (1974) *Language Origins*. Silver Spring, MD: Linstok Press.

Wynn, T. (1988). Tools and the evolution of human intelligence. In *Machiavellian Intelligence*, ed. R. Byrne & A. Whiten. Oxford: Clarendon Press.

Part I

Word, sign and gesture

Part I Introduction

Relations between visual–gestural and vocal–auditory modalities of communication

TIM INGOLD

The human hand is a restless organ. For the greater part of our waking lives it is constantly on the move, touching and feeling, holding and manipulating, or merely describing configurations in the air. It moves not of its own accord but under intentional control, yet that intentionality seems to be embodied in the movement itself – only from time to time is it preceded or interrupted by a consciously articulated plan. Moreover, we move with our eyes open, and the eyes, too, never stop still. Even in those artificial situations, such as at the barber's, when we are required to keep the head steady, the eyes strain to their rotational limits in our efforts to see what is going on around us. Most of the time, however, we can scan the world visually by moving not just the eyes themselves but also the head and indeed the whole body, and this movement, like that of the hands, carries an embodied intentionality (Gibson 1979: 218–19). Together, visual perception and manual kinaesthesis constitute aspects of one indivisible and continuous flow of action. For example, the orchestral 'cellist does not first form a percept of the conductor's gesture and then, in a separate, sequential act, respond with a renewed application of the bow to his instrument. For his motor response is part and parcel of the same, total process of intentional action as is his visual attention to the conductor. Were this not the case, the orchestral performance would dissolve into a cacophony (I return to this point in the Epilogue).

The orchestral player is, of course, also listening intently, both to the sound of his own instrument and to the sounds produced by his colleagues. Similarly, in social life, we attend both by watching and by listening to our own and others' actions. Orchestral performance does indeed provide a powerful metaphor for human social life (Leach 1976:43–5), if we grant that everyone is at least potentially a player, and with the provisos that social life normally affords far greater scope not only for improvisation but also for misunderstanding and breakdown in the flow of action and discourse. We attend (or fail

35

to attend) aurally to the vocal utterances of others, along with all the other kinds of noises they produce, just as we attend visually to their manual gestures, facial expressions, and so on, and the responses we make – both vocal and gestural – are integral to the same process as is our attention. However, the vocal–auditory modality of perception and action differs from the visual–gestural modality in two critical respects. First, whilst humans talk a lot they do not talk – in the way that they move their hands and other parts of their bodies – *almost all of the time*. Thus, every bout of vocal activity is framed within a context of gesture that is going on before we begin to speak, that proceeds throughout the utterance and continues after it has ended. Gesture flows, as it were, in and out of vocal discourse, serving (as in a conversation over dinner) at one moment to reinforce or illustrate an utterance, and at another to manipulate utensils such as knife and fork. Secondly, we can only hear someone or something as long as that person or thing is actively engaged in sound production (which is to say that we do not, like bats, make use of echolocation). By contrast, the things we see do not produce light energy, they reflect it, and therefore need not themselves be doing anything at all in order to be seen. Consequently, the vocal–auditory modality is inherently *interactive* in a way that the visual–gestural modality is not.

These reflections are prompted by the observation that the relation between using and making tools, on the one hand, and speech on the other, is an instance of the more general relation between manual and vocal activity, and between visual and auditory perception. In discussions of the evolution of human capacities in these respects it is customary to begin with the known attributes and abilities of contemporary non-human animals, to move on to those species – namely the great apes – which are most closely related to our human selves, and thence to the capacities of modern human beings. The chapters that make up this section, however, have been deliberately arranged in the reverse order. For there is a sense in which we can only understand what speech *is*, or grasp the character of the phenomenon whose evolution we are trying to explain, by seeing how it is used and how it works in the everyday context of communication among humans who have normal use of their hands and fully functioning organs of touch, sight and hearing. For such humans, as Wittgenstein famously put it (1953, para.23), the speaking of language is but part of a way of living, a 'form of life', that also includes all kinds of non-linguistic activities whose crucial characteristic is that they are carried out communally, or are in some sense shared by the members of a human group. If language owes its specific properties, and its effectiveness as a medium of communication, to the manner in which it is embedded in or relates to these other social activities, it follows that an account of the evolution of language must, *ipso facto*, be an account of the evolution of this relation of embeddedness.

Unfortunately for those afflicted, but perhaps fortunately for science, many people have to make do with organs of perception that are more or less defective. In their vocal communication, blind people have no use for manual gesture, yet they rely on their hands much more than do the sighted for finding their way in the world, through an enhanced sense of touch. Deaf people, by contrast, often rely on gesture for their communication, but this in turn can 'tie their hands up' so that they are not simultaneously free for touching, feeling and object manipulation. In principle, studies of how both blind and deaf people cope with their respective disabilities should shed light on the relation between manual and linguistic skills, and hence – indirectly – on the connections, in human evolution, between language and tool use. In practice, however, only the deaf have been the focus of attention in this regard. So what can we learn from their 'ways of coping'?

The most obvious and striking fact is that among communities of the deaf, sign languages have emerged, apparently quite spontaneously, which have all the formal properties of natural spoken languages. This, in itself, is proof – if any were needed – that language is not necessarily confined to the vocal–auditory channel, and can function just as well, if required to do so, in the visual–gestural modality. Whether, however, this fact can be adduced in support of the theory that language *originated* in the latter modality is a more contentious matter. Advocates of the theory (e.g. Hewes 1973, 1976) argue that the first language consisted of manual signs, imitating the operations of tool use and capitalising on the enhanced manual dexterity and cortical asymmetry resulting from natural selection for tool-using skills. They claim that the 'transfer' of language from the visual–gestural to the vocal–auditory modality was a later development. The arguments for and against the gestural theory of language origins are finely balanced; implicated in them are questions concerning the respective contributions of the right and left hemispheres of the brain and the neural circuitry involved in both vocal and gestural function – questions that may be better addressed through studies of people suffering from various forms of aphasia and apraxia (see Part III of this volume, especially chapters by Falk and Kempler). Of one thing, however, we can be certain – namely, that deaf people of today cannot be taken as models for our evolutionary ancestors. For the latter, so far as we know, could hear perfectly well, and were doubtless capable – as are contemporary non-human primates – of a varied repertoire of intentional vocalizations. They were not, as are modern deaf people, forced to elaborate a specialized gestural system in order to compensate for the deficiencies of the vocal–auditory channel in a world where linguistic communication is ubiquitous and a precondition for social incorporation.

However, what studies of the deaf do show, clearly and consistently, are the kinds of changes that movements of the body must undergo if they are to function as elements in a system of linguistic signs, and what the properties of these signs must be. And in this regard, what applies to manual gesture surely applies with equal force to vocal utterance, for speaking, too, is a form of intentional motor activity, and is experienced as such – as *movement rendered audible* (this is even more obviously true of song, or of 'cello-playing – in which the 'movement rendered audible' is of course manual rather than vocal). In this sense, as Merleau-Ponty put it, 'the spoken word is a genuine gesture, and it contains its meaning in the same way as the [manual] gesture contains its' (1962: 183). If the parallel between vocal and manual gesture is accepted, it is then possible to suggest that vocal utterances underwent similar changes, in the course of the evolution of speech, to those undergone by manual gestures as their role changes from accompanying speech to constituting, in themselves, a language of apparently arbitrary signs. But in the case of speech, the development would have been impelled not by the deficiency of the alternative modality, but by the relative *efficiency* of adopting for communication a channel whose use interfered only minimally with the conduct of everyday practical (usually manual) activities. One can get on with doing things whilst talking at the same time – indeed most human practical activities are also social, and are accompanied by conversation, if not song. And of course one can talk in the dark, which for ancestral populations whose only source of artificial light was fire must have been no mean advantage.

In his discussion of human manual gesture, Kendon distinguishes three forms, which he calls *gesticulations*, *emblems* and *signs*. Gesticulations are the manual movements that accompany all normal talk: indeed as Kendon shows with specific examples, they are not so much an accompaniment as an integral part of communicative action, providing a holistic, iconic and situation-specific imagery to fill out the more decontextualized, 'lexico-syntactic' representations of verbal speech. Whereas the latter are *constructed* by organizing units of meaning (words), according to rules of syntax, into strings, gesticulations are *presented* as total movements that cannot be decomposed into recombinable elements. Emblems, by contrast, are gestures that are less tied to the specific situation. They have conventional meanings, applicable in a range of circumstances, and can therefore be quoted. But although it is possible to establish catalogues of emblems in regular use in particular communities, the meaning of an emblem is established by virtue of iconic resemblance and does not depend upon systematic contrast with other emblems in the gestural repertoire. However, where we find gestures that are not only quotable, like emblems, but which may also be analysed as specific combinations of distinctive features – such that the meaning of each is given not by iconic resemblance but by its contrast with other gestures in the system – then we are dealing with signs. Gestural signs function, to all intents and purposes, like words and – like words – they are put together into displays that are constructed rather than presented. Kendon's claim, based on studies of the deaf and of societies in which ritual restrictions on the use of speech are imposed on certain sections of the community, is that the more limited the possibilities for using speech, the more the manual gestural system takes on the properties of a system of signs. But there is an underlying continuity from gesticulations, through emblems to signs. As a gesture becomes more 'sign-like', it gradually ceases to mimic the whole situation, becoming progressively attenuated so that only one element of the original depiction is retained – an element selected, moreover, not for its iconic aptness in epitomising the whole, but by virtue of the requirement that it should contrast with other forms. In this process, the semiotic status of gestures shifts from their 'standing in' for the real-world objects or actions to which the accompanying verbal discourse refers, to their establishment as the very elements out of which that discourse is built, discourse which must perforce be illustrated (if at all) by other means.

Kendon is, by and large, concerned with the speech and gesture of adults; Goldin-Meadow, by contrast, takes us back to that period of early childhood during which initial linguistic competence is acquired. This is of particular interest since at the onset of this stage of development, in children with normal hearing, the communicative functions assumed by manual gesture and voice are as yet not clearly differentiated (see Lock, this volume Chapter 13, for a fuller account). With time, in such children, the vocal–auditory modality comes to predominate, and any emergence of language-like structure in gesture is inhibited by the role it comes to play as a complement to speech. Goldin-Meadow's study, however, focuses on children who are deaf, yet reared by hearing parents without exposure to any conventional sign language. These children, she discovered, evolve a system of gestural communication that goes far beyond anything found among hearing children at an equivalent stage of development, and that possesses the same language-like properties as does the vocal communication of the latter. Still more remarkably, the quasi-linguistic structuring of the deaf children's gestural system was something they produced for themselves, quite independently of parental guidance. As normal, speaking and hearing adults, parents gestured a good deal to their children, but (to adopt Kendon's terms) these remained as gesticulations or perhaps emblems – never did they become signs. They were not arranged morphologically into a system of contrasts, nor were they combined syntactically into strings. The children's gestural communication, on the other hand, shared both these features, albeit at a level of complexity far below that of a fully developed vocal or sign language.

It might be said, then, that these children, each of his or her own accord, are 'inventing' language. Yet Goldin-Meadow wisely cautions against hasty comparisons between their situation and that of language's first emergence in ancestral populations. For although the children in her study, unlike hearing children of hearing parents or deaf children of deaf parents fully competent in sign language, have no conventional language model to follow, they are nevertheless immersed in an environment already replete with all kinds of artefacts, in which are inscribed the linguistically informed practices of their makers. They live, in short, in a cultural world, and draw for support on the objects of this world in their communicative performances. So, indeed, do the chimpanzees reared at the Yerkes Language Research Center, whose acquisition of language-like communicative skills is described in the chapter by Savage-Rumbaugh and Rumbaugh. Could they, perhaps, provide us with a better model for the emergence of language than the deaf children described by Goldin-Meadow? The star performer at Yerkes is a bonobo (pygmy chimpanzee), Kanzi, who – by the age of eight years – had already acquired a comprehension of English semantics and syntax virtually equivalent to that of a two-year-old human child reared in the same environment. Most importantly, Kanzi learned language *without training*, merely by attending to, and participating in, the talk of his human caregivers.

Chimpanzees reared in an environment of their conspecifics, in the 'wild', do not learn to speak. Thus Kanzi's achievement is, in a sense, the opposite of that of the deaf child. Children normally acquire language; the deaf child does so even in the absence of a model. Young chimpanzees normally do *not* acquire language; Kanzi did so in the presence of human models. What conclusions can we draw from the comparison of these two instances? In the light of their work with Kanzi and other human-reared apes, Savage-Rumbaugh and Rumbaugh claim that the normal presence of language in humans, as against its normal absence in apes, cannot be attributed to some specialized and innate mechanism or neurological component unique to human beings. Instead, they argue that language is a product of learning, of what an individual learns and how it learns. Yet Goldin-Meadow's evidence for the spontaneous 'invention' of language by deaf children seems, at first glance, to contradict this argument. The contradiction, however, is more apparent than real, for it results from the framing of the argument in terms of an artificial dichotomy between innate and learned structures (see Oyama 1985 for a critique). We do not have to suppose that language, or a template for its assembly, pre-exists either 'inside' the individual (as a genetic programme) or 'outside' the individual (as a feature of the environment), or even that it is a 'product' of the interaction of genetic and environmental factors. The syntactic and semantic structures that we identify as 'language' should, perhaps, be rather seen as *emergent properties* of a developmental process that cuts across the interface between the individual and his or her environment. This conclusion would be entirely consistent with Goldin-Meadow's findings with regard to deaf children growing up in an environment of speaking–hearing caregivers and artificial objects, and it is also the conclusion to which Savage-Rumbaugh and Rumbaugh move in closing their chapter. Their point is that where an intelligent creature is placed in a developmental context which imposes a situational need for complex communication with similarly intelligent creatures in the social environment, syntactic structures are *bound to emerge* as necessary solutions to the communicative problem. Thus language is no more given *in* the environment than it is *in* the organism; it emerges in the relational context of the organism in its environment, and is therefore a property of the developmental system constituted by these relations.

Now of course our evolutionary ancestors, on the threshold of language, were not surrounded and guided by caregivers with a pre-formed linguistic competence. To approximate their situation, and to model the context in which language initially emerged, we have to turn our attention to animals living in their natural environments, in the company of their own conspecifics. Snowdon's chapter is a review of studies of animal communication in natural settings. He is sceptical of many of the bolder claims that have been made for the linguistic properties of non-human communication systems, such as categorical perception, syntax and symbolic reference, showing that in many cases these claims rest on unwarranted inference or result from the ways in which experiments are designed or data recorded, rather than reflecting genuine features of the communicative phenomena themselves. Whilst recognizing that some of these properties may be found in rudimentary forms in the systems of communication of certain non-human species, Snowdon concludes that these species do not, in natural settings, display anything approaching the complexity of human language. Yet language should not be thought of as a unitary 'thing' that has undergone an evolutionary trajectory of its own; it is rather a particular and unique 'coming together' and integration of a number of distinguishable components or capacities, each of which may have shown up in diverse species, at different times in evolution, but in combination with other traits which form no part of 'language-as-we-know-it' (White 1985). Drawing on the principle of convergent evolution, by which similar features may develop in separate phyletic lines when the species concerned are faced with comparable challenges in their physical or social environments, Snowdon suggests that one way to approach the evolution of language is to attempt to specify the environmental conditions, above all those of the *social* environment, in which its various components have repeatedly emerged in the animal kingdom.

This, perhaps, is the principal contribution that studies of non-human animals, in natural settings removed from human linguistic influence, can make in helping us to resolve the question of language origins. Taken together, however, the chapters in this section demonstrate the effective complementarity between such studies and investigations of human communication in a human environment, and of the communicative abilities of non-human animals reared in a human cultural setting.

References

Gibson, J. J. (1979). *The Ecological Approach to Visual Perception*. Boston: Houghton Mifflin.

Hewes, G. (1973). An explicit formulation of the relationship between tool-using, tool-making and the emergence of language. *Visible Language*, 7, 101–27.

Hewes, G. (1976). The current status of the gestural theory of language origin. In *Origins and Evolution of Language and Speech*, ed. S. R. Harnad, H. D. Steklis and J. Lancaster, pp. 482–504. Annals of the New York Academy of Sciences, vol. 280.

Leach, E. R. (1976). *Culture and Communication*. Cambridge: Cambridge University Press.

Merleau-Ponty, M. (1962). *Phenomenology of Perception* (trans. C. Smith). London: Routledge & Kegan Paul.

Oyama, S. (1985). *The Ontogeny of Information: Developmental Systems and Evolution*. Cambridge: Cambridge University Press.

White, R. (1985). Thoughts on social relationships and language in hominid evolution. *Journal of Social and Personal Relationships*, 2, 95–115.

Wittgenstein, L. (1953). *Philosophical Investigations*. Oxford: Blackwell.

1

Human gesture

ADAM KENDON

In this chapter features of human gesture are reviewed. I begin with the movements of the body that accompany speech and then discuss gesture as it is used without speech, as an occasional substitute for speech, and in gesture systems or sign languages. My approach is comparative, with an emphasis on the continuities between these different forms. Since, as we shall see, there are also certain continuities between gesture and spoken language, it can be argued that processes shaping language are not specific to it, but operate in other forms of human expression as well. If this is so, it becomes easier to construct a model for the evolutionary origins of language and it is for this reason that the study of gesture is relevant to theories of language origins.

Gesticulation

Some movement of body parts such as the head, eyes, brows and mouth, as well as of the hands, commonly occurs when people speak. Such movements are patterned in relation to speech and contribute to the utterance in different ways. They can play a role in indicating to whom the utterance is addressed; contribute to the emotional aspects of the utterance; serve in various ways as markers of utterance structure; and, finally, they can represent aspects of the *content* of the utterances with which they are associated.

Most of the work on gesticulation has been concerned with hand and arm movements and my discussion will concentrate on this. The role of head movement in gesticulation has been dealt with by Birdwhistell (1970), Hadar (Hadar *et al.* 1983, 1984) and Calbris (1990). That of the face (apart from facial expressions of emotion), by Birdwhistell (1970), Ekman (1979), and Chovil (1989). The head and the face probably play somewhat different roles in gesticulation than the hands and arms. Head movements, closely involved with eye movements, are probably most important in turn regulation, in utterance address and discourse structure marking, although the head can also be used in pointing and in other ways, for the expression of content. Facial movements, specifically brow movements, have been shown to mark contrasting segments of discourse, and configurations of action patterned after facial affect displays are used as icons of emotional expression, as well as contributing to the emotional tone of the utterance. In general, it seems likely that the face and head play a greater role in the regulatory and affective functions of utterance than they do in the expression of referential content.

An early study of gesticulation is by Efron (1972 [1940]). He demonstrated stylistic differences between different cultural groups (East European Jews as compared to Southern Italians) and provided an influential discussion of the different ways in which gesticulation and speech may be related (cf. Ekman & Friesen 1969). However, this was impressionistic, and it was not until the 1970s, when sound film and then video-tape became more easily available, that systematic analyses of the gesticulation–speech relationship began. Pioneering studies include those of Freedman (1972) and Dittman (1972), however it was in Condon and Ogston's (1966) work and in my own early studies (Kendon 1972, 1980) that attempts were first made to look in detail at how speech and the movements of gesticulation are related. In my work I analyzed the phrasal organization of gesticulatory movements and sought to establish the temporal relationship between these phrases and the phrases of concurrent speech. In this, and subsequent work, it was shown that gesture phrases and speech phrases (defined as tone units following Crystal & Davy 1969) are matched. Phrases of gesticulation were found to precede or coincide with their associated tone units and, so long as each tone unit encodes but one "idea unit", only one gesticular phrase was found to be associated with it. The phrases of gesticulation, I concluded, are organized simultaneously with phrases of speech. Accordingly, it seemed that gesticulation and speech should be regarded as two aspects of a single process of utterance.

David McNeill, separately, began similar work (1979a), and arrived at similar conclusions. He has shown that the holistic, image-like, presentational character of expression in gesticulation is planned for and organized simultaneously with the planning and organization of speech. He argues that such gesticulations present aspects of the meaning of an utterance that are specific to the situation in which it is employed (its intrinsic meaning, in his terms), rather than the general meaning that can be attributed to it in purely linguistic terms. He argues that the representation of meaning within an utterance may be imagistic and lexical at one and the same time (McNeill 1987).

These points may be illustrated by some examples. Consider the following passage, from a recording of a telling of Little Red Riding Hood.[1] The speaker is recounting how the hunter rescued Little Red Riding Hood's grandmother from the Big Bad Wolf, who had just eaten her. The speaker says: "and he took his hatchet and with a mighty sweep, sliced the wolf's stomach open." The relationship between gesticulation and speech in this passage is as follows:

```
        and took his hatchet and with a mighty sweep sliced the wolf's stomach open
|~~~~|>>>>>>>>>|^^^^^^^^^^^^^^^^^^|===========|*******|==================|-----|
  [1]    [2]          [3]             [4]        [5]          [6]            [7]
```

As the speaker says "sliced" she brings both her hands down from an upraised position, sweeping them to her left as she does so, in what is clearly an enactment of the hunter's action [5]. However, to do this, the speaker had to have her hands positioned in advance of the word "sliced". She organizes her hands into an upraised position as she says "and he took his hatchet and with" ([2] and [3]), actually beginning this process before she begins to speak at all [1], and she holds her hands upraised while she says "a mighty sweep" [4].

The organization of the gesture phrase somewhat in advance of the speech with which it is associated observed here, is highly characteristic. It has been described in many examples by McNeill (1987), Schegloff (1984) and Moerman (1990), as well as by myself (Kendon 1972, 1980). It means that speech-concurrent gestures are planned prior to the spoken phrase with which they are concurrent. In the example given, for instance, it will be clear that the enactment of the hunter's action and the account given in speech must have been organized simultaneously. An image of the hunter's action must have been available before the speaker embarked on her speech. Note, further, that the speaker, having raised her arms in preparation for the slicing movement, does not perform it at once, but waits until she has reached the word "sliced". The placement of the stroke of the gesture phrase (in this case the "slicing" enactment) in relation to the verbal representation of the action described is thus an organized placement and must itself be seen as part of the utterance plan.

In the example just discussed, the speaker adjusted the performance of her gesture to fit with the structure of the concurrent speech phrase. In other instances we may observe the reverse of this: the speaker adjusts the production of speech to the performance of gesture.

For example, in a psychiatric case discussion film,[2] one of the participants, a social worker, is describing the manner of speech of the patient, in her first telephone encounter with her. The social worker says: "She talked very rapidly and this was all coming out very spontaneously." The relationship between gesture and speech in this utterance is as follows.

```
        She talked very rapidly and this was          all coming out very spontaneously
|~~~~|@@@@@@@@@@|^^^^^^^|~~~~~~~|>>>>>|---------------|
   [1]          [2]                [3]      [4]    [5]     [6]
```

Over "she talked very rapidly" the speaker, with forearm upraised, engages in an outward rotatory movement with her right hand [2]. This movement begins exactly as she begins to speak. Once again, we find that the hand is readied for the gesture before the speech begins [1]. Over "and this" she moves her hand forward in a short linear movement [3], evidently a kinesic deictic enacted in association with the spoken deictic. Then she moves her hand in a rapid outward sweep, completing this action just as she completes the word "all" [5].

In order to accomplish this outward sweep, however, the speaker first must have her hand close to her body. Thus time has to be found, between "this", when her hand is pointing forward, and before she says "all", for her hand to be moved from the outward pointing position to the starting position for the outward sweep. It is found between "was" and "all", for there is a pause in the speech here, during which the speaker moves her hand to the position from which it starts its outward sweep [4]. Evidently, the small pause in the speech here (which lasts 0.5 seconds) does not reflect an interference in the internal speech planning processes – as some pauses certainly do. Rather, it is a pause which permits the gesture to catch up with the speech.

Examples such as these show that the movements of gesticulation are organized at the same time as speech within a given utterance and that they and speech together are deployed in a unitary program of execution, adjusting one to the other. If we accept that the gesticular movements are representations in imagistic form of aspects of the content of what is being expressed (McNeill 1987, Calbris 1990), then we must agree that imagistic representation and lexico-syntactic representation of the same content can occur together.

But how does gesticulation express aspects of utterance content? Consider the following example taken from a video-tape of two couples eating dinner.[3] The conversation turns to the experiences of one of those present who has recently given up smoking. The speaker is explaining how he still has the habit of reaching for a cigarette, and that he sometimes does this without realizing it. Thus he says:

```
G: Jus like a reflex thing like las night we were watching some video tape
   +++++++++++++++++++++++++++++++++++++++++++++++++++++++++++++++
                               [1]
```

[....] of what uh [....] of whatever en I noticed at one point that my hand
+++|>>>>>|
[2]

jus reached for my pocket an I was not aware of it. at the time.
|~~~|*********|_____
[3]   [4]   [5]

It was really a kind of revealing thing as to its nature.

_____

As G says "reached" he moves his hand across to the front pocket of his shirt as he would if he were reaching for a cigarette [4]. During the first part of this utterance, he is spooning food from a bowl onto his plate [1]. In order to do the "reach" gesture he must first stop doing this. He releases the spoon into the bowl as he says "hand" [2], moves his hand up so that it is ready to descend to his pocket [3], and then lowers it to his pocket [4], where he lets it remain for a considerable period of time [5].

Here we may note that whereas the speaker says "reached for my pocket", which leaves unspecified the pocket he reaches for, by moving his hand to his left front shirt pocket we are shown *which* pocket he reached for. Further, in the manner in which he does this he conveys the automatic character of the action. He moves his hand into his front shirt pocket, but then he leaves it there, and he leaves it there not only throughout the rest of his speech, but even beyond the end of his utterance. By his own account, his hand had become an object with its own agenda, so to speak, and he demonstrates this in gesture by *dissociating* it from the speech with which its movement was, at first, coordinated. As he does this he says "I was not aware of it at the time." The objectification of his hand that he displays in his gesture is an imagistic representation of the experience of being unaware of the hand's action. He makes concrete and specific the report of his lack of awareness.

Consider also the first example. In this instance it would seem that the speaker produced a kinetic image of "slicing" which, as we might say, "illustrates" the action she describes in words. In providing an illustration in this manner, however, we may see that information beyond that contained in the words in the utterance is thereby provided. The speaker provides details in her gesture of just how the hunter did what he did, which are not given in her speech. It will thus be seen, first, that whereas the words "he took his hatchet" by themselves only tell us that in some way the hunter came to be in some kind of close association with his hatchet as a result of voluntary action on his part ("took" could mean "took the hatchet along" – e.g., by putting it in the back of his car or putting it in his belt), by adding the gestures, the speaker shows us in precise detail in just what way he "took" it. Second, with the action that follows, she shows us just what sort of an action it was that served to slice open the wolf.

The additional meaning provided by the "iconic" gesticulations described in the first and third examples arises as a consequence of the unavoidably specific character of any enactment. Although the communicative virtues of such specificity are not explicitly addressed by the speakers in these examples, we can also find examples where they are. For example, in the dinner party video-tape, a little later in the discussion of giving up smoking, another speaker, Y, is describing the behavior of someone she knows who tried to stop. She says:

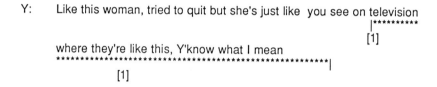

Y: Like this woman, tried to quit but she's just like you see on television
                                                                    |*********
                                                                     [1]

       where they're like this, Y'know what I mean
       ***|
                   [1]

As she begins to say "television" she moves both her hands forward, fingers spread, and moves them rapidly forward and back several times in a way that presents an image of someone in a frantic state of need [1]. Note how she says "where they're like this" – the referent of the demonstrative "this" is the kinesic demonstration which she provides at the same time.

The first and third examples discussed, and the one just presented, illustrate how, in gesticulation, a speaker may provide by manual action an image of something he or she is speaking of. In the three cases given, the images provided have been images of action or behavior of some sort, but images of the shapes of objects, spatial relations between objects and paths of movement through space can also be provided. The hands can assume shapes depictive of objects referred to, they can even be deployed as if they are objects and moved about and referred to as if they are independent of the speaker (cf. Streeck 1988). In such gesticulation, which McNeill (1987) has termed "iconic", we see how concrete representations of actions and objects talked about can be provided. As we have seen, when this is done, the speaker may, through such gesticulation, provide additional, more precise meanings and, as in the last example, such provision of additional meaning may be made an explicit component of the utterance.

However, gesticulation can also serve to present concrete images as metaphors for what is being referred to in the utterance (McNeill 1987, Calbris 1990). This is illustrated in the example from the psychiatric case discussion film, where the rapid circular movements of the speaker's wrist over "spoke very rapidly", and the outward sweep of the hand that just precedes "all coming out", provide images of rapid, ongoing action and of something extensive flowing outward. The forward movement of the hand over "this" is an example of a deictic movement, here pointing as if to the discourse of the patient which, evidently, is here being treated as a concrete object.

Now it is obvious that, if gesticulations encode some part of the referential content of the utterance, they do so in a way that is quite different from the way this is done in words. To represent content in words, one uses items which belong to a repertoire of standard forms, shared by others, and which refer to meaning units of great generality. For specific meanings to be represented, these standardized items must be organized in combinations in accordance with rules of syntax. The resulting units of meaning may be referred to as *constructed units*. In the gesticulation of the sort just illustrated, in contrast, there are no standardized items of general reference. A single configuration of action is organized which, though it develops in time as a phrase of movement, serves as a holistic depiction, like a picture or an enactment, presented in a single moment of time. This configuration of action is not built from recombinable elements according to rules of syntax. It is, thus, quite different from a linguistic construction. The unit of meaning in this case may be referred to as a *presentational unit*. Thus the gestural unit of the first example displays not only the meaning of "slice open" but also the mightiness of the action. It does this through the whole way in which the action is done. We cannot identify separate elements in the gesture that separately refer to the action and its manner of performance, as we can in the English expression it accompanies.

This apparently radical difference of meaning representation notwithstanding, presentational units of meaning and constructed units of meaning are produced simultaneously in utterance in a completely integrated fashion. This means that we cannot say that meaning representation in speakers is only *linguistic* or that linguistic representation is primary. That is to say, image-like representations cannot be said to derive from lexico-syntactic representations.

McNeill (1979b) has proposed that meaning is represented in the first place in the form of sensory–motor schemata. As soon as the individual is able to re-evoke a sensory experience in memory and as soon as the individual can execute a plan of action without actually carrying it out, he can be said to be operating in terms of sensory–motor schemata. Such schemata, by virtue of their status as virtual experiences and action sequences, are forms of representation. The ability of the individual to engage in patterns of movement that represent aspects of the meanings of his utterances, as manifested in gesticulation, is evidence that the meanings that are encoded and represented in utterances exist in the form of virtual experiences and action patterns.

Much evidence suggests that mental operations do not require linguistic representations and often do not involve them. Randhawa and Coffman (1978) and Kosslyn (1983) summarize numerous studies which demonstrate how thinking often appears as a process of internal manipulation of visual representations of perceptions and actions. The construction of overt models, as in architecture and engineering, and the employment of simulated actions in problem solving, provide evidence for the idea that mental activity, such as thinking, is far from being exclusively a linguistic process. As the evidence from the study of gesticulation makes clear, the expression of thought in utterance within interaction, which must be considered the primary locus of linguistic function, also is not exclusively linguistic.

Quotable gestures

Although it is subject to cultural regulation and the forms it assumes are not wholly idiosyncratic, gesticulation is not codified. There are no forms that are explicitly standardized. There is some evidence that different speakers may use similar gesticulatory forms when speaking about the same material (cf. Wiener *et al.* 1972, Calbris 1990) but these similarities appear to arise because of commonalities in strategies of graphic and pantomimic expression rather than from shared conventions. Nevertheless, in all speaker communities repertoires of codified gestures are to be found. These may be used within the context of spoken utterances, but they can also be used independently of speech, serving as kinesic utterances in their own right. Gestures of this sort Ekman (Ekman & Friesen 1969) termed "emblems", adapting Efron's (1972) term "emblematic gesture". Unlike gesticulations, these gestures have stable forms which are recognized as "correct", they may be cited out of context and within their community of use, when cited, they are usually given relatively similar glosses. Lists of such gestures for various parts of the world have been published, including Italy (Efron 1972, Munari 1963), France (Wylie 1977), Columbia and the United States (Saitz and Cervenka 1972), Spain (Green 1968), Iran (Sparhawk 1978), the Arab world (Barakat 1973), and East Africa (Creider 1977).

Despite these publications, surprisingly little is known about such gestures. The available lists rarely contain any information on how they were collected and how the glosses provided for them were arrived at (Sparhawk 1978 is a notable exception) and there are almost no studies of the contexts in which such gestures are employed. Their most salient characteristic, which sets them off clearly from gesticulations, is that they are *quotable*. That is, people can recall them when asked to do so and they can usually supply a gloss for them.

Comparative study of these glosses suggests that these gestures are used as the equivalent of complete speech acts (Kendon 1981), but that whereas some of them have relatively fixed meanings, functioning rather like cliches or interjections, others can function much more as if they are words, and can be employed to convey a range of different meanings, depending on context (Poggi 1981, 1983, Poggi & Zomparelli 1987).

Cultural groups differ considerably in the extent to which use is made of such gestures. In Europe, Neapolitans and Sicilians are known to have large vocabularies (De Jorio 1832, Pitrè 1877, Efron 1972), whereas in the more northerly parts, such as Holland and Scandinavia, these are much smaller (Morris *et al.* 1979). There are also differences in the use of gesticulation in these cultures. Efron (1972) showed that Southern Italians make much more use of pictorial or other kinds of illustrative gestures while speaking, as well as quotable gestures, than East European Jewish Yiddish speakers – who also were observed to gesture a great deal in conversation, but in quite a different way (both groups were studied in immigrant communities in New York City). In addition, two experimental studies suggest that Italians derive more information from speech-concurrent gesture than British and British-Australian people do (Graham and Argyle 1975, Walker and Nazmi 1979). If this is so, perhaps gesture also tends to be used more often as a complete alternative to speech, and this is why there is a larger vocabulary of quotable gestures among Italians.

Sign languages and quotable gestures

Where the use of speech in interaction is limited, or even impossible, on a routine basis, gestures may be used in place of speech and may assume properties similar in many respects to spoken language. A comparison between the gesture systems that have arisen in place of speech suggests that, the wider the range of circumstances of interaction in which speech cannot be used, the more elaborate the gestural system is found to be. Thus, in communities of deaf, where direct access to spoken language is impossible, gesture systems may arise which are as elaborated as spoken language and structurally fully analogous to it (Klima and Bellugi 1979, Kyle & Woll 1985, Wilbur 1987). Where, for ritual reasons, extended restrictions on the use of speech in all circumstances of everyday interaction are maintained, as among women of the Warlpiri and other groups of central Australia when in mourning, the gesture system that develops is also very elaborate, capable of serving all of the functions of speech, when necessary (Kendon 1988). Where speech use is prohibited or impossible in certain circumstances only, or only for certain periods each day, as in certain European monastic orders (Sebeok and Sebeok 1987, Kendon 1990) or in work environments such as sawmills (Meissner and Philpott 1975), the gesture systems employed have smaller vocabularies and are used for relatively more restricted sets of messages.

Nevertheless, in all of these circumstances the gestures employed in these systems, whether highly elaborate or not, develop certain properties analogous to those of structural units in spoken languages. They are, thus, different from the gestures of gesticulation and are commonly referred to as *signs*, instead. First, like quotable gestures or emblems, signs have standard or correct forms. If you vary the performance of a sign you may end up with a completely different meaning. For example, in Warlpiri sign language, to bring a flat hand into contact with the center of the upper chest is to sign "water". Should you move the hand laterally as you do so, however, you sign "man". To sign "crazy" you move a flat hand toward the ear. If you did this using a single index finger, however, you would either be signing "ear" or, possibly, depending upon the orientation of the hand, "understand" or "clever", the opposite of "crazy". To sign "ground" you lower a fist with palm upwards to make contact with the palm of the left hand. If you should do this with knuckles, you sign the verb "to press upon". If you hit the palm of the hand twice, with the back of the fisted right hand, you sign the color "brown" rather than "ground".

As might be gathered from these examples, there is a systematic character to the way signs contrast with one another. Indeed it is possible to show that, in a sign language, signs may be thought of as combinations of a limited number of contrastive features. Much as in spoken language, where words, or rather morphemes, are analyzable as combinations of contrastive sounds, or phonemes, so it can be shown that signs contrast in terms of handshapes, location of performance in relation to the body, and movement pattern (Stokoe 1978). That is, they are precisely specified in terms of their various features and a given sign language may be shown to employ a limited set of handshapes, movement patterns and locations of performance.

Much the same can be said for the gestural units in vocabularies of quotable gestures. Sparhawk (1978), for example, was able to show that the vocabulary of Iranian emblematic gestures she studied could be described in terms of formational properties analogous to those established in sign languages. There are also other ways in which emblems or quotable gestures are like the gestural units of sign languages. They are mostly one-handed and in sign vocabularies a substantial proportion of signs are one-handed. Among quotable gestures that are two-handed, a significant number are asymmetrical. That is, the left hand serves as a base or locus for the articulation of the other hand. This may be seen, for instance, in the *invito* gesture described by Munari (1963) for Italy, in which the tip of the index finger of the right hand is brought into contact with the palm of the left hand, held so it is oriented downward. Such asymmetry is also common for two-handed gestural units in sign languages. Finally, a high proportion of both signs and quotable gestures involve articulation in relation to some part of the head or face – against the ear, the eye, the cheek, and so on. Thus gestures, whenever they become codified, come to have certain formational characteristics. This is true whether they are items in a vocabulary of emblematic gestures or lexical units in a sign language.

Some further properties of sign languages

When gestures become codified, because they are stable in form they become, like words, capable of entering into combinations with one another, and in this way new codifed gestures may be created. This can be observed in sign languages, where new signs may be created from sign compounds in ways that are analogous to the creation of new words in spoken languages, by word (or part-word, i.e. morpheme) combinations. For example, in American sign language, to sign "home" you move a "tapered" hand first to touch the mouth, then to touch the cheek. This is in fact derived from a combination of two signs: the sign for "eat", in which you touch the mouth with a "tapered" hand, and the sign for "sleep," in which you place the palm of the hand against the side of the face. Notice that, in forming this sign, the handshape used in the first part of the sign has persisted into the second part, but the two distinct locations are retained. In this case, the two signs of the compound have tended to fuse together, and this is quite characteristic for compound signs in American Sign Language (Klima and Bellugi 1979).

In the Australian Aboriginal sign languages I have studied, compound signs also are found, but these show much less of a tendency to fuse. This is because, in these sign languages, signs tend to serve as representations of the morphemes of the spoken language. Signs for things which are referred to by a compound form in the spoken language tend to have a corresponding compound form in the sign language. Thus, in Warumungu, the sign for "scorpion" is a combination of the sign for "mouth" followed by the sign for "crab" and this corresponds directly to the spoken language word for "scorpion" which is *jalangartarta*, a compound of *jala* "mouth" and *ngartarta* "crab".

In a primary sign language, such as American Sign Language, movement and space are important both for morphology and syntax. For example, it has been found that a given handshape–location–movement pattern combination may be varied in the amplitude or speed with which it is performed, but that such variations function systematically in ways that are fully analogous to inflections in spoken languages. For example, in ASL there are many hand-shape–location–movement pattern combinations that function either as a noun or as a verb, according to whether the movement that the combination requires is restricted or extended in amplitude. Thus to sign the noun "chair" one lowers the extended index and second fingers of the dominant hand to make contact with the back of the similarly extended fingers of the other hand. Should this be done with an extended amplitude, however, one signs the verb "sit", instead (Supalla and Newport 1978). Likewise, the meaning of adjectival signs is intensified by increasing the speed with which the sign is done (Klima and Bellugi, 1979). This means that sometimes the way in which intensification in signing is achieved appears to go against what one might expect. Thus if one is to sign "very slow", one actually performs the sign for "slow" (drawing the finger of one hand along the back of the other) *more quickly*. Since this speed increase applies quite systematically as an intensifying device it may be abstracted as an intensifier morpheme. In the same manner, one may abstract contrast in amplitude as a morpheme that determines the grammatical status of a sign. If we take this view, we may say that ASL is a highly inflected language, and that, further, it is highly "fusional" in the sense that morphemes may overlay one another in a sign so that, according to the morphemes that are present, the performance of a sign can vary considerably.

Space is widely employed as a means of expressing grammatical relation-ships. In ASL there are many verb signs in which the movement pattern follows a definite path. For many of these the arguments of the verb (subject and object) may be directly incorporated into the verb sign in virtue of the direction of movement (Fischer and Gough 1978). A simple example would be if I were to sign "I see you." Instead of signing "I", "see" and "you" as a sequence of separate signs, it is possible for me simply to move the hand, shaped for the sign "see" in a direction that implies travelling from me to you. Prepositional relations are also expressed by means of space. To cite an example from Liddell (1980), consider a signed rendition of "A cat is sitting on the fence." Here the relationship between the cat and the fence is displayed directly as a spatial relationship. Furthermore, the act of "sitting" is displayed through a modification of a pronominal form that stands for "cat". Thus, in signing this, the signer first signs "fence" and then signs "cat". While "cat" is being signed, the left hand remains in position, now serving as a pronominal form (a so-called "classifier") for "fence". Following the signing of "cat", the right hand now assumes a V shape, which is a pronominal form, or "classifier", meaning "legged creature' which, because the sign "cat" has just preceded it, is now understood to mean "cat". The two fingers of the V hand are bent, however, and this means now "seated legged creature" and the hand, in this form, is brought and placed upon the left hand, thus showing the relationship between "cat" and "fence". Note that any other sort of relationship could be displayed in the same way. "Beside the fence" could have the "seated creature" hand placed to one side of the "fence" hand, "under the fence" one would place the "seated creature" hand under the "fence" hand, and so on.

From pictures to words

I have stressed the differences between gesticulation, quotable gestures and signs. Now I wish to draw attention to some similarities. First of all, consider again the "cat on fence" example. In the ASL rendition of this the way space is used to show how the cat and fence are related is really very "picture-like". This kind of picture building is actually very widespread in primary sign languages (DeMatteo 1977, Kyle 1983). However, the way in which it is done becomes highly simplified and standardized to the point where we have to say that signers are not really making pictures but are constructing sentences according to syntactic rules (cf. Newport 1982). However, the pictorial origin of these constructions seems obvious.

Many individual signs, likewise, can be seen to have a pictorial origin. Thus, even in Warlpiri sign language, where the spoken language has such a pervasive influence that picture-making as a sentence construction strategy is little exploited, the individual signs themselves can often be seen to have originated as a rendition of a concrete object or pattern of action, taken to stand for the concept symbolized by the sign. For example, the sign for "man" mentioned above almost certainly derives from a movement that serves to sketch the horizontal cicatrices which are cut on a man's chest as an indication that he is fully initiated. The sign for "crazy", in which a flat hand is moved forward and back beside the ear, can be understood as a gestural representation of a blocked ear. In central Australian culture the ear is the channel of understanding and persons who behave stupidly or crazily are said to have blocked ears. So also, a pointing finger, moved horizontally toward the ear, serves a sign for "wise", "understanding", etc., because what we now have is a gestural rendition of something going into the ear.

Many quotable gestures, like many signs, also often display a motivated relationship to their meaning of this sort. The "drink" gesture used in Italy, in which an extended thumb is directed toward the mouth, the other fingers being closed in a fist, clearly derives from modelling the shape of a bottle combined with an enactment of tipping it back to drain it. The gesture in which one pulls at one's collar to widen it slightly, which is glossed to mean either "that man is a fool" or "I am not a fool" (depending upon how it is combined with gaze direction, head movement and facial expression) is said to be a representation of the idea of widening one's throat to make room for something big to be swallowed, a metaphor for gullability (De Jorio 1832). It seems, in short, that for many such gestures we can recognize that they have been derived from a gestural representation of an image of some object or action. One has the strong impression, thus, that wherever gesture is used on its own, the forms employed started out as representations of images. These forms then undergo a transformation such that they end up no longer as image representations but as symbols of their referents. Thus pictorial gestures can, in certain circumstances, become transformed into elements that function as if they are words.

Finally, in regard to gesticulation, we saw that this often involves a frankly pictorial use of gesture, the "pictures" serving either as concrete representations of the things referred to in the speech or as concrete metaphors for them. As we saw, such gesticulations present, in imagistic form, parts of the intrinsic content of the utterance which have not been verbally articulated. Often this is implicit in the way the gesticulation and speech are combined. Sometimes, however, as we saw from one of the examples presented, the speaker may make explicit reference to this. In either case it seems clear that an imagistic representation of an "idea unit" is present at the same time as, or even somewhat in advance of, the speaker's formulation of a spoken representation of it.

Although gestural representation of meaning which is holistic and imagistic is sometimes employed apart from speech, if this is done repeatedly the forms employed tend to become codified and, as we have seen, this process follows certain principles with the result that codified gestures tend to share a number of formational properties. Codification also results in a shift in semiotic status. Codified gestures cease to function as pictures or enactments and become lexical symbols.

Let us look at this aspect of the process in a little more detail. One way in which one can attempt to convey a referential meaning to another is to show the other an example of what one means. This may be done either by presenting the object in question or by demonstrating the action that is meant. This may be called *reference by actual ostention.* However, if the object referred to is not to hand, or if the requisite circumstances for the action are not present, one may nevertheless attempt to create an image of the object or action in question. One might draw a picture of it. One might sketch it, model it by bodily posturing, imitate its sound (if it has one), or pantomime it. Here one engages in what may be called *reference by virtual ostention.* In "iconic" gesticulation, as we have suggested, this is, in effect, what speakers are doing, and in the "addicted smoker" example that we gave, this becomes quite explicit.

Where *only* gesture is available for communication, however, then the image that is presented may become established as the device by which the referent is referred to. Something that was created first as an image in lieu of the real thing, now becomes a symbol for the concept of that thing. We may see this in the process by which deaf people may develop signs. For example, Carolyn Scroggs (1981) describes the communication strategies of a nine-year-old deaf boy who had had no training in sign language. She describes how he told a story involving a motorcycle. When "motorcycle" was first introduced into his discourse he gave an elaborate pantomime of mounting the cycle, starting it and revving it up, using hand motions to indicate the twisting of the throttle on the handlebar. In subsequent references to the motorcycle, however, just this hand motion was used. Thus a hand action derived from a pantomime of twisting the throttle came to serve as a symbol for the concept of "motorcycle". In other words, the boy first created *representations* in gesture of the things he wished to refer to, and he then used *elements* from these representations as signs for these things.

In this case, apparently, these signs were not stable forms, and they were not shared by others, but it is easy to see that they could very quickly have become so. Another investigator, Tervoort (1961), also working with deaf children, has described how this may happen. He also describes the process by which an element from an elaborate pantomimic form comes to be used as a sign for the referent. He further describes, however, how, with repeated use, such elements become stripped of their complexities until all that is left of the gesture is a handshape and movement that serves only as something to contrast the gesture with other gestures in the system. Thus a complex depiction of a Bishop's miter (as a way of referring to Santa Claus, who is a Bishop in Holland) became an index finger pointed upward at the side of the head. An elaborate modelling of the facial appearance of a new teacher as a way of referring to her, became simply the point of a finger to the cheek. Similar instances of the simplification of complex depictions or pantomime, as signs for new concepts are developed, have also been described by Klima and Bellugi (1979). Very similar processes have been described for hearing speakers who, in an experimental situation, are asked to use gestures only for communication, for example to tell stories (Bloom 1979, as mentioned by McNeill 1987).

A consideration of the changes that complex depictions undergo as they become signs has suggested that certain quite regular processes are involved, as we have already indicated. Two-handed pantomimes tend to become one-handed, there is a tendency for movement patterns to become simplified to one- or two-phrase movements and there is a tendency for the hand movements to be performed within a relatively restricted, centralized space, typically immediately in front of the person. Furthermore, there is a tendency for features of the gesture to be retained only if they remain in contrast to features of other gestures within the sign system.

Three things are to be noticed from these accounts. First, as we have seen, an element from an elaborate pantomime gets selected for repeated use and comes to stand for the whole concept. Second, as a result of economy of action, the element becomes simplified to such a degree that its image-like or iconic character is no longer apparent. It turns into an arbitrary form as it comes to be shaped by the requirement that it be a distinctive form within a system of other forms. Third, however, as it undergoes this process, it is freed of the require-ment that it be a "picture" of something and it thus becomes free, also, to take on a general meaning. It thus becomes available for recombination with other forms and so may come to participate in compound signs or sentences.

Thus elaborate presentations of images, which serve as stand-ins for what the person wishes to refer to, can quite rapidly become transformed into highly simplified forms that serve as symbols for referents. They thereby come to be transformed into words and, like words, they can be combined in sequences to create units of meaning through discourse – constructed meanings, that is, rather than the presentational units of pictorial gesturing. The visual represen-tational character of the kinesic medium is transcended and a system of symbols that can operate in a quite abstract way becomes established.

The word-like character of signs and the picture-like character of gesticula-tions notwithstanding, we argue that there is a continuity between them. The picture making we see in gesticulation, we maintain, is a manifestation of the way meanings are represented before fully articulate symbolic vehicles are available. When used with speech, gesture can be brought in to provide image-like expression to those aspects of the meaning of an utterance that cannot be provided through the linear formulation of speech. When used without speech, gestures are again pictorial, but very quickly their character shifts and they become transformed into lexical symbols and serve as components of *con-structed* units of meaning.

Conclusion

In the foregoing I have surveyed some of the more salient features of gesture use as these appear in the light of recent research. Three points deserve emphasis.

1.   As the discussion of gesticulation showed, in the production of utterance, representation of meaning through pointing, depiction and enactment is as

primary a process as is its representation through spoken linguistic forms. When speech is used, speech and gesture are often employed in a complementary fashion, the meaning representation being apportioned between the two modes in ways that suggest that neither derives from the other.

2.   In considering examples where gestures are employed as the sole vehicle of utterance, we can show how they tend to become codified: they have a standard form and stable meanings. They are no longer illustrations but texts in their own right. In the study of how coded gestures become established, whether as elements of a sign language or not, we can observe how representations that are not unlike those observed in gesticulation come to be stabilized and transformed. This supports the view that actions that serve referentially in communication all ultimately derive from iconic representations and that arbitrariness can be shown to be the product of an historical process.

3.   The extent to which gesture exhibits systemic, code-like properties that are analogous to spoken language expands in proportion to the range of communicative demands it is employed to meet. Where gesture is employed as the sole means of utterance in all circumstances, as in its use by the deaf, it may come to have all of the features of a fully articulate language. This supports the view that the properties of a code that we recognize as "linguistic" are properties which emerge in response to communicative function.

The facts about human gesture which we have reviewed are relevant for theories of language origins because they support the view that there is continuity between language and other modes of expression. They do this in two ways. On the one hand, the study of how speaking humans use gesture in relation to speech supports the view that linguistic productions can be looked upon as involving derivations from specialized elaborations of much more general processes of representation and enactment. It is here that we can envisage a connexion between tool use and language. On the other hand, from the study of what happens to gesture when it is used on its own, apart from speech, we are able to gain insight into the way in which general principles of economy of effort combined with requirements for disambiguation can operate to produce structural features which are usually regarded as central to any notion of a "language."

Notes

1 From a video-tape recording for classwork at Connecticut College by Scott Pollack.
2 Van Vlack, J., Birdwhistell, R.L. & Hoffman, C.L. *Teaching Psychotherapy, V.* ISP00161, 16 mm film produced by Eastern Pennsylvania Psychiatric Institute, Philadelphia, 1961.
3 Video recording known as *Chinese Dinner*. I am indebted to Charles Goodwin for letting me use this material.

References

Barakat, R. A. (1973). Arabic gestures. *Journal of Popular Culture*, 6, 749–92.

Birdwhistell, R.L. (1970). *Kinesics and Context*. Philadelphia: University of Pennsylvania Press.

Bloom, R. (1979). Language Creation in the Manual Modality: A Preliminary Investigation. Unpublished Paper, Department of Behavioral Sciences, University of Chicago.

Calbris, G. (1990). *Semiotics of French Gesture*. Bloomington: Indiana University Press.

Chovil, N. (1989). Communicative Functions of Facial Displays in Conversation. Unpublished Ph. D. thesis, University of British Columbia.

Condon, W. C. & Ogston, W. D. (1966). Soundfilm analysis of normal and pathological behavior patterns. *Journal of Nervous and Mental Disease*, 143, 338–47.

Creider, C. (1977). Toward a description of East African gestures. *Sign Language Studies*, 14, 1–20.

Crystal, D. & Davy, D. (1969). *Investigating English Style*. Bloomington, Indiana: Indiana University Press.

De Jorio, A. (1832). *La Mimica degli antichi investigata nel gestire napoletano*. Naples: Fibreno.

DeMatteo, A. (1977). Visual imagery and visual analogues in American Sign Language. In *On the Other Hand: New Perspectives on American Sign Language*, ed. L. Friedman, pp. 109–36. New York: Academic Press.

Dittman, A. T. (1972). The body movement–speech rhythm relationship as a cue to speech encoding. In *Studies in Dyadic Communication*, ed. A. Seigman & B. Pope, pp. 135–52. Elmsford, New York: Pergamon Press.

Efron, D. (1972). *Gesture, Race and Culture* [=reprint from edition of 1940, with additional material] The Hague: Mouton & Co.

Ekman, P. (1979). About brows. In *Human Ethology: Claims and Limits of a New Discipline*, ed. M. Von Cranach, K. Foppa, W. Lepenies & D. Ploog, pp. 169–202. Cambridge: Cambridge University Press.

Ekman, P. & Friesen, W. (1969). The repertoire of non-verbal behavior: Categories, origins, usage and coding. *Semiotica*, 1, 49–98.

Fischer, S. & Gough, B. (1978). Verbs in American Sign Language. *Sign Language Studies*, 18, 17–48.

Freedman, N. (1972). The analysis of movement behavior during clinical interviews. In *Studies in Dyadic Communication*, ed. A. Siegman & B. Pope, pp. 153–75. Elmsford, NY: Pergamon Press.

Graham, J. A. & Argyle, M. A. (1975). A cross cultural study of the communication of extra verbal meaning by gestures. *International Journal of Psychology*, 10, 56–67.

Green, J.R. (1968). *A Gesture Inventory for the Teaching of Spanish*. Philadelphia: Chilton Books.

Hadar, U., Steiner, T. J., Grant, E. C. & Rose, F. C. (1983). Head movement correlates of juncture and stress at sentence level. *Language and Speech*, 26, 117–29.

Hadar, U., Steiner, T. J., Grant, E. C. & Rose, F. C. (1984). The timing of shifts of head posture during conversation. *Human Movement Science*, 3, 237–45.

Kendon, A. (1972). Some relationships between gesture and speech: an analysis of an example. In *Studies in Dyadic Communication*, ed. A. Siegman & B. Pope, pp. 177–210. Elmsford, NY: Pergamon Press.

Kendon, A. (1980). Gesticulation and speech: two aspects of the process of utterance. In *The Relation Between Verbal and Nonverbal Communication,* ed. M.R.Key, pp. 207–27. The Hague: Mouton & Co.

Kendon, A. (1981). Geography of gesture. *Semiotica*, 37, 129–63.

Kendon, A. (1988). *Sign Languages of Aboriginal Australia.* Cambridge: Cambridge University Press.

Kendon, A. (1990). Signs in the cloister and elsewhere. *Semiotica*, 79, 307–29.

Klima, E. & Bellugi, U. (1979). *The Signs of Language*. Cambridge, MA: Harvard University Press.

Kosslyn, S. M. (1983). *Ghosts in the Mind's Machine: Creating and Using Images in the Brain*. New York: W. W. Norton & Co.

Kyle, J.G. (1983). Looking for meaning in sign language sentences. In *Language in Sign*, ed. J.G.Kyle & B.Woll, pp. 184–94. London: Croom Helm.

Kyle, J. G. & Woll, B. (1985). *Sign Language: The Study of Deaf People and Their Language*. Cambridge: Cambridge University Press.

Liddell, S. (1980). *American Sign Language Syntax*. The Hague: Mouton & Co.

McNeill, D. (1979a). *The Conceptual Basis of Language*. Hillsdale, NJ: Erlbaum.

McNeill, D. (1979b). Language origins. In *Human Ethology: Claims and Limits of a New Discipline*, ed. M. Von Cranach, K. Foppa, W. Lepenies & D. Ploog, pp. 715–28. Cambridge: Cambridge University Press.

McNeill, D. (1987). *Psycholinguistics: A New Approach*. New York: Harper & Row.

Meissner, M. & Philpott, S.B. (1975). The sign language of sawmill workers in British Columbia. *Sign Language Studies*, 9, 291–308.

Moerman, M. (1990). Studying gestures in context. In *Culture Embodied*, ed. M. Nomura & M. Moerman, pp. 5–52. Senri Publications.Osaka: National Museum of Ethnology.

Morris, D., Collett, P., Marsh, P. & O'Shaughnessy, M. (1979). *Gestures: Their Origins and Distribution*. London: Jonathan Cape.

Munari, B. (1963). *Supplemento al Dizionario Italiano*. Milan: Muggiani.

Newport, E. (1982). Task specificity in language learning? Evidence from speech perception and American Sign Language. In *Langauge Acquisition: The State of the Art*, ed. E. Wanner & L. R. Gleitman, pp. 450–86. Cambridge: Cambridge University Press.

Pitrè, G. (1877). Gesti ed insegne del popolo Siciliano. *Rivista di letteratura popolare*, 1, 32–43.

Poggi, I. (1981). Le analogie fra gesti e interiezioni: alcune osservazioni preliminari. In *Comunicare nella vita quotidiana,* ed. F. Orletti, pp. 117–33. Bologna: Società Editrice il Mulino.

Poggi, I. (1983). La "mano a borsa": analisi semantica di un gesto emblematico olofrastico. In *Comunicare senza parole.* ed. G. Attili & P. E. Ricci-Bitti, pp. 219–38. Rome: Bulzoni.

Poggi, I. & Zomparelli, M. (1987). Lessico e grammatica nei gesti e nelle parole. In *Le parole nella testa. Guida a un'educazione linguistica cognitivista*, ed. I. Poggi, pp. 291–327. Bologna: Società editrice il Mulino.

Randhawa, B. S. & Coffman, W. E. (1978). *Visual Learning, Thinking and Communication*. New York: Academic Press.

Saitz, R. L. & Cervenka, E. J. (1972). *Handbook of Gestures: Columbia and the United States*. The Hague: Mouton & Co.

Schegloff, E. A. (1984). On some gestures' relation to talk. In *Structures of Social*

Action: Studies in Conversation Analysis, ed. J. M. Atkinson & J. Heritage, pp. 266–96. Cambridge: Cambridge University Press.

Scroggs, C. (1981). The use of gesturing and pantomiming: the language of a nine year old deaf boy. *Sign Language Studies*, 30, 61–77.

Sebeok, D. J. & Sebeok, T. A., eds. (1987). *Monastic Sign Languages.* Berlin: Mouton De Gruyter.

Sparhawk, C.M. (1978). Contrastive-identificational features of Persian gestures. *Semiotica,* 24, 49–86.

Stokoe, W.C. (1978). *Sign Language Structure,* 2nd edn, revised. Silver Spring, Maryland: Linstok Press.

Streeck, J. (1988). The significance of gesture: how it is established. *Papers in Pragmatics*, 21, 25–59.

Supalla, T. & Newport, E. (1978). How many seats in a chair? The derivation of nouns and verbs in American Sign Language. In *Understanding Language through Sign Language Research*, ed. P. Siple, pp. 91–132. New York: Academic Press.

Tervoort, B. T. (1961). Esoteric symbolism in the communication behavior of young deaf children. *American Annals of the Deaf,* 106, 436–80.

Walker, M., & Nazmi, M. K. (1979). Communicating shapes by words and gestures. *Australian Journal of Psychology*, 31, 137–47.

Wiener, M., Devoe, S., Rubinow, S. & Geller, J. (1972). Nonverbal behavior and nonverbal communication. *Psychological Review,* 79, 185–214.

Wilbur, R. (1987). *American Sign Language: Linguistic and Applied Dimensions.* Boston: Little-Brown & Co.

Wylie, L. (1977). *Beaux Gestes: A Guide to French Body Talk.* Cambridge, MA: The Undergraduate Press.

2

When does gesture become language? A study of gesture used as a primary communication system by deaf children of hearing parents

SUSAN GOLDIN-MEADOW

Perhaps the clearest example of the resilience of language comes from the fact that language is not tied to the mouth and ear but can also be processed by the hand and eye. Sign languages of the deaf have been found to take over all of the functions and to assume the structural properties characteristic of spoken languages (Klima & Bellugi, 1979). Moreover, when exposed to a conventional sign language such as American Sign Language, deaf children acquire the language as effortlessly as hearing children acquiring spoken language (Newport & Meier, 1985). Thus, the manual modality can serve as a medium for language, suggesting that the capacity for creating and learning a linguistic system is modality independent.

The manual modality is exploited even by those who use spoken language. Hearing adults and children frequently use gesture along with their speech. However, unlike conventional sign languages, the spontaneous gestures of hearing individuals do not stand on their own and must be interpreted in the context of the speech they accompany (McNeill, 1987). Moreover, although spontaneous gestures may reflect the ideas of the speaker (cf., Church & Goldin-Meadow, 1986; Perry, Church & Goldin-Meadow, 1988), they do so in a form that is distinct from the form assumed by speech and sign (McNeill, 1987). Thus, while the manual modality can assume all of the formal and functional properties of language in the conventional sign languages of the deaf, it does not appear to do so in the spontaneous gestures of hearing speakers.

The purpose of this chapter is to explore one condition under which gesture appears to take on both the form and the function of language. The children who are the focus of my work are deaf with hearing losses so severe that they cannot naturally acquire spoken language. In addition, these children are born to hearing parents who have not yet exposed them to a conventional sign language. Despite their lack of usable linguistic input, either signed or spoken, these deaf children develop gestures which they use to communicate. My colleagues and I have found that these gestures, which comprise the children's sole means of communication, take on many of the formal and functional properties found in the early communication systems of children learning conventional languages. Moreover, the deaf children's gestures are structured in ways that the spontaneous gestures of their hearing parents are not. These observations suggest that gesture will assume language-like properties when used as a primary communication system (but not when used as an adjunct to speech), and that language-like properties can develop in the absence of a conventional language model. I will consider these findings in terms of the light they may shed on the effects (or non-effects) of the environment on language development in an individual child, and on the circumstances compatible with the creation of language-like structure.

1. Background on deafness and language-learning

The sign languages of the deaf are autonomous languages which are not based on the spoken languages of hearing cultures (Klima & Bellugi, 1979). A sign language such as American Sign Language (ASL) is a primary linguistic system passed down from one generation of deaf people to the next and, like spoken language, is structured at syntactic, morphological, and "phonological" levels of analysis.

Deaf children born to deaf parents and exposed from birth to a conventional sign language such as ASL have been found to acquire that language naturally; that is, these children progress through stages in acquiring sign language similar to those of hearing children acquiring a spoken language (Newport & Meier, 1985). Thus, in an appropriate linguistic environment, in this case, a signing environment, deaf children are not handicapped with respect to language learning.

However, 90% of deaf children are not born to deaf parents who could provide early exposure to a conventional sign language. Rather, they are born to hearing parents who, quite naturally, tend to expose their children to speech (Hoffmeister & Wilbur, 1980). Unfortunately, it is extremely uncommon for deaf children with severe to profound hearing losses to acquire the spoken language of their hearing parents naturally, that is, without intensive and specialized instruction. Even with instruction, deaf children's acquisition of speech is markedly delayed when compared either to the acquisition of speech by hearing children of hearing parents, or to the acquisition of sign by deaf children of deaf parents. By age 5 or 6, and despite intensive early training programs, the average profoundly deaf child has only a very reduced oral linguistic capacity (Conrad, 1979).

In addition, unless hearing parents send their deaf children to a school in which sign language is used, these deaf children are not likely to receive conventional sign language input. Under such inopportune circumstances, these deaf children might be expected to fail to communicate at all, or perhaps to communicate only in non-symbolic ways. This turns out not to be the case.

Previous studies of deaf children of hearing parents have shown that these children spontaneously use gestures (referred to as "home signs") to communicate even if they are not exposed to a conventional sign language model (Lenneberg, 1964; Moores, 1974). Given a home environment in which family members communicate with each other through many different channels, one might expect that the deaf child would exploit his accessible modality (the manual modality) for the purposes of communication. However, given that no language model is present in the child's accessible modality, one might not expect that the child's communication would be structured in language-like ways.

My work has focused on the structural aspects of deaf children's gestures and, in particular, has attempted to determine whether any of the linguistic properties found in natural child language can also be found in those gestures. My colleagues and I have analyzed the gestures of ten deaf children of hearing parents, and found that these gestures consistently served many of the functions typical of child language and, in addition, were structured on several levels, as is child language. I will focus here on both the functions of the deaf children's gestures and on three aspects of their structure: lexicon, syntax, and morphology.

The ten children in my sample ranged in age from 1;4 (years;months) to 4;1 at the time of the first interview and from 2;6 to 5;9 at the time of the final interview. The children were videotaped in their homes during play sessions with their hearing parents or an experimenter every 2 to 4 months for as long as each child was available (the number of observation sessions per child ranged from two to 16). Six of the children lived in the Philadelphia area and four in the Chicago area. The children were all born deaf to hearing parents and sustained severe (70–90 dB) to profound (> 90 dB) hearing losses. Even when wearing a hearing aid in each ear, none of the children were able to acquire speech naturally. In addition, none of the children in the sample had been exposed to conventional sign language.

2. Functional uses of gesture in deaf children of hearing parents

All of the children used their gestures as "tools" for communication – to convey information about current, past, and future events, and to manipulate the world around them. Like children learning conventional languages, the deaf children requested objects and actions from others and did so using their gestures; e.g., a pointing gesture at a book, a "give" gesture, and a pointing gesture at the child's own chest, to request mother to give the child a book; or a "hit" gesture followed by a pointing gesture at mother, to request mother to hit a tower of blocks. Moreover, like children learning conventional languages, the deaf children commented on the actions of objects, people, and themselves, both in the past (e.g., a "high" gesture followed by a "fall" gesture to indicate that the block tower was high and then fell to the ground) and in the future (e.g., a pointing gesture at Lisa with a head-shake, an "eat" gesture, a pointing gesture at the child himself, and an "eat" gesture with a nod, to indicate that Lisa would not eat lunch but that the child would). Gestures were also used to recount events which happened some time ago; e.g., one child produced an "away" gesture, a "drive" gesture, a "beard" gesture, a "moustache" gesture, and a "sleep" gesture to comment on the fact that the family had driven away to the airport to bring his uncle (who wears a beard and a moustache) home so that he could sleep over.

Moreover, in addition to the major function of communicating with others, some of the deaf children used their gestures for other functions typically served by language. For example, the children used their gestures when they thought no one was paying attention, as though "talking" to themselves. In addition, one of the children used gesture to refer to his own gestures. For example, to request a Donald Duck toy that the experimenter held behind her back, the child pursed his lips to imitate Donald Duck's bill, then pointed at his own pursed lips and pointed toward the Donald Duck toy. When offered a Mickey Mouse toy, the child shook his head, pursed his lips and pointed at his own pursed lips. The point at the lips is roughly comparable to the words "I say," as in "I say 'Donald Duck bill'." It therefore represents a communicative act in which gesture is used to refer to a particular act of gesturing and, in this sense, is reminiscent of a young hearing child's quoted speech (cf., Miller & Hoogstra, 1989). The deaf child appeared able to distance himself from his own gestures and treat them as objects to be reflected on and referred to, thus exhibiting in his self-styled gesture system the very beginnings of the reflexive capacity that is found in all languages and that underlies much of the power of language (cf., Lucy, 1992).

In sum, the deaf children were able to use their gestures for many of the major functions filled by hearing children's words and deaf children's signs. The next three sections explore the form of the deaf children's gestures, and show that those gestures were structured at different levels as are the words and signs of children learning conventional languages.

3. Lexical structure in the gestures of deaf children of hearing parents

The deaf children produced three types of gestures that differed in form. *Pointing* gestures maintain a constant kinesic form in all contexts and were used predominantly to single out objects, people, places, and the like in the surroundings. In contrast, *characterizing* gestures were stylized pantomimes whose forms varied with the intended meaning of each gesture (e.g., a fist pounded in the air as someone was hammering; two hands flapping in the presence of a pet bird). Finally, *marker* gestures were typically head or hand gestures (e.g., nods and headshakes, one finger held in the air signifying "wait") which are conventionalized in our culture and which the children used as modulators (e.g., to negate, affirm, doubt). Markers are not included in the analyses presented here).

3.1. Pointing gestures

At the outset, it is important to note that pointing gestures and words differ fundamentally in terms of the referential information each conveys. The pointing gesture, unlike a word, serves to direct a communication partner's gaze toward a particular person, place, or thing; thus, the gesture explicitly specifies the location of its referent in a way that a word (even a pro-form such as "this" or "that") never can. The pointing gesture does not, however, specify what the object is, it merely indicates where the object is. That is, the pointing gesture is "location-specific" but not "identity-specific" with respect to its referent. Single words, on the other hand, can be identity-specific (e.g., "lion" and "ball" serve to classify their respective referents into different sets) but not location-specific, unless the word is accompanied by a pointing gesture or other contextual support.

Despite this fundamental difference between pointing gestures and words, the deaf children's pointing gestures were found to function like the object-referring words of hearing children in two respects. First, the referents of the points in the deaf children's gestured sentences encompassed the same range of object categories (in approximately the same distribution) as the referents of nouns in hearing children's spoken sentences (Feldman, Goldin-Meadow & Gleitman, 1978). Secondly, the deaf children combined their pointing gestures with other points and with characterizing gestures; if these points are considered to function like nouns and pronouns, the deaf children's gesture combinations turn out to be structured like the early sentences of children learning conventional languages (see below). Thus, the deaf children's pointing gestures appear to function as part of a linguistic system.

In addition, the deaf children used their pointing gestures in ways that went beyond merely directing gaze toward a particular object. The children primarily used their pointing gestures to refer to real-world objects in the immediate environment (e.g., the child pointed at a jar of bubbles, followed by a "blow" characterizing gesture, to request that the bubbles be blown). However, the children also used their pointing gestures to refer to objects that were not present in the here-and-now, and did so by pointing at a real-world object that was similar to the (absent) object they intended to refer to (e.g., the child pointed at an empty jar of bubbles, followed by a "blow" gesture, to request that the absent, full jar of bubbles be blown). We have examined pointing gestures in detail in one of our deaf subjects, and found that this child could extend his use of points even further beyond the here-and-now by pointing at an arbitrary location in space set up as a place-holder for an absent, intended referent (e.g., the child pointed at a spot on his own gesture – a "round" gesture representing the shape of a Christmas tree ball – to refer to the hook typically found at that spot on Christmas tree ornaments). This child was found to use points to indicate objects in the immediate context when he was first observed at age 2;10; he first used his points to indicate objects that were not present in the here-and-now at age 3;3, and began using points to indicate arbitrary locations set up as place-holders for objects at age 4;10 (Butcher, Mylander & Goldin-Meadow, 1991). Hoffmeister (1978) reports a similar developmental pattern from points at real-world objects, to "semi-real-world" objects, to arbitrary loci, in deaf children who have been exposed to a conventional sign language (ASL) from birth.

3.2. Characterizing gestures

The characterizing gesture is the lexical item the deaf children used to denote actions and attributes. It differs somewhat from the words or signs typically used by young language learners exposed to conventional language models. The form of the deaf children's iconic characterizing gesture captures an aspect of its referent and, in this respect, is distinct from the far less transparent verb and adjective word forms hearing children use to denote actions and attributes. It also differs from the early sign forms of deaf children acquiring ASL, most of which are not iconic (Bonvillian, Orlansky & Novack, 1983) or, if iconic from an adult's point of view, are not recognized as iconic by the child (Schlesinger, 1978). Note, however, that in contrast to their location-specific pointing gestures, the deaf children's characterizing gestures resemble hearing children's words in that the characterizing gesture (via its iconicity) can specify the identity of its referent.

We used the form of the children's gestures as the basis for assigning a lexical meaning to each characterizing gesture. As an example of an action form, one child held a fist near his mouth and made chewing movements to comment on his sister eating snacks; this gesture was assigned the meaning "eat". Another child moved his hand forward in the air to describe the path of a moving toy, and this gesture was assigned the meaning "go". Similarly for attribute forms, one child formed a round shape with his hand to describe a Christmas tree ornament; basing the meaning of the gesture on its form, we assigned the meaning "round" to this gesture.

The characterizing gestures that the deaf children produced showed considerable stability of form throughout our observations; that is, the children tended to use the same form to convey the same meaning over time. For example, 91% of the 170 different forms one child produced over a two-year period were used to convey a consistent meaning throughout that period; conversely, 99% of the 188 different meanings the child conveyed were conveyed by a consistent form. Thus, the child's system appeared to be characterized by standards of form, although those standards were idiosyncratic to him or her and not shared by a community of language users.

4. Syntactic structure in the gestures of deaf children of hearing parents

4.1. Predicate structure

The deaf children in our studies combined their gestures into strings that functioned in a number of respects like the sentences of early child language. First, the children's gesture sentences expressed the semantic relations typically found in early child language (in particular, action and attribute relations), with characterizing gestures representing the predicates and pointing gestures representing the arguments playing different thematic roles in those semantic relations (Goldin-Meadow & Mylander, 1984). For example, one child produced a pointing gesture at a bubble jar (representing the argument playing the patient role) followed by the characterizing gesture "twist" (representing the act predicate) to request that the experimenter twist open the bubble jar. Another child produced a pointing gesture at a train (representing the argument playing the actor role) followed by the characterizing gesture "circle" (representing the act predicate) to comment on the fact that a toy train was circling on the track.

In addition, the predicates in the deaf children's sentences were comparable to the predicates of early child language in having underlying frames or structures composed of one, two, or three arguments. For example, all of the children produced "transfer" or "give" gestures with an inferred predicate structure containing three arguments – the actor, patient, and recipient (e.g., you/sister give duck to her/Susan). The children also produced two types of two-argument predicates: transitive gestures such as "eat" with a predicate structure containing the actor and patient (e.g., you/Susan eat apple), and intransitive gestures such as "go" with a predicate structure containing the actor and recipient (e.g., you/mother go upstairs). Finally, the children produced gestures such as "sleep" or "dance" with a one-argument predicate structure containing only the actor (e.g., you/father sleep).

We attributed these one-, two- and three-argument predicate structures to the deaf children's gestures on the basis of the following evidence (see Goldin-Meadow, 1979, 1985, for further types of evidence for these constructions). We found that each child, at some time during our observations, produced gestures for all of the arguments associated with a particular predicate structure. For example, one child produced the following different two-gesture sentences, all conveying the notion of transfer of an object: "cookie–give" (patient–act), "sister–David" (actor–recipient), "give–David" (act–recipient), "duck–Susan" (patient–recipient). By overtly expressing the actor, patient, and recipient in this predicate context, the child exhibited knowledge that these three arguments are associated with the transfer predicate (although few children ever explicitly gestured all of the semantic elements required for three-argument predicates within a single sentence).

4.2. *Ordering and production probability rules*

The deaf children's gesture sentences were structured on the surface as are the sentences of early child language (Goldin-Meadow & Feldman, 1977; Goldin-Meadow & Mylander, 1984). The sentences the children produced were found to conform to regularities of two types: ordering regularities and production probability regularities. Moreover, the particular structural regularities found in the children's sentences showed considerable consistency across the ten children in the sample.

Ordering regularities were based on the position a gesture for a particular thematic role tended to occupy in a sentence. The children tended to order gestures for patients, acts, and recipients in a consistent way in their two-gesture sentences. The following three ordering patterns were found in many, but not all, of the children's two-gesture sentences: *patient–act* (e.g., the gesture for the patient, cheese, preceded the gesture for the act, eat), *patient–recipient* (e.g., the gesture for the patient, hat, preceded the gesture for the recipient, cowboy's head), and *act–recipient* (e.g., the gesture for the act, move-to, preceded the gesture for the recipient, table).

Production probability regularities were based on the likelihood that a particular thematic role would be gestured in a sentence. If the children were randomly producing gestures for the thematic roles associated with a given predicate, they would, for example, be equally likely to produce a gesture for the patient as for the actor in a sentence about eating. We found, however, that the children were not random in their production of gestures for thematic roles – in fact, they used likelihood of production in such a way as to distinguish among thematic roles. We found, in particular, that all ten of the children were more likely to produce a gesture for the patient, e.g., cheese in a sentence about eating, than to produce a gesture for the actor, mouse. Note that this particular production probability pattern tends to result in two-gesture sentences that preserve the unity of the predicate: i.e., patient + act sentences (akin to object–verb in conventional systems) were more frequent in our deaf children's gestures than (transitive) actor + act sentences (akin to subject–verb in conventional systems).

In addition, nine of the ten children produced gestures for the intransitive actor (e.g., the mouse in a sentence describing a mouse running to his hole) as often as they produced gestures for the patient (e.g., the cheese in a sentence describing a mouse eating cheese), and far more often than they produced gestures for the transitive actor (e.g., the mouse in a sentence describing a mouse eating cheese). This production probability pattern is analogous to the structural case-marking patterns of ergative languages in that the intransitive actor is treated like the patient rather than like the transitive actor (note, however, that in conventional ergative systems it is the transitive actor which is marked, whereas in the deaf children's gesture systems the transitive actor tends to be omitted and, in this sense, could be considered unmarked; cf., Silverstein, 1976). In addition to an ergative-like pattern in production probability, the one child who produced a sufficient number of sentences with transitive actors to allow us to determine a pattern also showed an ergative pattern in the way he ordered his gestures. He tended to produce gestures for the patient and the intransitive actor *before* gestures for the act in his two-gesture sentences, but gestures for the transitive actor *after* gestures for the act. This one child thus treated patients and intransitive actors alike, and distinct from transitive actors, not only with respect to production probability but also with respect to gesture order.

The ergative pattern found in the deaf children's gestures could reflect a bias on the part of the child toward the affected object of an action. In an intransitive sentence such as "you go to the corner", the intransitive actor "you", in some sense, has a double meaning. On the one hand, "you" refers to the goer, the actor, the effector of the going action. On the other hand, the "you" refers to the gone, the patient, the affectee of the going action. At the end of the action, "you" both "have gone" and "are gone", and the decision to emphasize one aspect of the actor's condition over the other is arbitrary. By treating the intransitive actor like the patient, the deaf children appear to be highlighting the affectee properties of the intransitive actor over the effector properties.

4.3. Complex sentences

We determined the boundaries for a string of gestures on the basis of gesture form (using relaxation of the hand as the criterion) and then determined the number of propositions conveyed within that gesture string. We found that all ten of the deaf children in our sample generated complex sentences containing at least two propositions (Goldin-Meadow, 1982). The propositions conjoined in the children's complex sentences often had a temporal relationship to one another; these sentences either described a sequence of events or requested that a sequence of events take place. For example, one child pointed at a tower, produced a "hit" gesture and then a "fall" gesture to comment on the fact that he had hit [act$_1$] the tower and that the tower had fallen [act$_2$]. The children also produced complex sentences conveying propositions which were not ordered in time. For example, one child pointed at Mickey Mouse, produced a "swing" gesture and then a "walk" gesture to comment on the fact that Mickey Mouse both swings [act$_1$] on the trapeze and walks [act$_2$].

5. Morphological structure in the gestures of deaf children of hearing parents

5.1. Derivational morphology

At this point in our studies, we have completed our investigation of morphological structure in the gestures of only one deaf child in our sample (we do, however, have extensive preliminary evidence from two other children suggesting that the gesture systems of these children are also characterized by morphological structure; data from the remaining seven children in our sample have not yet been coded for morphological structure). We found that the corpus of characterizing gestures the child produced over a two-year period (from age 2;10 to 4;10) could be regarded as a system of handshape and motion morphemes (Goldin-Meadow & Mylander, 1991). The gestures were composed of a limited and discrete set of five handshape and nine motion forms,

Table 2.1. *Examples of hand and motion morphemes in the deaf child's gestures*

Motions	Handshapes		
	Fist-hand (handle a small, long object)	O-hand (handle a small object of any length)	C-hand (handle a large object of any length)
Short Arc motion (reposition)	Reposition a small, long object by hand (e.g., scoop utensil)	Reposition a small object of any length by hand (e.g., take out bubble wand)	Reposition a large object of any length by hand (e.g., pick up bubble jar)
Arc To and Fro motion (move to and fro)	Move a small, long object to and fro by hand (e.g., wave balloon string back and forth)	Move a small object of any length to and fro by hand (e.g., move crayon back and forth)	Move a large object of any length to and fro by hand (e.g., shake salt shaker up and down)
Circular motion (move in a circle)	Move a small, long object in a circle by hand (e.g., wave flag pole in circle)	Move a small object of any length in a circle by hand (e.g., turn crank)	Move a large object of any length in a circle by hand (e.g., twist jar lid)

each of which was consistently associated with a distinct meaning and recurred across different gestures. For example, the Fist handshape (meaning "handle a small, long object") combined with a Short Arc motion (meaning "reposition" in place) formed a gesture which meant "reposition a small, long object by hand" (e.g., scoop a spoon at mouth). Table 2.1 presents examples of this same Fist handshape combined with the Short Arc motion and other motions – the Arc To and Fro motion (meaning "move to and fro") and the Circular motion (meaning "move in a circle") – as well as examples of other handshapes – the O-hand (meaning "handle a small object of any length") and the C-hand (meaning "handle a large object of any length") – combined with these three motions. As the table illustrates, the meaning of each gesture is predictable from the meaning of its handshape component and its motion component. Note that the motions in the gestures presented in Table 2.1 all represent transitive actions, with the handshapes of these gestures representing the hand of the actor as it is shaped around the patient. These handshape morphemes are comparable to Handle classifiers in ASL which combine with motions to convey transitive actions (McDonald, 1982).

As in ASL, various handshapes were used not only to represent the handgrip around objects of varying sizes and shapes, but also to represent objects themselves; for example, our deaf child also used the C-hand to mean "a curved object". These object handshape components similarly combined with motion components to create paradigms of meanings; for example, the C-hand, when combined with a Linear motion (meaning "change location"), formed a gesture which meant "a curved object changes location" (e.g., a toy turtle moves forward) and, when combined with an Open and Close motion (meaning "open and/or close"), formed a gesture which meant "a curved object opens and/or closes" (e.g., a bubble expands). As these examples suggest, the object handshapes were typically combined with motions representing intransitive actions, with the handshape representing the size, shape, or semantic class of the actor. These object handshapes are comparable to Semantic-Class and Size-and-Shape classifiers in ASL which combine with motions to create intransitive verbs of motion (Supalla, 1982).

The deaf child in our study, at times, also produced his object handshapes with motions representing transitive predicates; in these gestures, the handshape represented the size, shape, or semantic class of the patient – omitting any representation of the actor entirely. For example, to represent placing a toy cowboy on a horse, the child produced a C-hand with his fingers pointed downward (meaning "a curved object") combined with a Short Arc motion (meaning "reposition"), thereby focusing attention on the curved legs of the cowboy as they are placed around the horse. Gestures of this sort are comparable to Size-and-Shape classifiers in ASL which combine with motions typically to represent instruments of transitive actions (Schick, 1987).

The morphemes in the deaf child's gestures were thus organized into a framework or system of contrasts. When the child generated a gesture to refer to a particular object or action, the form of that gesture was determined not only by the properties of the referent object or action, but also by how that gesture fitted with the other gestures in the lexicon. Thus, the child's gestures appeared to reflect a morphological system, albeit a simple one, akin to the system that characterizes the productive lexicon in ASL.

5.2. Inflectional morphology

Analyses of the deaf child's gestures suggest that the system also has inflectional morphology. In conventional sign languages such as ASL, inflectional systems have been described in which spatial devices are used to modify verbs to agree with their noun arguments (e.g., the sign "give" is moved from the signer to the addressee to mean "I give to you", but from the addressee to the signer to mean "you give to me"; Padden, 1983). The deaf child in our study could vary the placement of his characterizing gestures, producing gestures either in neutral space (e.g., a "twist" gesture performed at chest level) or oriented toward particular objects in the room (e.g., a "twist" gesture produced near a jar). In the latter case, the placement of the gesture served to identify an entity playing a particular thematic role in the predicate represented by the gesture and, as such, served to modify the predicate to agree with one of its arguments. As an example, for transitive predicates, the characterizing gesture was typically displaced toward the object playing the patient role – the jar in the above example – thereby marking the jar as the patient of the predicate. In contrast, for intransitive predicates, the characterizing gesture was typically displaced toward the object playing the recipient role; for example, the child moved his "go" gesture toward the open end of a car-trailer to indicate that cars go into the trailer, thereby marking the trailer as the recipient of the predicate. Gestures were very rarely displaced toward the actor of either transitive or intransitive predicates.

As in ASL (cf., Hoffmeister, 1978), it was not necessary that an object be in the room for the deaf child in our study to mark that object morphologically via displacement. The child could produce his gestures near an object that was similar to the object he wished to refer to (e.g., a "twist" gesture produced near an empty jar of bubbles to indicate that he wanted the full jar of bubbles in the kitchen twisted open). Or, if the object the child wanted to indicate were animate, the child could indicate the object by producing his gestures on his own body (e.g., a "twist" gesture produced on the side of the child's body to indicate that he wanted the experimenter to twist a key on the side of a Mickey Mouse toy). Note that, in this example, the child is representing one individual with his hand (the experimenter) and a different individual with his body (Mickey Mouse); thus, as is frequently the case in ASL, the child appears to be using his body as a stage for his own gestures.

In a developmental analysis, we found that the child first began to displace his gestures toward objects that were similar to his intended-but-absent referents between the ages of 3;3 and 3;5 – the age at which this same child began producing points at objects in the room to refer to objects that were not in the room (Butcher *et al.*, 1991). Thus, this child's morphological marking system began to be freed from the here-and-now situation at about the same moment in development as was the child's system of pointing gestures.

6. The role of parental gestures in guiding the deaf child's system

The deaf children in our studies were found to elaborate gestural communication systems characterized by a lexicon, a simple syntax, and a simple morphology without the benefit of a conventional language model. It is possible, however, that the children's hearing parents spontaneously generated their own structured gesture systems which their children saw and learned. The parents – not the children – would then be responsible for the emergence of structure in the children's gestures.

The hearing mothers of the deaf children in our studies all produced gestures as they spoke to their children. Indeed, five of the six mothers whose gestures we analyzed in detail produced single gestures (as opposed to gesture strings) more often than their children. Moreover, the mothers produced both pointing and characterizing gestures, and produced them in approximately the same proportions as their children. However, the mothers produced fewer different types of characterizing gestures than their children, and their lexicons of characterizing gestures were different from their children's, overlapping no more than 33% and as little as 9%. Thus, the deaf children and their mothers both produced lexicons containing characterizing and pointing gestures, although the lexical items themselves did differ.

Despite the fact that the mothers were prolific producers of single gestures, they were not prolific producers of gesture strings: Five of the six mothers produced gesture strings less often than did their children. In addition, the mothers' gesture strings did not show the same structural regularities as their children's. The mothers showed no reliable gesture order patterns in their strings. Moreover, the production probability patterns in the mothers' gesture strings were different from the production probability patterns in the children's strings. Finally, the mothers began conveying two propositions in their gesture strings later in the study than their children, and produced proportionately fewer sentences with conjoined propositions than their children (Goldin-Meadow & Mylander, 1983, 1984).

With respect to morphology, the mother of the deaf child whose gestures were shown to be characterized by a morphological system was found to produce the same five handshape and nine motion forms as her child. In terms of meanings, however, only 50% of the mother's handshapes and 51% of her motions conformed to the child's system; in contrast, 95% of the child's handshapes and 90% of his motions conformed to the system. Moreover, the fit between mother's and child's meaning systems did not improve over the two-year period during which the pair was observed. In addition, the child appeared to have generalized beyond his mother's gestures in two respects: 1. The child produced almost all of the different types of handshape/motion combinations that his mother produced (20 of his mother's 25) but, in addition, produced another 34 combinations that were not found in his mother's repertoire. In order to go beyond his mother's gestures as he did, the child must have isolated the handshape and motion dimensions and used them as a basis for generating his novel combinations. 2. The mother used her gestures to refer to individual events (e.g., she used the C-hand combined with a circular motion only to refer to opening a jar and to no other types of actions or objects), while the child used his to refer to classes of related events (Goldin-Meadow & Mylander, 1990).

Thus, if a source for the handshape and motion components in the deaf child's gestures could be found in his mother's gestures, the child would have had to search through considerable noise in order to arrive at those components. Moreover, the child appeared to treat whatever structure he might have found in his mother's gestures as a starting point, using it to generalize to novel combinations and to novel referential uses.

With regard to the input issue in general, it is important to note that we are not claiming that the deaf child develops his gesture system in a vacuum. It is clear that the child receives input from his surroundings which he undoubtedly puts to good use. The crucial question, however, is: How close is the mapping between this input and the child's output? We have looked for isomorphic patterns between mother's gestures and child's gestures on the assumption that the child might have been inclined to copy a model that was easily accessible to him. We found that the gesture systems developed by the deaf children in our studies had some obvious similarities to the gestures produced by their hearing mothers: Both the children and their mothers produced pointing and characterizing gestures which they used to express the action and attribute relations typical of early mother–child conversations. However, the children consistently surpassed their mothers by organizing these gestural elements into productive systems with consistent patterns on at least two linguistic levels – the level of the sentence and the level of the word. All of the deaf children regularly combined the gestural elements into linear strings characterized by an, albeit simple, syntactic structure. The one child studied thus far analyzed the gestural elements into component parts characterized by a productive morphologic structure. Thus, our deaf children had, indeed, gone beyond the input, contributing linearization and componentialization to the gestures they received as input from their hearing mothers.

7. Gesture as a primary communication system versus gesture as an adjunct to speech

7.1. Comparison to conventional sign languages

The deaf children's gestures exhibited formal structuring at many of the same levels as a conventional sign language such as ASL, and exhibited similar kinds of organizational principles, in particular, constrained systems of components, rules based on underlying forms, and recursive processes (cf., Bellugi *et al.*, 1988). However, the deaf children's gestures formed a linguistic system that was far less complex than the linguistic system of ASL, a conventional language with a long history and shared by a wide community of signers. For example, ASL makes use of many more handshape and motion forms than the limited set described for the deaf children's gestures (cf., Wilbur, 1987); moreover, deaf children acquiring ASL from their deaf parents have already begun to acquire many of these handshape and motion forms at ages comparable to those at which we have observed our deaf children (cf., Supalla, 1982).

The simplicity of the deaf children's system relative to ASL highlights the importance of a community in generating and maintaining complexity in a linguistic system. Our present study of deaf children is a study of the kind of language system an individual (more specifically, an individual child) can create without the participation of a second language-user. We suggest that at least two language-users are likely to be required in order to introduce arbitrariness into a language system. Moreover, it may well be necessary for language to be passed on from one generation of users to the next (that is, for a group of fresh minds to learn the language as a whole) in order for language to undergo the sort of reorganization necessary for complex linguistic structures to develop (cf., Singleton, 1989).

7.2. Comparison to gestures in hearing children and adults

It is important to note that despite the simplicity of the deaf children's gestures, their gestures did exhibit structural regularities and, in this sense, went beyond the gestures typically produced by hearing children learning spoken language at the same age. Hearing children in the early stages of spoken language development do indeed gesture, and certain communicative functions may even appear in gesture before they appear in speech (Volterra & Caselli, 1986; Goldin-Meadow & Morford, 1985). Not surprisingly, however, speech comes to dominate over gesture in the hearing child and this domination typically occurs before the child's gestures become complex. For example, hearing children rarely produce their pointing gestures in combination with other gestures, even other points (Masur, 1983), and tend not to produce strings of characterizing gestures (Petitto, 1988).

In fact, young hearing children produce very few motor acts that would even meet our criteria for characterizing gestures (i.e., motor acts that do not involve direct manipulation of objects and that are used for communication rather than symbolic play). Even when hearing children produce the same characterizing gestures as the deaf children in our studies, they use those gestures differently. For example, one of the most common characterizing gestures hearing children produce is the "give" gesture – open palm extended as though to receive an object. Hearing children use this gesture almost exclusively to request objects for themselves (Petitto, 1988), while the deaf children in our studies used the "give" gesture across a variety of semantic situations to request the transfer of objects to other people and locations as well as to themselves. In general, hearing children tend to use their characterizing gestures as names for particular objects (often non-transparent names developed in the context of interactive routines with parents, e.g., index fingers rubbed together to refer to a spider, Acredolo & Goodwyn, 1988) and, as such, their gestures do not appear to have the internal handshape and motion structure characteristic of the deaf child's gestures. Unlike the deaf child's gestures, the gestures produced by hearing children do not appear to be organized in relation to one another and, thus, do not form a system of contrasts.

Overall, McNeill (1992) has described the gestures that characteristically accompany speech in hearing individuals (adults as well as children) as less clear, less disciplined, less reproduceable, and less schematic than the gestures used by the deaf children in our studies. Unlike the deaf children's gestures, which tend to be linear and segmented, the gestures that accompany speech in hearing individuals are "*global* in that the symbol depicts meaning as a whole (noncompositionally) and *synthetic* in that the symbol combines into one symbol meanings that in speech are divided into segments" (McNeill, 1987:18).

McNeill (1987) has shown that the gestures which accompany speech in hearing individuals form an integrated system with the speech they accompany. This fact may explain why the hearing mothers of our deaf subjects produced gestures which were organized so differently from their deaf children's gestures. Since almost all of the mothers' gestures were accompanied by speech, it is likely that the mothers' gestures (like those of all hearing speakers) were influenced by the spoken utterances with which they occurred. It is worth noting that this influence is a powerful one. One might have expected, over the two-year period during which we observed one of our deaf subjects and his mother, that the mother would have adapted her gestures to her child's, since gesture was essentially the child's only means of communication. However, the fit between the child's gesture system and the mother's remained at the same low level throughout the two-year period. In fact, the mother's gestures were not dramatically different from those that hearing mothers typically produce when they converse with their hearing children (Bekken, Goldin-Meadow & Dymkowski, 1990). Thus, it appears that the gestures the deaf child's mother produced – because they formed an integrated system with speech – were not "free" to take on the language-like structure that characterized the deaf child's gestures.

7.3. *When does gesture become language?*

The study of gesture provides a unique window into the conditions that foster language-like structure. The fact that the gestures of hearing individuals do not exhibit inter-gesture and intra-gesture structure suggests that communication in the manual modality does not inevitably result in structure at the sentence and word levels. Thus, language-like structure is not forced by the manual modality.

A priori, one might have thought that language-like structure would arise whenever information is conveyed. The gestures that hearing individuals produce along with their speech *do* convey information – information that is interpretable not only to experimenters (cf., McNeill, 1992; Church & Goldin-Meadow, 1986; Perry *et al.*, 1988) but also to individuals who have not been trained in coding gesture (Goldin-Meadow, Wein & Chang, 1992, e.g., adults, both trained and untrained, are able to observe a child who demarcates the width of a container with her hands in a Piagetian conservation task and infer that the child is, at some level, aware of this dimension of the task object). Nevertheless, these gestures do not exhibit language-like structure.

What then is the difference between the gestures produced by hearing individuals, which do not exhibit language-like structure, and the gestures produced by the deaf children in our studies, which do? I suggest that the function gesture serves in these two situations differs, and that this difference may contribute to the observed variations in structure (see Kendon, this volume, for a similar view). Gestures produced by hearing individuals serve as an adjunct to speech, which itself assumes the primary burden of communication. Unlike words, which are organized into combinations according to rules of syntax and morphology, gestures which accompany those words are rarely combined (each spoken clause being accompanied by a single gesture, McNeill, 1987) and are not themselves decomposeable (each gesture serving as a holistic depiction, like a picture or an enactment, presented in a single moment of time, Kendon, this volume). This holistic representation is adequate simply because gesture is framed by the speech it accompanies; that is, speech supplies the focus and context that allows interpretation of the accompanying gesture.

In contrast to the gestures of hearing and speaking individuals, the gestures produced by the deaf children in our studies assume the burden of a primary communication system and thus, in a sense, must frame themselves. To understand this distinction better, consider how holistic gesture of the type that typically accompanies speech might fare as a primary communication system. It is possible to depict an event, for example, "eating an apple", by enacting that event (i.e., one might move a hand shaped as though holding an apple toward one's open mouth). However, given this holistic representation, how would one request someone else to eat the apple, or comment on the fact that the apple had been eaten in the past, or warn a hopeful eater that this apple is wormy? It becomes increasingly difficult to fulfill the diversity of communicative functions that language typically serves without being able to isolate certain elements of the event and comment on those elements specifically. It appears as if gesture must be both decomposeable and combinatorial (i.e., it must be composed of "constructed units" in Kendon's terms) in order to function as a primary "linguistic" communication system. We have shown that the deaf children's gestures do indeed serve as elements in gesture strings (forming a simple syntax) and are themselves composed of recombineable elements (forming a simple morphology). It is precisely this combinatorial system which appears to be necessary for language to fulfill the range of functions it typically serves and which gives the deaf children's gesture its language-like quality.

In sum, the diversity of communicative uses to which the deaf children put their gestures have, in a sense, forced the children to go beyond holistic representation, requiring them to break their gestures into parts and to use those gestures as parts of larger wholes. Nevertheless, it is important to realize that, in order for those parts to form a combinatorial *system*, the deaf children must have been capable of, and inclined toward, creating that system.

8. The resilience of language

In general, the phenomenon of gesture creation in deaf children is a testament to the robustness of language in humans. However, children can be raised in circumstances which are not compatible with the development of language. For example, children raised under conditions of extreme deprivation, lacking not only linguistic input but also the social supports of typical human existence, do not develop language during their periods of deprivation (cf., Skuse, 1988). Thus, language-learning is not infinitely robust and, although it may not be necessary to have a language model to develop the rudiments of a linguistic system, it does appear to be essential to have another human to communicate with.

I have previously referred to the language-like properties found in the deaf children's gestures as "resilient" (Goldin-Meadow, 1982) – properties that appear in children's communication despite extensive variation of the learning conditions (such as no exposure to an established language). Properties displayed under such extreme conditions are evidently among the most basic and indispensible for a structured system of human communication, and they should spontaneously appear in any deliberate communication of meaning (cf., McNeill, 1992). That these same resilient properties are not systematically found in the spontaneous gestures accompanying the speech of both hearing children and hearing adults underscores (and continues to clarify by contrast) the language-like nature of the deaf children's gestures.

In sum, we have shown that a child who is not exposed to a usable conventional language model can create a communication system that is indeed language-like. This situation of language creation is quite clearly not a simulation of the situation in which language was created for the first time, simply because the deaf children are developing their communication systems in a world in which language and its consequences are pervasive. Thus, although it may not be necessary for a child to be exposed to a language model in order to create a communication system with language-like structure, it may be necessary for that child to experience the human cultural world. It is very likely that, as language evolved, the cultural artifacts that characterize our world evolved with it. Indeed, Hockett (1977:149) argues that the ability to carry artifacts (in particular, tools) and the ability to refer to objects that are not visible (communication beyond the here-and-now) developed side-by-side, each developing in small increments furthered by the already-achieved increments of itself and of the other. The deaf children in our studies, while lacking conventional language, nevertheless had access to the artifacts whch evolved along with language and which could have served as supports for the child's invention of a language-like system for communicating both within and beyond the here-and-now.

Thus, the techniques necessary to communicate in language-like ways appear to be fundamental to human interaction – so fundamental that they can be reinvented by a child who has access to the artifacts of the modern world but not to a culturally-shared linguistic system.

Acknowledgements

This research was supported by Grant No. BNS 8407041 from the National Science Foundation and Grant No. RO1 NS26232 from the National Institutes of Health.

References

Acredolo, L. P., & Goodwyn, S. W. (1988). Symbolic gesturing in normal infants. *Child Development*, 59, 450–66.

Bekken, K., Goldin-Meadow, S., & Dymkowski, T. (1990). *Dissociation of maternal speech and gesture to deaf children of hearing parents.* Paper presented at the 15th Annual Boston University Conference on Language Development, Boston, MA.

Bellugi, U., van Hoek, K., Lillo-Martin, D., & O'Grady, L. (1988). The acquisition of syntax and space in young deaf signers. In *Language Development in Exceptional Circumstances*, ed. D. Bishop & K. Mogford, pp. 132–49. New York: Churchill Livingstone.

Bonvillian, J. D., Orlansky, M. D., & Novack. L. L. (1983). Developmental milestones: Sign language acquisition and motor development. *Child Development*, 54, 1435–45.

Butcher, C., Mylander, C., & Goldin-Meadow, S. (1991). Displaced communication in a self-styled gesture system: Pointing at the non-present. *Cognitive Development*, 6, 315–42.

Church, R. B., & Goldin-Meadow, S. (1986). The mismatch between gesture and speech as an index of transitional knowledge. *Cognition*, 23, 43–71.

Conrad, R. (1979). *The Deaf Child*. London: Harper & Row.

Feldman, H., Goldin-Meadow, S., & Gleitman, L. (1978). Beyond Herodotus: The creation of language by linguistically deprived deaf children. In *Action, Symbol, and Gesture: The Emergence of Language*, ed. A. Lock., pp. 351–414. New York: Academic Press.

Goldin-Meadow, S. (1979). Structure in a manual communication system developed without a conventional language model: Language without a helping hand. In *Studies in Neurolinguistics* (Vol. 4), ed. H. Whitaker & H. A. Whitaker, pp. 125–209. New York: Academic Press.

Goldin-Meadow, S. (1982). The resilience of recursion: A study of a communication system developed without a conventional language model. In *Language Acquisition: The State of the Art*, ed. L. R. Gleitman & E. Wanner, pp. 51–77. New York: Cambridge University Press.

Goldin-Meadow, S. (1985). Language development under atypical learning conditions: Replication and implications of a study of deaf children of hearing parents. In *Children's Language* (Vol. 5), ed. K. Nelson, pp.198–245. Hillsdale, New Jersey: Lawrence Earlbaum Associates.

Goldin-Meadow, S., & Feldman, H. (1977). The development of language-like communication without a language model. *Science*, 197, 401–3.

Goldin-Meadow, S., & Morford, M. (1985). Gesture in early child language: Studies of deaf and hearing children. *Merrill-Palmer Quarterly*, 31(2), 145–76.

Goldin-Meadow, S., & Mylander, C. (1983). Gestural communication in deaf children: The non-effects of parental input on language development. *Science*, 221, 372–4.

Goldin-Meadow, S., & Mylander, C. (1984). Gestural communication in deaf children: The effects and non-effects of parental input on early language development. *Monographs of the Society for Research in Child Development*, 49, 1–121.

Goldin-Meadow, S., & Mylander, C. (1990). The role of parental input in the development of a morphological system. *Journal of Child Language*, 17, 527–63.

Goldin-Meadow, S., & Mylander, C. (1991). Levels of structure in a language developed without a language model. In *Brain Maturation and Cognitive Development*, ed. K. R. Gibson & A. C. Peterson, pp. 315–44. Hawthorn, New York: Aldine Press.

Goldin-Meadow, S., Wein, D., & Chang, C. (1992). Assessing knowledge through gesture: Using children's hands to read their minds. Cognition and Instruction, in press.

Hockett, C. F. (1977). *The View from Language: Selected Essays 1948–1974*. Athens, Georgia: The University of Georgia Press.

Hoffmeister, R. (1978). The development of demonstrative pronouns, locatives and personal pronouns in the acquisition of American Sign Language by deaf children of deaf parents. Unpublished doctoral dissertation, University of Minnesota.

Hoffmeister, R., & Wilbur, R. (1980). Developmental: The acquisition of sign language. In *Recent Perspectives on American Sign Language*, ed. H. Lane & F. Grosjean, pp. 61–78. Hillsdale, New Jersey: Lawrence Earlbaum Associates.

Klima, E. & Bellugi, U. (1979). *The Signs of Language*. Cambridge, MA: Harvard University Press.

Lenneberg, E. H. (1964). Capacity for language acquisition. In *The Structure of Language: Readings in the Philosophy of Language*, ed. J. A. Fodor and J. J. Katz, pp. 579–603. New Jersey: Prentice-Hall.

Lucy, J. A. (1992). Reflexive language and the human disciplines. In *Reflexive Language: Reported Speech and Pragmatics*, ed. J. Lucy. New York: Cambridge University Press, in press.

Masur, E. F. (1983). Gestural development, dual-directional signaling, and the transition to words. *Journal of Psycholinguistic Research*, 12(2), 93–109.

McDonald, B. (1982). Aspects of the American sign language predicate system. Unpublished doctoral dissertation, University of Buffalo.

McNeill, D. (1987). *Psycholinguistics: A New Approach*. New York: Harper & Row.

McNeill, D. (1992). *Hand and Mind: What Gestures Reveal about Thought*. Chicago: University of Chicago Press, in press.

Miller, P. J. & Hoogstra, L. (1989). *How to Represent the Native Child's Point of View: Methodological Problems in Language Socialization*. Paper presented at the annual meeting of the American Anthropological Association, Washington, DC.

Moores, D. F. (1974). Nonvocal sytems of verbal behavior. In *Language Perspectives: Acquisition, Retardation, and Intervention*, ed. R. L. Schiefelbusch & L. L. Lloyd, pp. 377–418. Baltimore: University Park Press.

Newport, E. L., & Meier, R. P. (1985). The acquisition of American Sign Language. In *The Cross-Linguistic Study of Language Acquisition, Volume 1: The Data*, ed. D. I. Slobin, pp. 881–938. Hillsdale, New Jersey: Lawrence Earlbaum Associates.

Padden, C. (1983). Interaction of morphology and syntax in American sign language. Unpublished Ph.D. dissertation, University of California at San Diego.

Perry, M., Church, R. B., & Goldin-Meadow, S. (1988). Transitional knowledge in the acquisition of concepts. *Cognitive Development*, 3, 359–400.

Petitto, L. A. (1988). "Language" in the pre-linguistic child. In *The Development of Language and Language Researchers*, ed. F. Kessel, pp. 187–221. Hillsdale, New Jersey: Lawrence Earlbaum Associates.

Schick, B. S. (1987). The acquisition of classifier predicates in American sign language. Unpublished doctoral dissertation, Purdue University.

Schlesinger, H. (1978). The acquisition of bimodal language. In *Sign Language of the Deaf: Psychological, Linguistic, and Sociological Perspectives*, ed. I. Schlesinger, pp. 57–96. New York: Academic Press.

Silverstein, M. (1976). Hierarchy of features and ergativity. In *Grammatical Categories in Australian Languages*, ed. R. M. W. Dixon, pp. 112–71. Canberra: Australian Institute of Aboriginal Studies.

Singleton, J. L. (1989). Restructuring of language from impoverished input: Evidence for linguistic compensation. Unpublished doctoral dissertation, University of Illinois, Champaign-Urbana.

Skuse, D. H. (1988). Extreme deprivation in early childhood. In *Language Development in Exceptional Circumstances*, ed. D. Bishop & K. Mogford, pp. 29–46. New York: Churchill Livingstone.

Supalla, T. (1982). Structure and acquisition of verbs of motion and location in American sign language. Unpublished doctoral dissertation, Universtiy of California at San Diego.

Volterra, V. & Caselli, M. C. (1986). First stage of language acquisition through two modalities in deaf and hearing children. *Italian Journal of Neurological Sciences*, 5, 109–15.

Wilbur, R. (1987). *American Sign Language: Linguistic and Applied Dimensions*. 2nd edn. Boston, MA: Little, Brown & Co.

3

The emergence of language

E. SUE SAVAGE-RUMBAUGH AND DUANE M.
RUMBAUGH

Introduction

The issue to be dealt with in this chapter is: Can an account of what an organism learns and how it learns it adequately explain the phenomenon of language – or must any such account be insufficient because of innate mechanisms which determine both the capacity for reference and the capacity for syntax? A satisfactory answer to this question must precede any discussion of the relationship between tool use and language, as tool use is not viewed as innate, in the strong sense that has been claimed for grammatical processes (Chomsky, 1988a,b).

The position to be argued here is based on the premise that there is *no* neurological component which is unique to linguistic processes and independent of general cognitive function. The biological capacities which *Homo sapiens* bring to language are the anatomical specialization of the speech apparatus, a very large brain, and a species tendency to live in groups where resources are shared. The anatomical reorientation of the laryngeal cavity, coupled with the ability to produce voluntary controlled exhalation (accompanied by vocal cord modulation) make it possible for *Homo sapiens* to produce lengthy vocal sequences with ease. This skill possibly permitted early hominids to take advantage of extant cognitive capacities for representation and intentional communication. The tendency to share resources and to cooperate in their procurement set the stage for the use of sounds to coordinate interactions.

It will also be argued that syntax, rather than being biologically predetermined, is a skill which arises naturally from the need to process sequences of words rapidly. As overall intelligence increased, spurred by the ever-increasing use of language for planning future activities, communications became increasingly complex and increasingly independent of context. When complex ideas began to require groups of words for their expression, it became essential to devise a means to specify which of the words in a group modified (or were related to) which other words. Syntactical rules were developed to solve this dilemma. Such rules were the inevitable outgrowth of complex symbolic communication involving multiple symbols. Because the medium of transmission was a rapidly fading auditory system, rules appeared which grouped words together, permitted rearrangements of word groups (transformations), and allowed a great deal of redundancy and predictability.

In support of this perspective is the fact that the major difference between the brain of human and ape is that of size rather than structure or organization (Passingham, 1982). However, increased size alone cannot account for language as there are many cases in which persons whose brains are far below the normal size display no intellectual deficit with regard to language (Seckel, 1960). Humans are in the awkward position of having to acknowledge that the known physical basis of the dissimilarity between themselves and apes does not justify the perceived degree of behavioral dissimilarity.

This raises the possibility that more of the apparent intellectual gap between *Pan* and *Homo* may be attributable to learning than is commonly acknowledged. A small physical difference, such as one which conferred the ability to control the expiration of air, could have permitted a complex ape-like intelligence to take advantage of this new found physical ability by inventing a crude vocal language. Such an invention, perhaps primitive at first, would have been culturally transmitted across generations and could have exerted its own pressure for biological improvement. The invention of even a simple language would place severe pressures upon the members of the species that produced the invention. Any members of the group that did not acquire the new communicative skill would rapidly be excluded from the larger community – just as such persons are today. It is surely self-evident that a behavioral–cultural invention of the order of a primitive language would allow any species having this capacity to outpace other hominids very rapidly.

Attempts to teach apes language

Understanding the bio-linguistic substrate of language, particularly as it is manifested by apes, has been the focus of the research program at the Georgia State/Yerkes Language Research Center since the early 1970s. The center now houses eleven apes, including five common chimpanzees (*Pan troglodytes*), four bonobos or pygmy chimpanzees (*Pan paniscus*) and two orangutans (*Pongo pygmaeus*), all of whom are subjects in ongoing studies of linguistic, numeric, attentional, and hemispheric processes. Studies of numeric, attentional, and learning capacities in rhesus monkeys are also under way (Rumbaugh *et al.*, 1989).

The first ape subject at this laboratory, Lana (*Pan troglodytes*), demonstrated that apes could readily discriminate among geometric symbols and could sequence symbols to form differentiated symbol strings for the purposes of receiving various rewards, for example, "Please machine give M&M" (Rumbaugh 1977; Rumbaugh, Gill & von Glasersfeld, 1973). Her abilities illustrated the value of providing an ape with a graphic communication system. Because the system made it easy for a chimpanzee to *produce* symbols, and even to combine them in sequence, interest in the fact that the ape could construct something that "looked like sentences" was quickly superseded by the theoretical question of how one could legitimately draw parallels between human and ape communications. When, how, and under what conditions was it appropriate to equate symbol presses with words and sentences?

Clearly, questions of communicative competence could not be answered by determining whether or not chimpanzees could learn to answer questions or associate certain movements with different stimuli, as early work implied (Premack, 1970; Gardner & Gardner, 1969). The essence of human language lies in the capacity to use symbols to tell others something that they do not already know. The more complex the message, the greater the need for structural rules regarding the units of which the message is composed. Did Lana understand the relationships between the symbols she used? Was she using key presses to convey messages or were these key presses simply conditioned responses performed for the purpose of obtaining food? Even more importantly, upon what basis could a legitimate determination be made of why and how Lana, or other apes, used symbols?

These questions were addressed with the addition of four more *Pan troglodytes* chimpanzees (Ericka, Kenton, Sherman and Austin) to the Language Center's research program. They, too, learned to differentiate and use symbols. However, unlike Lana, they were not taught to "name" things, but rather to *ask* for things of interest to them such as foods and tools. As with children, these skills were acquired through a process which stressed the development of communicative intentionality and joint regard (Savage-Rumbaugh, 1986). Communications were negotiated first at the nonverbal level and secondarily at the verbal level. For example, when a chimpanzee needed a particular tool to solve a problem, it first began to express this desire nonverbally, by pointing to the tool, as would a child. Only later was pointing superseded by use of a symbol to express which tool was needed when none were visible.

For research purposes, a measure of "concordance" was devised to determine whether or not the chimpanzees abided by the "rule of correspondence". This rule, essential to any representational system, states that a correspondence of some reasonable nature must exist between what is said and what is intended. In other words, it is not possible to utter any randomly chosen symbol or string of symbols to convey a specific idea or goal. There must exist a correspondence such that the words which are selected must overlap, in some meaningful way, with the words which others would have selected to express a similar idea or goal. Without this rule, any language system would collapse upon itself in a muddle of miscommunication (Savage-Rumbaugh, 1990a).

Such correspondence was determined for Sherman and Austin by checking to see if what they said agreed with what they did, when given a choice. For example, if they asked for a key and were shown ten tools, did they select a key? And was a key really the tool they needed or just a word that was used to gain access to a group of tools? Although this idea of looking for such a concordance seems simple, it has not been used in other ape-language studies.

This approach permitted "reference" to be objectified as a verifiable inter-individual process, rather than a postulate of internal mental structure. It put into place an empirically testable means of determining whether or not a chimpanzee knew what it had said. It also revealed that an ape could make statements which expressed its intended future actions. For example, the announcement of the intent to "tickle" followed by tickling, or the announcement of the intent to "go outdoors" followed by a trip outside, was noted because of the emphasis on concordance.

Other language trained apes have not evidenced such skills of announcement. They have asked for things they wanted and have named things when told to do so, but they had not used symbols for the purpose of making statements or conveying intended actions.

The use of a concordance measure "legitimized" the ape's linguistic capacity; it established that apes could use symbols in many of the same ways that human beings do, and that they were *not* engaging in a series of tricks or conditioned routines while lacking cognizance of the significance of their utterances (Darley, Glucksberg & Kinchla, 1986). What this work had not done yet, however, was to address one of the most significant differences between human and ape – the fact that nearly all children acquire language skills, not only spontaneously, but effortlessly. Indeed, it is hardly possible to prevent children from learning language if they are reared in anything approaching a normal environment. Why did apes need tutoring, and fairly constant tutoring at that?

The addition of five more subjects, four of whom were of a different species, (the bonobo or pygmy chimpanzee, *Pan paniscus*) permitted studies at the Language Research Center to address this question (Savage-Rumbaugh *et al.*,

1986; Savage-Rumbaugh, 1988; Savage-Rumbaugh *et al.*, in press). These subjects included a wild-caught adult bonobo female (Matata) and her offspring; a male (Kanzi) born in 1980, two females (Mulika and Panbanisha) born in 1983 and 1985, respectively, and a female common chimpanzee (Panpanzee) born in 1985 and co-reared with Panbanisha.

Observational learning in apes

Studies with these subjects began with attempts to replicate the findings described above with Sherman and Austin. Unlike them, the adult bonobo, Matata, showed great difficulty in discriminating between symbols and in sequencing symbols. She was also deficient in other areas such as match-to-sample, sorting, and tool use. Significantly, Matata could not display stable concordances between her symbolic utterances and her nonverbal behaviors, suggesting that her lexigram usage should not be characterized as "symbolic" or "referential", at least at the level typified by the *Pan troglodytes* subjects Sherman and Austin (Savage-Rumbaugh, 1986). While Matata did poorly on these tasks, her overall social behavior suggested that she was quite intelligent. Presumably her early experiences in the wild did not prepare her for laboratory tests, although they did result in excellent social skills.

All of Matata's offspring (Kanzi, Mulika, and Panbanisha) acquired large vocabularies. The most important finding with these young ape subjects was that *it was not necessary to train language.* Simply by observing and listening to their caretakers' input (as children observe and listen to those around them) Kanzi, Mulika, Panbanisha, and Panpanzee learned to use symbols appropriately. Kanzi's acquisition of these skills has been described in detail elsewhere (Savage-Rumbaugh *et al.*, 1986; Savage-Rumbaugh, 1988); however, a brief summary is relevant here.

Kanzi was exposed to caretakers who pointed to keyboard symbols while speaking. Caretakers talked to him about daily routines, events, and about comings and goings at the laboratory. Regular events included trips to the woods to search for food, games of tickle and chase, trips to visit other primates at the laboratory, play with favorite toys such as balloons and balls, visits from friends, watching preferred TV shows, taking baths, helping pick up things, and other numerous simple activities characteristic of daily life. Instead of requiring symbol production, caretakers talked about what they were doing and what was to be done next. Conversations were rarely, if ever, repetitive in the sense that the same word or sentence was uttered over and over. Nonetheless, across days, routines evidenced basic structures or schemata that tended to repeat themselves, just as do the routines of preschool children (Savage-Rumbaugh, 1991.)

Table 3.1 *First lexigrams acquired by infant apes*

The lexigrams acquired by three different subjects are listed in the order in which they were acquired. The differences reflect the individual interests of the different subjects in the topics to which the different vocabulary words referred.

Kanzi	Mulika	Panbanisha
Orange	Milk	Milk
Peanut	Key	Chase
Banana	T-room	Open
Apple	Surprise	Tickle
Bedroom	Juice	Grape
Chase	Water	Bite
Austin	Grape	Dog
Sweet potato	Banana	Surprise
Raisin	Go	Yogurt
Ball	Staff office	Soap

Unlike Sherman, Austin and Lana, whose symbol vocabularies were assigned by the experimenter, Kanzi and his siblings "selected" the symbols they were ready to acquire from the hundreds used around them each day. Like children, Kanzi's first words were not the same as those of his siblings, though there was some overlap (see Table 3.1). Also unlike earlier subjects, Kanzi learned to associate a spoken English word with its world referent. Indeed, only after they learned the relationship between a *spoken* word and its referent did Kanzi and his siblings begin to connect the word and the *geometric symbol*.

Kanzi's mode of symbol acquisition was therefore very different from that of Lana, Sherman and Austin. Kanzi first evidenced comprehension by responding to sentences uttered in context. For example, when he began to associate the light switch with the word "light", he would rush over and flip the switch anytime someone mentioned "light". Later, he simply looked over at the switch, as though visually forming the pairing. He then began to watch as people pointed to the light lexigram when they said "light". Next he began to touch the symbol at times when he heard the word. Occasionally he placed a flashlight on the light lexigram. Although "light" was one of the first words Kanzi comprehended he rarely used it except to indicate that he wanted to play with the light switch or a flashlight.

By contrast "ball", another early word, was used in many ways which reflected Kanzi's fascination with balls. Kanzi commented "ball" when he observed children playing ball on TV, while holding his ball, and apparently to indicate he was thinking of his ball. He could remember where he left various balls for many days and would lead people to these places later. He invented games with balls, such as "slap ball", "chase ball", "grab ball", and "tickle ball". However, his competence with the word "ball" did not differ from that with the word "light", in terms of test performances on single words or comprehension of novel combinations; suggesting that he understood both words, but had more interest in employing the word "ball" than the word "light" (Savage-Rumbaugh, 1988).

Kanzi's linguistic skills continued to increase, though at a much slower pace than those of a normal child. He became increasingly able to negotiate miscommunications and to make his communicative intentions clear. His comprehension skills continued to outpace his production skills to a significant degree. As he grew older, it became possible to administer a variety of formal tests that were independent of context. By age five, Kanzi was able to listen, through headphones, to a prerecorded tape of spoken words and identify the words he had heard by selecting a photograph or lexigram (Figure 3.1). Because the experimenter did not utter the word and did not know which word Kanzi heard, this procedure eliminated the possibility of inadvertent cuing. (Savage-Rumbaugh, 1988). It is important to note that Kanzi was *not* trained to listen and identify words through headphones, nor was he rewarded for correct symbol selections during the test period. Just as a five-year-old child could point to pictures of words heard through headphones at once if it knew the words, Kanzi was also able to do this.

Tests of competencies

Through such tests it was determined that, at age six, Kanzi could identify 150 lexigram symbols upon hearing their spoken words. In addition, he could understand Votrax synthesized speech, though not as well as human speech. After listening to synthesized words he identified 100 lexigram symbols. Synthesized speech lacks the intonation and stress patterns characteristic of spoken speech. Kanzi's ability to understand synthesized monotonic speech indicated that he could segment words and decode their phonemic component. This meant that his English comprehension was not based singularly upon intonation or stress pattern. Further evidence for this view came from his success in tests of the ability to distinguish between many pairs of words that differed in only one or two phonemic components, such as clover and collar, peas and peaches (Savage-Rumbaugh, 1988).

During the first three to four years of Kanzi's life, his comprehension of spoken English was limited to individual words, or to phrases that were generally uttered as a whole in the same way each time. During the fifth year, he

Fig. 3.1. The test in which Kanzi must listen to single words through headphones and then select the photograph or lexigram that matches the English word.

began to respond appropriately to more complex utterances, even novel requests that were elaborate in character and required him to integrate meanings and relationships appropriately across groups of words.

In order to gain an appreciation of the types of complex communications that Kanzi could understand, he was presented with a large number of novel sentences under controlled conditions. A two-year-old child (Alia) was presented with the same sentences. Both Kanzi and Alia were exposed to a similar linguistic environment from birth and neither was trained to use language (Savage-Rumbaugh *et al.*, 1986). Both had similar experiences with lexigrams (i.e. geometric symbols that serve as words in our laboratory). Additionally, both subjects shared a primary caregiver, Jeannine Murphy, who was the mother of the child and whose language input and caretaking behavior was similar for both subjects. For purposes of the current study both subjects were presented with 660 novel sentences to determine the extent to which they comprehended utterances based upon speech cues alone. The sentences were presented in spoken English in a normal voice. Relative clauses were utilized as well as word-order reversals to investigate comprehension of specific syntactical devices.

The test of comprehension skills took place across nine months, from May 1988 to February 1989 for Kanzi, and across six months, from January 1989 to July 1989 for Alia. Sentences were generated by assembling a random array of objects, locations, and agents, and forming requests from this array. In the majority of instances, only requests that the subjects were unlikely to have encountered in their daily environment were used as test items. The speaker who uttered the sentences sat behind a one-way mirror to prevent any unintentional cuing of the subjects (blind condition). At first both subjects were uncomfortable in responding to a speaker they could not see, so some preliminary sentences were administered with the speaker visible (nonblind condition).

Coding of responses was done by viewing the video record of each sentence to determine whether the subject correctly completed all parts of the sentence, whether they partially executed the sentence, or whether they were completely wrong. Inter-observer reliability was computed for 386 trials for each subject and was 0.98 for Kanzi and 0.89 for Alia. In addition, a verbatim transcription was made of *all* trials from the video tape. (See Savage-Rumbaugh *et al.*, in press.)

Kanzi's and Alia's overall performances, as well as their performances on different sentence types are shown in Table 3.2. Overall, Kanzi was correct on 72% of all sentences and 74% of the blind sentences. Alia was correct on 66% of all sentences and 65% of the blind sentences. There were no significant differences between the blind and nonblind conditions in terms of overall performance for either subject. The overall high performance of both subjects provides strong evidence of their ability to comprehend all sentence types and subtypes.

Comprehension of word order and recursion

Sentences which entailed word-order reversals were intentionally presented during data collection, generally as subsets of some of the sentence types described in Table 3.2. To ensure that the subjects responded to reversed sentences without bias, reversed presentations of given sentences were usually separated by intervals of days or even weeks.

The pairs of reversed sentences are presented in Table 3.3. Eighty-eight sentences, forming 44 pairs of reversed sentences occurred. They were of two types. Group 1 was fully reversible. For Group 2, X and Y were reversed, but

Table 3.2. *Percentage of correct responses to different sentence types*

Sentence type	Kanzi (%)	Alia (%)
All sentences presented in the blind condition	74	66
1-A. *Put object X in moveable object Y.* Put the rubber band on your ball. Put the sparklers in the shoe	63	73
1-B. *Put object X in nonmoveable object Y.* Put the shoe in the potty. Take your ball to the table	77	71
2-A. *Give object X to animate A.* Go get a coke for Rose. Give a knife to Kelly.	78	84
2-B. *Give object X & object Y to animate A.* Give the apple and the hat to Rose. Show Sue the toothpaste and the milk.	37	57
2-C. *(Do) action A on animate A.* Go vacuum Liz. Hammer the doggie.	91	91
2-D. *(Do) action A on animate A with object X.* Tickle Rose with the bunny. Go put some soap on Liz.	76	61
3. *(Do) action A on object X.* Vacuum your ball. Bite the stick	82	63
4. *Announce information.* Kanzi is going to tickle Liz with the bunny. There is a new ball hiding at Sherman's.	67	83
5-A. *Take object A to location Y.* Take the doggie to the colony room. Take the toothpaste outdoors.	78	71
5-B. *Go to location Y and get object A.** Go to the potty and get the sparklers. Go to the microwave and get a shoe.	82	45
5-C. *Go get object A that's in location Y.** Go get the snake that's outdoors. Go get the collar that's in the refrigerator.	77	52
6. *Make pretend animate A act on person B.* Can you make the bunny eat the sweet potato? Make the toy orang bite Rose.	67	56
7. *All others*	78	33

* Indicates that Kanzi performed significantly better than Alia on these sentence types.

Table 3.3. *Examples of sentences with word-order reversals*

The word order of key words is reversed. In groups one and two, the subject must carry out the action in the same sequence as the sentence itself. In the final group, the subject must carry out different actions for each sentence even though the order of the key words is reversed.

Group 1 sentences: Word order reversed – same verb

Pour coke in the lemonade.
Pour the lemonade in the coke.

Make the doggie bite the snake.
Make the snake bite the doggie.

Put the pine needles on the ball.
Put the ball on the pine needles.

Kanzi	Alia
32/42	26/39

Kanzi made 2 inversion errors, Alia 5

Group 2 sentences: Word order reversed – different verb

Take the potato to the bedroom.
Go to the bedroom and get the potato.

Take the melon to the refrigerator.
Go to the refrigerator to get the melon.

Take the carrots outdoors.
Go outdoors and find the carrots.

Put the sparklers in the potty.
Go to the potty and get the sparklers.

Kanzi	Alia
34/46	26/44

Neither made any inversion errors

Group 3 Sentences: Word order constant – action must differ

Put the doggie in the refrigerator.
Go get the doggie that's in the refrigerator.

Put the melon in the potty.
Get the melon that's in the potty.

Put your collar in the refrigerator.
Go get your collar that's in the refrigerator.

Kanzi	Alia
22/28	18/27

Neither made any inversion errors

the verb was different (generally "go get" versus "take"). Also included in Table 3.3 are Group 3 pairs in which the order of the actions in the response should have varied, even though the order of the location and object terms did not vary (i.e. "Put the melon in the potty" versus "Get the melon that's in the potty").

Thus the subjects were presented with the double challenge of the English language. On the one hand, some sentences required that the order of X and Y be treated as a signal about the sequence that ensuing actions should take. In other cases, the order of X and Y was to be ignored.

Kanzi was correct on 80% of the sentences in which different orders of X and Y were to be followed by different actions. He correctly executed *both sentences of a given pair* for 66% of the pairs. Alia was correct on 63% of the sentences and on 57% of the pairs. When *only fully reversible object–object pairs were considered*, Kanzi was correct on 32 of the 42 possible sentences and on both sentences in 12 of the 21 pairs. Alia was correct on 26 of 39 possible sentences and on both sentences in 7 of the 19 pairs.

Overall, on fully reversible (Group 1) sentences, Kanzi made only three errors of inversion and Alia made only seven, indicating that *even when they were incorrect on reversal sentences, their errors generally were not generated by an inability to interpret word order*, but instead were semantic errors and errors of inattention. For example, when Kanzi heard "Pour the milk in the cereal" he poured the milk in the mushrooms. He thus acted with the milk, rather than the cereal (as the sentence specifies), but poured the milk onto the wrong item. Similarly, when Alia was asked to put the peaches in the tomatoes, she responded by putting the peaches in the yogurt.

The ability of the subjects to understand the syntactical device of re-cursion was tested by contrasting sentences such as "Go get the tomato that's in the microwave" (5-C) with "Go to the micro-wave and get the tomato" (5-B) when the object (in this case a tomato) was present in the array of objects in front of the subject, as well as in the announced location. To carry out 5-C sentences correctly, the subjects had to ignore the mentioned object that was in front of them and travel to the correct location to find it.[1] How-ever, 5-B sentences employed a linear construction and were ambiguous, in that they did not specify which X was to be obtained (that is one could interpret them as a request to do two separate things, to go to the microwave, and to get a tomato.) By contrast, 5-C sentences were nonlinear, but did indicate, through the device of recursion, that the tomato to be retrieved was in a different location.

Because the embedded phrasal structure of 5-C sentences removed the semantic ambiguity inherent in linear 5-B sentences, better performance on 5-C control trials should indicate that subjects processed the phrasal modifier appropriately. Kanzi's data indicate that the syntactical relationship expressed in 5-C sentences was helpful to him. When given type 5-B sentences, Kanzi acted on the object in front of him in some manner on 50% of the trials. That is, the ambiguity inherent within the sentence structure was reflected by ambiguity in Kanzi's behavior. However, in the 5-C sentences, Kanzi acted on the object in the near array on only 9% (or 2) of the trials and even these errors indicated comprehension of the syntactical structure of 5-C sentences.[2]

More impressive was the manner in which Kanzi responded to the 5-C sentence format. When Kanzi heard 5-C sentences, he typically did not even glance at the array in front of him, as he did when presented with the ambiguous 5-B sentences. Instead, he headed directly for the location, suggesting that he had deduced from the structure of the sentence itself that there was no need to search for the object in the array in front of him. Alia's data followed a similar pattern. She was correct on 25% of the 5-B control trials and on 63% of the 5-C control trials, suggesting that the syntactical structure of 5-C sentences functioned to clarify ambiguity for her as well.

The overall performance of both subjects, on 5-A, 5-B and 5-C sentences, regardless of whether the object to be retrieved was present in the array or not, indicated they were able to comprehend the syntactical relationships among word units, not just the units themselves. The ability to respond correctly to a suite of sentences such as "Take the tomato to the microwave", "Go to the microwave and get the tomato", and "Go get the tomato that's in the microwave", demonstrated an understanding of the fact that such sentences reflect an intended relationship between all words (the action, the object, and the location).

The clear outcome of this study was that two normal individuals of different ages and different genera (*Homo* and *Pan*) were remarkably closely matched in their ability to understand spoken language. One, a two-year-old human female, the other an eight-year-old bonobo male, demonstrated that under relatively similar rearing conditions and virtually identical test conditions, they could comprehend both the semantics and the syntactical structure of quite unusual English sentences.

Both subjects appeared to process the experimenters' words at the sentence level. The meaning which they assigned to a word was based on its role in the sentence rather than upon a dictionary-like set of referents. Both subjects also responded appropriately to very unusual sentences. For example, the opposing requests "Make the doggie bite the snake" and "Make the snake bite the doggie" required the subjects to imbue inanimates with action and cause them to pretend to do something that they had never seen themselves. In both

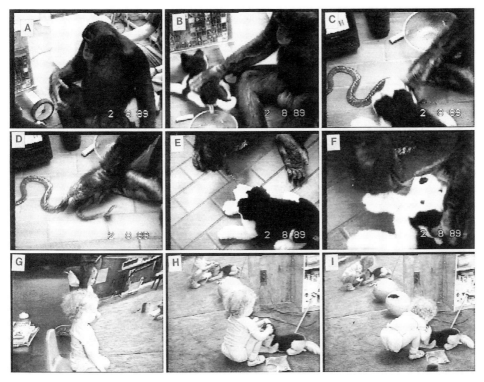

Fig. 3.2. Kanzi and Alia respond to the sentence "Can you make the doggie bite the snake?" (a) Kanzi listens to the sentence. (b) Kanzi picks up the dog. (c) Kanzi places the dog near the snake. (d) Kanzi picks up the snake. (e) Kanzi moves the snake toward the dog. (f) Kanzi puts the snake's head in the dog's mouth, and using his thumb, closes the dog's mouth on the snake's head. (g) Alia listens to the sentence. (h) Alia approaches and looks at the dog. (i) Alia bites the dog herself.

instances, Kanzi picked up the agent first and moved the agent toward the recipient. Alia, in this sentence, misinterpreted the agent and bit the doggie herself (see Figure 3.2)

Combinations by Kanzi

The communicative effect produced by a combination of words, with or without syntax, is different from that which can occur with single words in a way that is not often discussed. The production of novel combinatorial utterances is a powerful communicative process that characterizes all languages. The appearance of "sequenced words" antedates the emergence of syntax proper, and serves, like syntax, to create new meanings which are not simply an additive result of the separate words.

For example, two-word utterances, such as "Car trailer" or "Grouproom Matata," convey novel meanings that the individual components, if uttered alone, could never generate. Additionally, these sentences can convey their novel meanings regardless of the order of their components or whether or not this order is fixed according to any rule-based syntactical system. For example, when Kanzi produced the combination "Car trailer" he was in the car and employed this utterance as a means of indicating that he wanted the car to be driven to the trailer rather than to walk (Savage-Rumbaugh, 1990b). He followed the utterance with a gesture toward the trailer and the steering wheel of the car. Had Kanzi said "car" alone, this single symbol utterance would have been interpreted as a comment about being in the car and would have simply been acknowledged. Had he said "trailer" alone, the caretaker would probably have simply gotten out of the car and walked with Kanzi to the trailer, since it was a very short distance to drive. However, by saying "car trailer" Kanzi produced a novel meaning and brought about a set of events that otherwise would not have been likely to occur (i.e., taking the car to the trailer).

Kanzi similarly produced the combination "Grouproom Matata" to convey something different from what either symbol could convey alone. Kanzi was in the grouproom when he produced this combination and had just heard Matata (his mother), who was in another room, vocalize. Typically, if Kanzi wanted to visit Matata, he would simply say "Matata" and gesture toward her room. However, on this occasion, by producing the combination, he indicated that he wanted Matata to come to the grouproom where he was, instead of going to visit her. In response to this utterance, the caretaker asked, "Do you want Matata to come to the grouproom?" Kanzi answered with loud positive vocal noises, directed first to the experimenter, then to Matata. He seemed to be announcing something about his intent to Matata. Had Kanzi said only "grouproom", his utterance would have been interpreted as a comment on his location, just as "car" would have been in the preceding example. However, because it is not possible to take a room somewhere (a piece of real-world knowledge known and assumed by both Kanzi and the caretaker), it was surmised that Kanzi wanted Matata to come to the grouproom. Kanzi's vocalization in response to the caretaker's inquiry affirmed his intent.

These sorts of combinatorial processes characterized Kanzi's utterances even though the majority of his productions were still single words. Table 3.4 presents a categorical breakdown (with examples), of Kanzi's combinations at five years of age (Greenfield & Savage-Rumbaugh, 1991).

The majority of Kanzi's multi-symbol utterances were novel and most of his messages could not have been conveyed by single words. Unlike Nim or Washoe, many of Kanzi's utterances referred to events or objects which were

Table 3.4 *Distribution of two-element semantic relations in Kanzi's corpus*

Relation	No.	Example (of dominant order)
Action–action	92	"*Tickle bite*," then positions himself for researcher/caregiver to tickle and bite him.
Action–agent	120	"*Carry*" *person (gesture)*, gesturing to Phil, who agrees to carry Kanzi.
Agent–action	13	
Action–object	39	"*Keepaway balloon*," wanting to tease Bill with a balloon and start a fight.
Object–Action	15	
Object–agent	7	"*Balloon*" *person (gesture)*, Kanzi gestures to Liz; Liz gives Kanzi a balloon.
Agent–object	1	
Entity–demonstrative	182	"*Peanut*" *that (gesture)*, points to peanuts in cooler.
Demonstrative–entity	67	
Goal–action	46	"*Coke chase*", then researcher chases Kanzi to place in woods where coke is kept.
Action–goal	10	
Entity–entity	25	"*M&M Grape*." Caregiver/researcher: "You want both of these foods?" Kanzi vocalizes and holds out his hand.
Location–location	7	"*Sue's-office childside*," wants to go to those two places.
Location–entity	19	"*Playyard Austin*," wants to visit Austin in the playyard.
Entity–location	13	
Entity–attribute	12	"*Food blackberry*", after eating blackberries, to request more.
Attribute–entity	10	
Miscellaneous relations	37	These include low frequency (less than seven) such as attribute of action, attribute of location, affirmation, negation, and those involving an instrument.
Two-mode paraphrase	4	"*Chase*" *chase (gesture)*, trying to get staff member to chase him in the lobby.
No direct relation	6	"*Potato oil*." Kanzi commented after researcher had put oil on him as he was eating a potato.
Total	724	

absent at the time of the utterance. As illustrated in Table 3.4, Kanzi employed a wide variety of semantic relations and he tended to order symbols according to semantic function. The variety of lexically distinct combinations was so great as to make it impossible for the combinations to have been formed on the basis of lexical position preference.

The evolution of syntactical capacities

Let us now move from the issue of reference to that of syntax. Is it possible to account for the presence of these powerful structure-dependent rules in language, given the overwhelming evidence that adults do not explicitly teach children syntactical rules, nor do they correct syntactical errors?

As long as accounts of language acquisition concentrate upon what children are saying, the biologically predetermined component of language will loom large. This is because the kinds of things that children are able to say change very rapidly, with few intervening errors which can be used to illuminate the learning process. However, to the extent that accounts of language concentrate upon what children are able to comprehend, the learning process begins to lose its mystery. Unfortunately, because what children say is easier to measure and record than what they understand, very few child linguists focus upon comprehension.

The ability to understand sentences begins very early, at least by age $\frac{1}{2}$, and perhaps earlier (Golinkoff *et al.*, 1987). Indeed, the ability to parse sentences in some manner must predate the understanding of single words since the child cannot even know which sounds are words and which are sentences prior to some parsing. How does the child learn to do this? Do children come equipped with a speech-parsing mechanism?

Almost all communications directed toward children are simple, present tense, and emphasize a single noun and a single verb. The same verbs are used over and over and the same nouns are used over and over, in many different combinations, for example;

> Can you *get* the *ball*?
> Why don't you *throw* the *ball*?
> *Catch* the *ball*.
> *Put* the *ball* down now.
> *Leave* the *ball* outside.
> *Kick* the *ball*.
> *Give* your brother a *ball*.
>   or
> Why don't you *get* yourself a *cookie*?
> You must not *throw cookies*.
> Please *catch* the *cookie* that's about to fall off your plate.
> Could you *put* the *cookie* down while you drink?
> *Leave* the *cookie* on your plate.
> Please don't *kick* the *cookies* that fell on the floor.
> *Give* your friend some *cookies*.

Many other sentences that a child hears will use these verbs, and other nouns with these verbs. If the child does not know how to respond to the sentence, the caretaker will take whatever steps are needed to make sure that the response is properly carried out – not steps to ensure that the child understands, just steps to ensure that the child does what is needed or that at least whatever happens is what is needed. For example, if the child does not respond to "Put the ball down" the parent may point to the ground. If the child still does not understand, the parent may point to the ball then to the ground. If the child continues to hold the ball, the parent will ask for it with a give gesture and then place it on the ground usually with a remark such as "See, now it's down". Then the caretaker will follow with another comment such as "Now we can kick it".

Each different verb–noun combination requires a different response on the part of the child. As the number of such combinations rises, it quickly becomes neurologically inefficient to treat each new sentence that is heard as a whole, and to try to learn the relationship between that string of words and the action that the parent intends. It rapidly becomes more efficient to parse the sentence into units and abstract some commonality across units and across situations. For example, in all the sentences with "ball" above, the subject is performing some behavior towards, on, or with a large round object. Thus whenever the sound "ball" occurs, actions on that object are expected to follow. Once the sound can be parsed from the others, and once the single common element "the large round ball" is abstracted as the common element to all of these situations, the sound "ball" then becomes paired with the object "ball" in a manner that is not linked to any specific class of behaviors. Rather, its occurrence means that some specific class of behaviors is likely to be requested towards a ball. Having formed then some realization that "ball" is related in some manner to "large round objects", and that "hit", for example, is related to "the striking of one thing against another" the child may hear "hit ball" for the first time and respond appropriately by hitting the ball with his/her fist or even with some object.

This may sound like an extremely complicated task, but remember, children are usually exposed to such sounds daily for more than a year before any attempt is made even to say the first word. Far more complicated than such parsing would be a process of recalling each string of words and the behavior that was to go with them as a separate unit.

In recognizing that a child must identify both sounds and the intended meanings associated with those sounds, it is important to observe that in all the communications which the child observes or hears, it will be the case that neither actions nor objects will stand alone, they will always be in some type of relationship with one another. This is because it is not possible to produce a functional symbolic communication without constructing a relationship between, at minimum, action and object, either explicitly or implicitly.

From the perspective of early language users (be they hominids, *Homo*, or *Homo sapiens*), as long as the idea that needed to be expressed could be gotten across by a single noun and a single verb, the relationship between them was unambiguous and did not need to be specified in any way. For example, hit–ball, throw–ball, is–ball, see–ball, are as easily understood regardless of order (ball–hit, ball–throw, ball–is, ball–see). As long as such utterances are directed to a single recipient identified by direction of gaze, there is no ambiguous interpretation of the relationship between such single nouns and single verbs – the only relationship that can be manifest is inherent in the concept of the given noun, the given verb and the possible ways in which they can relate. Thus in the case of hit–ball, one might ask what or who is to hit the ball, has the ball already been hit, where was it hit, how hard was it hit, etc., yet such questions only add greater specificity to the hit–ball relationship, they do not alter it in any way.

Once the concept that one wishes to express is sufficiently complex that one or more nouns, one or more verbs, and one or more modifiers occur together, however, syntax is needed. This is because as the number of words in an utterance increases, so does the number of possible relationships among these words. Thus a nonsemantic device such as order or inflection now becomes mandatory.

Such a device *must in fact be nonsemantic* for if it were semantic it could alter the intended content carried by the "content" words. The semantic content must remain constant while the nonsemantic device declares which nouns go with which verbs, which modifiers go with which verbs, and which nouns are related to each other.

The central debate of linguistics centers around whether or not there exist some general species-specific commonalities to the ways in which different languages have come to solve the problem of ambiguity caused by having more than one noun and one verb co-occur. If such a commonality can be found across languages, it will be taken as evidence that something about the process of how to form relational rules between content words is innate.

Language learning in *Homo sapiens*

Whether or not something can be found to be common among all languages, it is already clear that specific word-sound classes are learned, specific order rules are learned, and that even specific concepts are learned in the sense that a syntactic concept extant in one culture may be completely absent from another.

It is also clear that syntax will not be needed by symbol using creatures until their form of symbolic communication becomes complex enough to generate more than a single noun and a single verb. For example, consider the sequence, modifier–noun–verb. Does the modifier modify the noun or the verb? In most cases this is solved semantically, without recourse to rules of organization because the same type of things in the real world usually cannot modify both verbs and nouns. For example, a ball can be green, but "pushes" are not green. However, this is not true for all nouns and verbs. "Kicks" can be up or down, for example, and trees can be up or down as well.

In this case the problem of ambiguity is typically solved by introducing the special verb, "to be", as in "The tree *is* down", so that "down" becomes a comment on the state of the tree. Here, "down" cannot be a comment on the state of the verb, because actions have no states, they are processes. Thus, whenever the "to be" verb is used with a modifier, the modifier is directed to the noun simply because the "to be" verb itself cannot be modified. The purpose of the copula is not one of semantic content, but rather to permit the typical noun–verb format to occur in expressions where the only action is one of existence. This is necessary because the act of formally noting existence cannot in and of itself be modified and still retain its status as a denotator of existence.

The quintessential situation which causes a need for syntactical rules arises when there is more than one noun and the verb itself is insufficient to specify the direction of this relationship, as in "put the box on the hat". "Put hat box" is ambiguous because it can be enacted in a number of different ways. In a sentence containing words such as "put", "hat" and "box," the relational problem can be solved by assigning words to all possible positional relation-ships between hat and box (in, on, under, etc.). However, in "Tickle Jane Sue" the number of potential spatial relations that can be assumed between Jane and Sue is too numerous to specify what it is that the speaker intends to have occur during a tickling bout. Some means is needed to specify agent and recipient and role reversal of these classes. In English, the chosen device is word order. In Latin, the chosen device is a case system in which nouns take on different endings indicating their grammatical roles in the sentence. In other languages, devices such as pitch may be used to accomplish the same goal.

The point of commonality here is not really that all languages use a device – leading sometimes to the erroneous conclusion that grammar must be innate. Rather, the situation of uttering multiple nouns with a single verb creates a situation to which the listener cannot properly respond knowing only the general concept typically associated with each noun and verb. The listener must be told more about the way in which the nouns are to relate via the verb in this case. A syntactical device *must* be invented if it does not exist, or the speaker will forever have to show the listener his intent, for example, by taking Jane's hand and moving toward Sue, thereby indicating that Jane is the agent and Sue is the recipient in the tickling action.

Whatever commonalities there are among grammars may well exist because only a limited number of solutions to the same problems are workable, given the constraints placed on the problem itself (oral communication in a rapidly fading sound medium, the need to limit rules to the same modality as the noun–verb–modifier so that information will be attended to, etc.).

Similar constraints have been identified for the construction of tools. The intended purpose of the tool defines, to a large extent, its form and substance. For example, things that are intended to be missiles for hurling at prey tend to have different characteristics from implements intended to be used to retrieve food from the ground. Such implements in turn are quite different in form and substance from those used to carry water. Across cultures the implements used to carry water vary markedly, yet they share certain commonalities as a result of their function. These commonalities are so strong that it is possible to identify the intended use of objects in many cultures that are now long extinct and where there exists no written record of what the function of the implement was to be. It would hardly be correct to say that an innate template existed in the human brain for making water vessels. Rather it is the problem, and the constraints placed upon possible solutions by the demands of the real world, that have led to common solutions to the same problem on many different occasions.

One would assume that any brain capable of understanding the water transport problem, whether dolphin, chimpanzee, or human, would be forced to come up with a similar solution to the problem of how to transport water on dry land – unless special features of their anatomy rules out those solutions (as in the case of the dolphin) or made possible radically different solutions (as in the case of the camel). It may similarly be argued that any brain capable of understanding the problem of determining which noun is to be linked to which verb in a complex utterance would have to come up with syntax, and that syntax would have to have certain features (for example, verb–noun units, noncontent words or morphemes that serve only to mark relationships, means of indicating past, present and future, etc.) Moreover, given that the output channel is rapidly fading and auditory, and that the human vocal tract is limited to a certain phonemic set, it would seem inevitable that completely *independent* non-innate solutions to the problem would necessarily show some overlap. By viewing language as the inevitable outcome of the social interactions of intelligent creatures, humankind may lose some sense of uniqueness, but gain in return a deeper understanding of itself.

Notes

1 In all cases where objects were placed in different locations, three or more objects were utilized. Thus the subjects could not simply go to a location and return with the first item they saw, instead they had to remember the item they were asked to retrieve.

2 One error occurred because Kanzi appropriately looked for, but did not see the object in the distal location, thus his interpretation of the sentence structure was still syntactically correct. His other error occurred when the object of the modifier "that's" was not in a distal array, but in the array immediately in front of him in the sentence "Take the potato that's in the water outdoors." Both potatoes were side by side, but only one was in a bowl of water. In this case Kanzi took both potatoes outdoors.

    Since Kanzi's performance on 1-A sentences indicated that he did *not* differentiate "in" from "next to", it seems reasonable to attribute this error to a lack of understanding of the terms of spatial contiguity (in, on, under, and next to) rather than as a misreading of the syntactical structure of the sentence.

References

Chomsky, N. (1988a). *Language and Problems of Knowledge: The Managua Lectures.* Cambridge: The MIT Press.

Chomsky, N. (1988b). *Lectures on Government and Binding: The Pisa Lectures (5th edn.).* Dordrecht: Holland.

Darley, J. M., Glucksberg, S., & Kinchla, R. A. (1986). *Psychology.* Englewood Cliffs, NJ: Prentice-Hall.

Gardner, R. A., & Gardner, B. T. (1969). Teaching sign language to a chimpanzee. *Science* 165, 664–72.

Golinkoff, R. M., Hirsch-Pasek, K., Cauley, K. M., & Gordon, L. (1987). The eyes have it: Lexical and syntactic comprehension in a new paradigm. *Journal of Child Language,* 14, 23–46.

Greenfield, P., & Savage-Rumbaugh, E. S. (1991). Imitation, grammatical development, and the invention of protogrammar by an ape. In *Biobehavioral Foundations of Language Development,* ed. N. Krasnegor, D. M. Rumbaugh, M. Studdert-Kennedy, & D. Scheifelbusch pp. 235–58. Hillsdale, NJ: Lawrence Earlbaum.

Passingham, R. (1982). *The Human Primate.* San Francisco: W. H. Freeman and Company.

Premack, D. (1970). The education of Sarah: A chimp learns language. *Science* 170, 54–8.

Rumbaugh, D. M. (1977). *Language Learning by a Chimpanzee: The Lana Project.* New York: Academic Press.

Rumbaugh, D. M., Gill, T. V., & von Glaserfeld, E. C. (1973). Reading and sentence completion by a chimpanzee. *Science,* 182, 731–33.

Rumbaugh, D. M., Richardson, W. K., Washburn, D. A., Savage-Rumbaugh, E. S., & Hopkins, W. D. (1989). Rhesus monkeys (*Macaca mulatta*), video tasks, and implications for stimulus-response spatial contiguity. *Journal of Comparative Psychology,* 103, 32–8.

Savage-Rumbaugh, E. S. (1986). *Ape Language: From Conditioned Response to Symbol.* New York: Columbia University Press.

Savage-Rumbaugh, E. S. (1988). A new look at ape language: Comprehension of vocal speech and syntax. In *Nebraska Symposium on Motivation,* vol. 35, ed. D. Leger. Lincoln, NE: University of Nebraska Press.

Savage-Rumbaugh, E. S. (1990a). Language as a cause-effect communication system. *Philosophical Psychology,* 3, 55–76.

Savage-Rumbaugh, E. S. (1990b). Language acquisition in a nonhuman species:

Implications for the innateness debate. *Developmental Psychobiology*, 23, (7), 599–620.

Savage-Rumbaugh, E. S. (1991). Language learning in the bonobo: How and why they learn. In *Biobehavioral Foundations of Language Development*, ed. N. Krasnegor, D. M. Rumbaugh, M. Studdert-Kennedy, & D. Scheifelbusch pp. 209–33. Hillsdale, NJ: Lawrence Earlbaum.

Savage-Rumbaugh, E. S., McDonald, K., Sevcik, R., Hopkins, B., and Rubert, E. (1986). Spontaneous symbol acquisition and communicative use by pygmy chimpanzees (*Pan paniscus*). *Journal of Experimental Psychology: General*, 115, 211–35.

Savage-Rumbaugh, E. S., Murphy, J., Sevcik, R., Brakke, K., Williams, S. & Rumbaugh, D. M. (in press). Language Comprehension in Ape and Child. *Monograph Series of the Society for Research on Child Development*.

Seckel, H. P. G. (1960). *Bird Headed Dwarfs*. Basel: Karger.

4

A comparative approach to language parallels

CHARLES T. SNOWDON

When we are interested in the evolutionary origins of a human trait such as toolmaking, language or cognition, we typically look to the archeological record or to the nearest relatives of humans, the great apes, for comparative evidence. Both sources contain pitfalls. Behavior is no more preserved in the archeological than in the fossil record, so that any conclusions about intelligence or language are highly speculative inferences based on a variety of assumptions that can probably never be confirmed or refuted. Debates concerning the validity of the assumptions and the logic of the inferences made from them will, no doubt, keep archaeologists busy for centuries without ever being conclusively resolved.

Data on tools, language and cognition from great apes are more tangible, but raise problems of their own. Which great apes shall we choose as models? How do we decide among models? Do we work with African apes since human origins are thought to have been in Africa? Do we choose animals which are closest to humans genetically or those that have the most similar social systems? How do we take into account the several million years since humans diverged from other apes? Finally, since speech is such an important part of human language, how can we justify as models species such as chimpanzees, bonobos or gorillas that apparently have minimal vocal communication skills (Marler & Tenaza, 1977)?

A third approach, which I will defend, seems at first glance even less likely to tell us about the origins of tools, language and cognition. This approach involves the naturalistic study (in the wild or in semi-naturalistic captive environments) of a wide range of species, not just those most closely related to humans. I argue that this approach provides us with information that complements the archeological record and the results of studies of great apes. Since we can never gain direct information on the evolution of human language and intelligence, we must approach the subject from several converging points of view. The naturalistic study of a variety of species other than the great apes is one important source of information.

In addition to justifying the study of non-hominoids, I also wish to discuss the importance of studying communication in a naturalistic environment and the importance of social influences on the expression of complex skills, and I will consider various standards for evaluating the performance of non-human animals. In this context I want to assess some of the recent claims for these animals' linguistic and cognitive accomplishments.

Why study non-hominoid animals?

For some phenomena there are simply no hominoid parallels to be found. Chimpanzees appear to be intractable to training in speech (see however Savage-Rumbaugh, this volume). Kellogg & Kellogg (1933) reared a chimpanzee with their own son and while both chimpanzee and son were given identical home environments and identical exposure to spoken language, only the son learned to speak despite the fact that the chimpanzee surpassed the son in sensory and motor tasks. Hayes & Hayes (1951) reared a chimpanzee as they would a human child and tried to teach her to speak. Although this chimpanzee displayed considerable cognitive competence (Hayes & Nissen, 1971), after seven years of training she produced only four intelligible words: *mama, papa, cup* and *up*. Lieberman (1975) has argued that chimpanzees lack the vocal tract structures to produce the full range of human phonemes. Nonetheless, chimpanzees possess, in theory, enough similar vocal tract structures to be able to produce several phonemes of human language, but still they do not speak. Observers of wild apes have commented on the relative lack of complex vocalizations in both chimpanzees and gorillas (Marler & Tenaza, 1977). The lack of evidence of complex vocalizations in wild great apes means that we must look to other species if we are interested in the evolution of the vocal aspects of language.

Converging versus diverging evolution

The difficulty of using the vocalizations of great apes as prototypical exemplars of vocal language leads us to consider an alternative evolutionary model. We generally think of evolutionary processes as inducing divergence. As two populations become isolated from each other they become more dissimilar both through genetic drift and through mutations that occur in one population that can no longer be transferred back to the other population. In addition, as populations become more isolated, each will respond via natural selection to different aspects of their environment and become increasingly divergent. Several studies have used differences in patterns of communication to make inferences about the pattern of phylogenetic differentiation or speciation among closely-related species (e.g. Lorenz, 1941 for anatid ducks; Smith, 1966 for Tyrannid flycatchers, Snowdon, in press, for Callitrichid primates). The logical outcome of this view of evolution is that populations of animals that are more similar will have diverged from one another more recently than populations that are more dissimilar. Therefore, if one wants to find the evolutionary origins of some aspect of human behavior, one might best look at great apes.

What do we do, however, when two species that we know are otherwise quite similar to each other differ in one critical respect such as complexity of vocalization? We have two alternatives. We can ask what might be different in the history of human beings as against great apes. This is discussed in several other chapters in this volume. Alternatively, we can deliberately seek out other species that do show apparent vocal complexity, document the degree of similarity of this vocal complexity to that of human speech, and then ask what is similar in the social and physical environments of humans and these other species that might have given rise to such convergence. For many traits, convergent evolution is already well documented. For example, color vision has appeared at different times in evolution, in bees, certain fishes, birds, and certain mammals. By documenting which species do and do not have color vision, we can then develop a theory of the processes that have favored selection for color vision. Similarly, if we examine a variety of other highly vocal species to see which features they do and do not share with human speech, we can then develop a theory concerning the evolution of these features. Simpson (1953) has noted that parallelism and convergence have been extremely common in evolution, and with behavior which leaves no fossil record it is extremely difficult to determine the ancestral origins of a trait. The best solution, it seems to me, is to consider both divergence and convergence hypotheses.

In the next sections I will provide data from a variety of avian and mammalian species with respect to ontogeny, categorical perception, syntax, and representational signalling in order to use the hypothesis of convergent evolution to develop ideas about the evolutionary origins of each of these phenomena.

Ontogeny

Marler (1970) argued that birds might provide better models for studies of the origins of speech and language since they are so much more vocal than the great apes and display highly complex vocal signals. He noted several parallels in the ontogenies of bird song and human language which further supported his arguments. In both birds and humans young must learn from adults, and dialect differences result from learning in different populations. In both, there is a critical period for vocal learning during which the capacity for such learning is at its peak, and species-specific dispositions guide the learning process in certain directions. Both birds and humans must have intact hearing to learn from adults and to monitor the development of their own vocalizations. Babbling has an important role in vocal practice and there is neural lateralization of vocal production. All of these characteristics of song learning by birds appear to parallel characteristics of language development in human beings (Lenneberg, 1967).

In contrast, little conclusive evidence has been found for the learning of vocal signals in non-human primates. In squirrel monkeys the deafening of animals does not disrupt vocal development (Talmage-Riggs *et al.*, 1972). Infants reared in isolation show normal vocal development (Winter *et al.*, 1973), and also respond with appropriate behavior and vocalizations on initial exposure to potential predators (Herzog & Hopf, 1984). There are population-specific vocal patterns in the isolation calls of young squirrel monkeys, but these dialect differences appear to be under genetic control rather than learned (Newman & Symmes, 1982). Playing back calls of each dialect type to adults of each population produced dialect-specific responses. Adults responded only to the calls of infants using their own dialect, and ignored the calls of isolated infants using a different dialect (Snowdon, Coe & Hodun, 1985).

In other primate species there is evidence that learning is important in the perception and appropriate use of calls, but still no evidence of vocal learning. For vervet monkeys Seyfarth & Cheney (1986) found evidence of observational and social learning in the comprehension and use of alarm calls. Hauser (1988) found that vervet monkeys having greater contact with superb starlings responded to the alarm calls of starlings at an earlier age than did infants in groups with less exposure to starlings. Similarly, he showed that infants from groups where intergroup "wrrs" were produced more frequently used these calls in appropriate contexts at an earlier age than in groups where adults gave these calls at low rates (Hauser, 1989). However, there is no evidence that experience is important in shaping the structure of these calls. Snowdon (1987) provided indirect evidence for perceptual learning of individual vocal signatures in pygmy marmosets, and Elowson, Sweet & Snowdon (1992) showed that pygmy marmosets increase the accuracy of their usage of trills with increasing age although they could find no consistent pattern in the development of adult vocal structure.

Several studies have demonstrated progressive changes in vocal production with increasing age (Seyfarth & Cheney, 1986 for vervet monkeys, Gouzoules and Gouzoules, 1989 for pigtail macaques), but only one study has provided definitive evidence of vocal learning in any non-human primate. Masataka and Fujita (1989) cross-fostered one Japanese macaque to a rhesus macaque group and two rhesus macaques to a Japanese macaque group. After examining a large number of calls from each individual, they concluded that the cross-fostered macaques imitated their adopted species on a single pitch parameter. However, since it has been so difficult to find evidence of vocal learning in any ape or monkey species, anyone interested in finding evolutionary parallels to the ontogeny of language in humans would be well-advised to stick with birds.

Why should there be such divergence in ontogeny between birds and non-human primates? Why should humans appear to be more similar in their vocal development to birds than to other primates which are much more closely related? One suggested explanation is based on a discussion by Marler (1987) of why some populations of birds show limited song learning while other populations of the same species show much more openness in song learning. Marler argued that relatively sedentary populations are unlikely to come into contact with birds of other dialects after the first few months of life, so that song learning is limited to the song types a young bird hears early in life. However, in migratory populations, birds are likely to make errors in navigation and end up in populations with dialects quite different from those with which they were reared. For these birds vocal learning must remain possible throughout their lives.

Almost all primates are much more sedentary than even the most sedentary of song birds. Dispersing primates rarely move more than a few home ranges away from their natal group, and often move to the nearest group. If, in fact, the selective pressures for vocal learning result from animals having contact with other populations with vocal variants, then primates which are highly unlikely to range far enough to join a population with a novel dialect may never have been subjected to selective pressures for vocal learning (Snowdon & Elowson, 1992).

Categorical perception

Despite the fact that non-hominoid primate models have not yet proven to be useful for understanding the ontogenetic development of speech and language, several other language-like phenomena do appear in monkeys. Categorical perception has been suggested as a unique feature of human speech (Liberman, 1982). If a continuum of sounds is synthesized between two phonemes, such as

/ba/ and /pa/, human subjects do not usually perceive a continuously varying set of sounds, but rather label the sounds as belonging to one or the other phonetic category. When asked to discriminate between two stimuli, subjects do no better than chance when the sounds both come from the same labeled category, but when two sounds, differing by the same degree of acoustic variation, are labeled as belonging to different phonetic categories the subjects do discriminate between them.

Several studies indicate that animals are capable of categorical perception of human speech sounds (Kuhl & Miller, 1975, with chinchillas; Morse & Snowdon, 1975; Kuhl & Padden, 1983; Sinnott *et al.*, 1976, Waters & Wilson, 1976, with macaques). Even birds can be trained to perceive speech sounds categorically (Kluender, Diehl & Killeen, 1987).

In addition, several studies have shown that animals can categorically perceive their own vocalizations. Snowdon & Pola (1978) demonstrated categorical labeling of trill variants by pygmy marmosets. Masataka (1983) showed categorical labeling in alarm calls of Goeldi's monkeys; May, Moody & Stebbins (1989) demonstrated categorical labeling and discrimination of coos of Japanese macaques. Ehret (1987) found that house mice could categorically perceive ultrasounds produced by pups, and Nelson & Marler (1989) demonstrated categorical perception of song elements in the swamp sparrow. Thus a wide variety of species demonstrate categorical perception of both human speech and of their own vocalizations, clearly showing that categorical perception is not unique to human beings, but has a long evolutionary history.

Syntax

Chomsky (1957) proposed that syntax rather than vocabulary size, phonetic structure or semantics was the unique characteristic of human language. Thus if non-human animals are to be useful models for studies of language origins, they must show some capacity for syntax. While many animals display some sorts of rule-based features in their utterances, Chomsky argued that a "generative" grammar was uniquely human. A generative grammar is one that can produce an infinite number of grammatical utterances, and this underwrites the complexity of human language.

Hailman, Ficken & Ficken (1985) studied the black-capped chickadee and suggested that a generative grammar was necessary to account for the sequences of the "chick-a-dee" call. This call is made up of four note types. Note A is followed by note D, or the call can begin with note B, which is followed by note C, which is followed by note D. Any of the note types can be repeated any number of times. Silence always follows the last D note. Nearly 3500 chick-a-dee calls were analyzed and 362 different sequences were found. All but 11 of these sequences followed the rules given above. Using information theory analysis Hailman *et al.* (1985) found that the chick-a-dee call contained 6.7 bits of information, compared with 11.7 bits in the average English word. Thus, the chick-a-dee call has the potential for encoding a great deal of information, and the fact that each note type can be repeated means that an infinite number of sequences can be generated. There is no evidence, however, that chickadees actually make use of the semantic potential of this system, i.e. that the variety of sequences produced is of any functional significance. Thus the semantics of this system are not known, but then Chomsky (1957) argued that semantics was not relevant to a discussion of syntax.

The chickadee provides us with the only known case of a generative grammar in non-human animals, but examples of simple grammars exist in several species. Using a bioassay of female copulation posture, Ratcliffe & Weisman (1987) showed that female cowbirds are sensitive to the sequence of ordering of phrases in song. They also found that chickadees could discriminate between normal and altered sequences of notes (Ratcliffe & Weisman, 1986). Robinson (1979) found that titi monkeys organized syllables into sequences with predictable orders, and when these orders were rearranged and played back to the monkeys, they could discriminate between normal and abnormal sequences.

Marler (1977) related syntax to semantics. He defined Phonetic Syntax as the combining of individual notes or syllables into a sequence where the sequence represents something different from the individual elements, analogous to forming a word from different phonemes. He defined Lexical Syntax as the combination of elements into a sequence where the sequence retained the meaning of the individual units, analogous to the formation of a phrase or sentence from individual words. Marler predicted that lexical syntax was a property of human language and would rarely be found in animals.

A few examples of lexical syntax have emerged. Robinson (1984) found that 38% of the vocalizations of wedge-capped capuchin monkeys were sequences of calls where the sequence was used in contexts intermediate between those in which the individual units would be given. Cleveland & Snowdon (1982) also found two types of vocal sequences in cotton-top tamarins that were used in situations intermediate between those where one or the other of the individual elements would be used.

However, although these examples indicate that simple – and even generative – grammars can be found in non-human animals, there is an enormous gap between these grammars and the syntax of human speech. Few parallels exist and therefore the complex syntax of human language must be of fairly recent origin.

Representational or symbolic communication

After syntax the most commonly sought parallel to human language has been symbolic communication. Do animals have in their natural communication anything equivalent to words? In studies with chimpanzees and other animals trained to use language analogues, the issue of whether animals can use arbitrary symbols or gestures to represent objects or actions has been resolved. Chimpanzees and bonobos can acquire arbitrary symbols and use these symbols to represent objects and actions both in making and in responding to requests (see Savage-Rumbaugh, this volume). It has proven to be more difficult to provide unambiguous evidence that animals use symbolic signals in their natural communication.

Many species use different types of vocalizations in response to aerial and ground predators respectively. The best known example of differential predator calls comes from vervet monkeys; Seyfarth, Cheney & Marler (1980) demonstrated calls specific to martial eagles, to snakes and to leopards. When these calls were played back through a hidden loudspeaker, in the absence of the actual predators, the vervet monkeys responded in an appropriate fashion. Thus after an eagle alarm was played, the monkeys in trees would run to the ground and take cover in the brush. When a leopard alarm was played, animals on the ground would climb into trees. These results indicate that listeners can infer what type of predator is present from the structure of the call alone. But is this evidence of referential or symbolic communication?

California ground squirrels also have separate calls for aerial and ground predators, and when these vocalizations are played back through hidden speakers animals take different actions. In response to a ground predator call animals become alert and scan the horizon. In response to the aerial alarm call, they rapidly take cover in their burrows. However, Owings & Hennessey (1984) have reinterpreted these calls as reflecting the urgency of response rather than direct reference to predator type. They observed that ground squirrels give the "aerial" alarm for a terrestrial predator who has approached close to the burrow without being noticed, and the "ground" predator call is often given for aerial predators sighted at a distance. These calls can be interpreted in nonlinguistic terms as indicating what actions the caller is going to take next. It is not necessary to argue that the calls "symbolize" a particular predator type.

The major method used to test for whether communication is referential or symbolic is the playback method, and this has only limited usefulness. When a stimulus is presented in a playback test, we can observe how the animals react. If they respond "appropriately", we can conclude that listeners are able to make good inferences from the information in the signal, but that in itself is not sufficient to determine that the signal was referential or symbolic. We can listen to the non-verbal utterances of other humans, and on the basis of learned associations between those utterances and specific contexts, we can make a good inference about what is happening even though the communicator is not using words. May not the same be possible with animals? A monkey, a squirrel or a bird hearing a call that indicates a degree of urgency to escape, or of fear, may have learned a connection to a specific predator type without the caller being able actually to symbolize the nature of the predator. This learned association between a specific predator and an emotionally driven signal elicited by its presence may be the basis on which symbolic communication eventually evolved.

Cheney & Seyfarth (1988) have developed a clever modification of the playback technique to circumvent some of these problems. They presented vervet monkeys with one of two forms of intergroup alerting calls given by an individual and then played back over the next few hours several examples of a different type of intergroup alerting call from the same individual. With repeated presentations responses to this latter call declined, showing habituation. When the listeners were again tested with the first intergroup alerting call, they continued to display the habituation displayed to the second form of call. In contrast, when eagle and leopard alarm calls were used, listeners habituated to the repeated presentation of one alarm call but did not transfer this habituation to the test with a different alarm call. Thus, semantic categories are distinct for different types of predators.

Another class of referential calls that has been well studied are those associated with food. Dittus (1984) reported that toque macaques gave certain calls upon the discovery of a high-quality food resource such as a tree of ripe figs and interpreted these calls as symbolizing high-quality food. However, on a few occasions the calls were also given in response to other pleasurable events, such as the first appearance of the sun after the monsoon or the first rain clouds at the end of the dry season. There is obviously an affective component to these calls. Marler, Dufty & Pickert (1986a) have described food calls in chickens, showing that the rate of calling varied with the quality and amount of food available. Hauser & Wrangham (1987) reported that captive chimpanzees emitted food calls at a rate that correlated directly with the amount of food available.

Recently, Elowson, Tannenbaum & Snowdon (1991) studied food-associated calls in cotton-top tamarins. Of these calls, 97% were given only in the presence of food. Non-food, manipulable objects, the size of food pieces, did not elicit calls. Each individual tamarin had its own ranking of food preferences, and for individual tamarins there was a close correlation between food preference and the number of calls given when that food was presented.

There has been a tendency to dichotomize referential and affective signals in studies of animal communication, but the results of studies of food-associated calls in several species indicate that these calls have both a referential and an emotive component. Food-associated calls do indicate the presence of a preferred food, but they also indicate an individual's interest in or desire for a particular food. Predator calls also have an emotive component to them, and part of the variation in calling to different types of predators and in response to the playbacks of these calls may reflect this emotive component. Animal signals can convey information about specific classes of object in the environment, but as yet there is no conclusive evidence for their having a purely symbolic function. Both affective and referential functions can coexist in the same signal.

Importance of social context

Most research on human speech and language as well as on animal communication has focused on the utterances and responses of single individuals. However, the proper metaphor for communication is not the utterance or sound but the conversation (Snowdon, 1988). Communication is a highly social process, and yet in attempts to simplify the analysis of communication and language we have all too often removed the social context. In the next sections I will discuss social influences affecting ontogeny, categorical perception, syntax and how signals are used with different social companions.

Social influences on ontogeny

Contrary to the expectation that animals should be hard-wired to detect and avoid predators, a variety of studies have shown the importance of social learning and observational learning in developing a fear of predators and appropriate predator responses. Seyfarth & Cheney (1986) showed that young vervet monkeys give alarm calls to a variety of objects in contrast to the specificity of adults. For example, eagle alarm calls might be given to starlings, to hawks and to falling leaves as well as to eagles. However, adult vervet monkeys responded only to those infant alarm calls given to the actual predator (a form of social reinforcement), and infants often waited to give alarm calls until after an adult had given one (observational learning). Hauser

(1988) described a gradual development of infant vervet monkeys' responses to the alarm calls of superb starlings. In addition, infants in groups with more exposure to alarm calls from starlings responded to these calls at an earlier age than did infants from groups with less exposure to starlings. Thus vervet monkeys can even learn about predators from other species.

We attempted to study predator alarm calls in our colony of captive cotton-top tamarins, but quickly discovered that captive born animals showed minimal fear and few responses to a natural predator, a boa constrictor (Hayes & Snowdon, 1990). When a live boa constrictor was presented to a group mild alarm responses were given, but these quickly habituated by the second day of testing. In no case were the responses as severe as the snake-mobbing responses that have been observed in encounters with snakes by wild tamarins (Bartecki & Heymann, 1987). Furthermore, when the captive tamarins were presented with a laboratory rat, they gave the same intensity of response as to the snake (Hayes & Snowdon, 1990). Even when a live hawk was brought into the colony many animals reacted with curiosity rather than alarm (unpublished observations). Thus, cotton-top tamarins require social learning of responses to predators just as vervet monkeys do.

Social aspects of categorical perception

Categorical perception is a puzzling phenomenon. A listener appears to ignore information that is present in a sound. Snowdon & Pola (1978) tested human subjects with synthesized pygmy marmoset trills that pygmy marmosets labelled categorically. Human subjects were able to make clear discriminations between sounds that pygmy marmosets treated as equivalent. Why should human subjects be able to make finer distinctions between pygmy marmoset calls than the marmosets themselves? If we think about the social functions of communication, these results are not so puzzling. When we first hear a speaker we may require broad perceptual categories in order to classify the speaker's sounds, but with a familiar voice, we can be much more precise in perceptual categories. Our need to recognize individual differences varies with the nature of the communication. If someone yells "Fire!" in a crowded theatre, it is important to attend to the phonetic information, but if we hear someone whisper "Kiss me, I love you" it is very important that we determine the age, gender and individual speaking before we take action. In a real social environment we attend to many non-phonetic cues.

To evaluate the role of individual differences in categorical perception we replicated our study with pygmy marmosets but made some important design changes (Snowdon, 1987). Instead of asking whether monkeys labeled trills as of one type or another we used a response measure that indicated whether monkeys recognized a specific individual or not. We synthesized trills to mimic those of specific individuals in the colony and then varied these calls systematically on the duration dimension which had been the dimension of categorical labeling. When specific calls from familiar individuals were used and animals were asked to identify a specific individual, there was clear within-category perception. Whether monkeys perceive a sound categorically or not depends on the testing paradigm used and on whether characteristics of familiar individuals are incorporated into the stimuli. These results cast doubt on the reality of categorical perception as a phenomenon. The results also provide evidence of perceptual learning for the vocal features of social companions. It seems probable that humans can also make fine within-category distinctions, but we are not yet able to synthesize such fine details in human speech.

Social aspects of syntax

One definition of syntax is any rule-based system that predicts patterns of utterances. In a social context syntax can refer to the ordering in patterns of communication between individuals. Examples of duetting are found in a wide variety of bird species (Farabaugh, 1982) and duetting is also common in some primates. Deputte (1982) has described the complex coordination of singing between male and female white-cheeked gibbons. Robinson (1979) and Kinzey & Robinson (1983) have described duetting patterns in two species of titi monkeys. Snowdon & Cleveland (1984) described turn-taking behavior in a group of three pygmy marmosets. There were significantly more sequences of each of the three animals calling in turn than there were sequences where an animal called two or three times in succession. One pattern of calling (animals 1, 2 and 3) was more often found than the other order of calling (animals 1, 3 and 2). McConnell & Snowdon (1986) simulated territorial encounters among captive groups of cotton-top tamarins and found that there was a structure to their vocal bouts. If the present call was answered by someone in the caller's own group, the subsequent call was one indicating a more intense level of aggression. On the other hand responses from the group other than the one from which the call originated tended to imitate the call that was given.

Responses to different audiences

Many species appear to alter their signals as a function of who is present as a possible audience. Sherman (1977) reported that ground squirrels give alarm calls more often when close kin are present than they do when only non-kin are present. Cheney & Seyfarth (1985) reported that vervet mothers give alarm calls more often when their own infants are present than when another infant is present. Bayart *et al.* (1990) have shown that infant rhesus macaques give structurally different forms of coo vocalizations when they are separated from their mothers yet housed so as to retain visual contact with them, and when they are separated and housed out of sight of their mothers. Recently, Kalin, Shelton & Snowdon (in preparation) have shown that infant rhesus macaques briefly separated from their mothers give different calls depending on whether the mother is visible, an adult male is visible, or no conspecific is visible at all during the separation. Different calls are also given upon reunion with the mother, and upon union with the male. Thus monkeys and ground squirrels are able to differentiate between the presence of kin and non-kin, and alter their calling accordingly.

Owings *et al.* (1986) found that California ground squirrels gave non-repetitive alarm calls more often after their pups had emerged from the den than prior to their emergence. Karakashian, Gyger & Marler (1988) found that chickens gave more alarm calls in the presence of a cock or a hen than they did when another species was present or when no animal was present. Marler, Dufty & Pickert (1986b) found that cocks gave food calls in the presence of their mates or a strange hen but rarely called in the presence of a rival cock or when alone.

In a very different type of study Snowdon & Hodun (1981) recorded trill vocalizations from wild pygmy marmosets in the Peruvian Amazon. Several variations of trills were described in captive pygmy marmosets which varied in the acoustic cues that might be used for sound localization. Snowdon & Hodun hypothesized that in the wild it would be adaptive for marmosets to use the most cryptic calls with the fewest localization cues when they were close to other group members, and that increasing cues for localization would be added as monkeys were further and further separated from their group. Recordings were made of several examples of each type of trill, and the distances between the caller and the nearest observable conspecific were calculated. There was a close relationship between cues for sound localization and distance to other group members. The most cryptic call forms were used at very close distances and the most easily localized forms were used at the greatest distances. The monkeys must be estimating how far they are from the rest of their group and adjusting the structure of their calls accordingly.

All of these studies show that animals do not call reflexively in response to a particular stimulus type, but rather modulate their signals according to the nature of the audience present or the distance to the nearest neighbor. We typically assume, but rarely bother to prove, that humans modulate their speech according to their intended audiences. That such a broad array of animal species also modulate their calls according to the audience present suggests that this phenomenon is quite widespread and phylogenetically quite old.

How should we evaluate the accomplishments of non-human animals?

Finding the appropriate criteria for evaluating what non-human animals can do has been quite complex. In some of the criticisms leveled against the chimpanzee language-training studies psychologists have tried to hold chimpanzees to a stricter criterion for language development than one would even hold for human children (e.g. Terrace *et al.*, 1979). We can consider two evaluative extremes. Are chimpanzees, non-human primates, ground squirrels or birds showing the full complexity of human language? The answer is obviously "No!" Even the most successful of the ape-language projects has not produced a chimpanzee or bonobo with all of the linguistic skills and accomplishments of a human child of the same age. Primates, other than great apes, other mammals, and birds are even less impressive by these standards.

The other evaluative extreme would be that non-human animals have no capacity for language or language-like phenomena. Is this any more true? Again the answer is "No". It is clear that non-human animals display some precursors of language. Many species are capable of producing complex sounds. In some species these sounds can be combined to form complex sequences, at least some of which are superficially analogous to simple phrases. Some animals, especially birds, learn how to produce sounds, and learn what sounds to produce from adult animals. Even primates, for which there is little evidence that the productive capacities of vocalizations are learned from adults, still display perceptual learning of sounds and rely on observational learning and social reinforcement to use sounds appropriately. Many animals are sophisticated communicators, altering the structure or amount of calls according to their distance from other group members or according to the type of conspecific present. Non-human animals are not simple automata which call reflexively in response to fixed environmental stimuli.

The approach to studying the origins of language that I have presented here is quite different from that of Savage-Rumbaugh (this volume). She and her colleagues have achieved remarkable success in demonstrating the linguistic potential of the bonobo Kanzi and other apes. Their success has been due to their creation of a highly social, interactive environment coupled with an objective method of recording the responses of the apes. They have demonstrated quite clearly what apes can achieve, given a high quality environment with caring social companions who work to elicit linguistic utterances and responses. However, the environment in which these apes are tested is not at all like the normal social environments in which these animals evolved. The differences between Savage-Rumbaugh's approach and the one I have presented is between determining what animals can do with explicit training in language and what they actually do in their natural environments without the benefit of training. The two approaches are complementary, each illuminating a different aspect of the evolution of language.

Although studies of natural communication systems in a wide array of species have not led to the stunning capacities that Kanzi has displayed, they are still of great value. First, by studying a diversity of species, we can discover which phenomena are widespread, and hence either of early phylogenetic origin or consequent upon convergent evolution. In the latter case, we can try to extract common environmental factors that might have led to the evolution of the same trait in several different species. We can learn more directly about the process of the evolution of language through the study of a large number of diverse species.

Second, by studying the communication of animals in their natural social groups, we can often discover new phenomena of interest that we might not have noticed had we only used a paradigm of training animals with human language analogues. The discovery of new phenomena enriches our notion of human language and can lead to new studies of human beings. For example, it is likely that we alter our speech according to whether our intended recipient is close or far away, yet I am aware of no studies of how human speech varies with distance that parallel those of Snowdon & Hodun (1981) on pygmy marmosets. The finding that monkeys must learn how to identify predators suggests that our own fear of snakes is not innate but culturally transmitted. Hundreds of studies have used synthesized speech sounds to explore categorical perception, and numerous authors have claimed this to be a unique attribute of human speech. However, the demonstrations that monkeys, chinchillas and even quail can categorize human speech suggests that this is hardly a uniquely human phenomenon. The finding that pygmy marmosets will categorize calls if tested with synthesized calls that mimic no particular individual, but will respond to subtle within-category differences if tested with calls synthesized to mimic familiar social companions, suggests that much of categorical perception may be an epiphenomenon of the type of stimuli and testing instructions used. Here, animal studies can suggest methodological changes to improve studies of humans. Finally, the animal studies reviewed here have shown the importance of social context in communication. This may seem to be an obvious point, but much research on human speech and language has avoided its social aspects. We will have a more complete understanding of human speech and language when we include a consideration of the importance of social influences in our research.

Despite the value of studying a wide array of animal species this approach also suffers from some limitations. Many of the species which appear so impressive in one or another aspect of communication show striking limitations in other aspects. Cheney & Seyfarth (1990), in a summary of their 15 years of research on vervet monkeys, conclude that despite the ability of vervets to refer to specific predators, to form semantic categories, to detect unreliable signallers and to understand which animals are related to their friends and which to their enemies, they are unable to act as though they can understand what information other animals have. Thus despite a very sophisticated and complex communication system, vervet monkeys are mentally egocentric. At least some descriptions of the language-using apes suggest that apes are also egocentric and unable to understand the minds of their companions. Marmosets and tamarins, which have a complex vocal communication system, do not appear to have the same social sophistication as vervet monkeys. The complex vocal syntax of the black-capped chickadee is not accompanied by any corresponding semantic complexity.

By studying the communication of a broad array of species in their natural or naturalistic environments we can better understand what animals are able to do in these environments, and this in turn can provide new insights for looking at our own human language. Given the range of findings, in relation to diverse species, no simple, linear model can adequately account for the evolution of language and intelligence. Many of the components of human language and intelligence have appeared at different times in evolution, but only in human beings have all of the components that define our linguistic abilities come together in a single species.

Acknowledgments

Preparation of this chapter and the author's research has been supported by USPHS Grant MH 29,775 and a National Institute of Mental Health Research Scientist Award. I thank Bertrand Deputte, A. Margaret Elowson, Jack P. Hailman, Keith Kluender, W. John Smith and Karen B. Strier for discussion of the ideas presented here.

References

Bartecki, U. & Heymann, E. W. (1987). Field observations of snake-mobbing in a group of saddleback tamarins, *Saguinus fuscicollis nigrifrons*. *Folia Primatologica*, 48: 199–202.

Bayart, F., Hayashi, K. T., Faull, K. T., Barchas, J. D. & Levine, S (1990). Influence of maternal proximity on behavioral and physiological responses to separation in infant rhesus monkeys (*Macaca mulatta*). *Behavioral Neuroscience*, 104: 98–107.

Cheney, D. L. & Seyfarth, R. M. (1985). Vervet monkey alarm calls: manipulation through shared information. *Behaviour*, 94: 150–66.

Cheney, D. L. & Seyfarth, R. M. (1988). Assessment of meaning and the detection of unreliable signals by vervet monkeys. *Animal Behaviour*, 36: 477–86.

Cheney, D. L. & Seyfarth, R. M. (1990). *How Monkeys See the World*. Chicago: University of Chicago Press.

Chomsky, N. (1957). *Syntactic Structures*. The Hague: Mouton.

Cleveland, J. & Snowdon, C. T. (1982). The complex vocal repertoire of the adult cotton-top tamarin (*Saguinus oedipus*). *Zeitschrift fuer Tierpsychologie*, 58: 231–70.

Deputte, B. L. (1982). Duetting in male and female songs in the white-cheeked gibbon (*Hylobates concolor leucogenys*). In *Primate Communication*, ed. C. T. Snowdon, C. H. Brown & M. R. Petersen, pp. 67–93. New York: Cambridge University Press.

Dittus, W. P. J. (1984). Toque macaque food calls: Semantic communication concerning food distribution in the environment. *Animal Behaviour*, 32: 470–7.

Ehret, G. (1987). Categorical perception of speech sounds: Facts and hypotheses from animal studies. In *Categorical Perception*, ed. S. Harnad, pp. 301–31. New York: Cambridge University Press.

Elowson, A. M., Sweet, C. S. & Snowdon, C. T. (1992). Ontogeny of trill and J-call vocalizations in the pygmy marmoset (*Cebuella pygmaea*). *Animal Behaviour*, 43: 703–15.

Elowson, A. M., Tannenbaum, P. L. & Snowdon, C. T. (1991). Food associated calls correlate with food preferences in cotton-top tamarins. *Animal Behaviour*, 42, 931–7.

Farabaugh, S. M. (1982). The ecological and social significance of duetting. In *Acoustic Communication in Birds, Vol. 2: Song Learning and its Consequences*, ed. D. E. Kroodsma & E. H. Miller, pp. 85–124. New York: Academic Press.

Gouzoules, H. & Gouzoules, S. (1989). Design features and developmental modification of pigtail macaque, *Macaca nemestrina* agonistic screams. *Animal Behaviour*, 37: 381–401.

Hailman, J. P., Ficken, M. S. & Ficken, R. W. (1985). The "chick-a-dee" call of *Parus atricapillus*: A recombinant system of animal communication compared with written English. *Semiotica*, 56: 191–224.

Hauser, M. D. (1988). How infant vervet monkeys learn to recognize starling alarm calls: the role of experience. *Behaviour*, 105: 187–201.

Hauser, M. D. (1989). Ontogenetic changes in the comprehension and production of vervet monkey (*Cercopithecus aethiops*) vocalizations. *Journal of Comparative Psychology*, 103: 149–58.

Hauser, M. D. & Wrangham, R. W. (1987). Manipulation of food calls in captive chimpanzees. *Folia Primatologica*, 48: 207–10.

Hayes, K. J & Hayes, C. (1951). The intellectual development of a home-raised chimpanzee. *Proceedings of the American Philosophical Society*, 95: 105.

Hayes, K. J. & Nissen C. H. (1971). Higher mental functions of a home-raised chimpanzee. In *Behavior of Nonhuman Primates: Modern Research Trends, Vol. 4*, ed. A. M Schrier & F. Stollnitz, pp. 59–115. New York: Academic Press.

Hayes, S. L. & Snowdon, C. T. (1990). Predator recognition in the cotton-top tamarin (*Saguinus oedipus*). *American Journal of Primatology*, 20: 283–91.

Herzog, M. & Hopf, S. (1984). Behavioral responses to species-specific warning calls in infant squirrel monkeys reared in social isolation. *American Journal of Primatology*, 7: 99–106.

Karakashian, S. J., Gyger, M. & Marler, P. (1988). Audience effects on alarm calling in chickens (*Gallus gallus*). *Journal of Comparative Psychology*, 102: 129–35.

Kellogg, W. N. & Kellogg, L. A. (1933). *The Ape and the Child*. New York: Hafner Publishing.

Kinzey, W. G. & Robinson, J. G. (1983). Intergroup calls, range size and spacing in *Callicebus torquatus*. *American Journal of Physical Anthropology*, 60: 539–44.

Kluender, K. R., Diehl, R. L. & Killeen, P. R. (1987). Japanese quail can learn phonetic categories. *Science*, 237: 1195–7.

Kuhl, P. K. & Miller, J. D. (1975). Speech perception in the chinchilla: voiced–voiceless distinction in alveolar plosive consonants. *Science*, 190: 69–72.

Kuhl, P. K. & Padden, D. M. (1983). Enhanced discriminability at the phonetic boundary for the place feature in macaques. *Journal of the Acoustical Society of America*, 73: 1003–10.

Lenneberg, E. (1967). *Biological Foundations of Language*. New York: Wiley.

Liberman, A. M. (1982). On finding that speech is special. *American Psychologist*, 37: 148–67.

Lieberman, P. (1975). *On the Origins of Language*. New York: Macmillan.

Lorenz, K. (1941). Comparative studies of the motor patterns of Anatinae. Translation from the German printed in K. Lorenz (1971). *Studies in Human and Animal Behavior, Vol. 2* pp. 14–114. Cambridge, MA: Harvard University Press.

Marler, P. (1970). Birdsong and human speech: Can there be parallels? *American Scientist*, 58: 669–74.

Marler, P. (1977). The structure of animal communication sounds. In *Dahlem Workshop on the Recognition of Complex Acoustic Signals*, ed. T. H. Bullock, pp. 17–35. Berlin: Dahlem Konferenzen.

Marler, P. (1987). Sensitive periods and the roles of general and specific sensory stimulation in birdsong learning. In *Imprinting and Cortical Plasticity*, ed. J. P. Rauschecker & P. Marler, pp. 99–135. New York: Wiley.

Marler, P., Dufty, A. & Pickert, R. (1986a). Vocal communication in the domestic chicken: I. Does a sender communicate information about the quality of a food referent to a receiver? *Animal Behaviour*, 34: 188–93.

Marler, P., Dufty, A. & Pickert, R. (1986b). Vocal communication in the domestic chicken: II. Is a sender sensitive to the presence and nature of a receiver? *Animal Behaviour*, 34: 194–8.

Marler, P. & Tenaza, R. (1977). Signalling behavior of apes with special reference to vocalizations. In *How Animals Communicate*, ed. T. E. Sebeok, pp. 965–1033, Bloomington: Indiana University Press.

Masataka, N. (1983). Categorical responses to natural and synthesized alarm calls in Goeldi's monkeys (*Callimico goeldii*). *Primates*, 24: 40–51.

Masataka, N. & Fujita, K. (1989). Vocal learning of Japanese and rhesus monkeys. *Behaviour*, 109: 191–9.

May, B., Moody, D. B. & Stebbins, W. C. (1989). Categorical perception of conspecific communication sounds by Japanese macaques, *Macaca fuscata*. *Journal of the Acoustical Society of America*, 85: 837–47.

McConnell, P. B. & Snowdon, C. T. (1986). Vocal interactions among unfamiliar groups of captive cotton-top tamarins. *Behaviour*, 97: 273–96.

Morse, P. A. & Snowdon, C. T. (1975). An investigation of categorical speech discrimination by rhesus monkeys. *Perception and Psychophysics*, 17: 9–16.

Nelson, D. S. & Marler, P. (1989). Categorical perception of a natural stimulus continuum: Birdsong. *Science*, 244: 976–8.

Newman, J. D. & Symmes, D. (1982). Inheritance and experience in the acquisition of primate acoustic behavior. In *Primate Communication*, ed. C. T. Snowdon, C. H. Brown & M. R. Petersen, pp. 259–78, New York: Cambridge University Press.

Owings, D. H. & Hennessey, D. F. (1984). The importance of variation in sciurid visual and vocal communication. In *The Biology of Ground Dwelling Squirrels*, ed. J. O. Murie & G. L. Michener, pp. 169–200. Lincoln: University of Nebraska Press.

Owings, D. H., Hennessey, D. F., Leger, D. W., & Gladney, A. B. (1986). Different functions of "alarm" calls for different time scales: A preliminary report on ground squirrels. *Behaviour*, 99: 101–16.

Ratcliffe, L. & Weisman, R. G. (1986). Song sequence discrimination in the black-capped chickadee (*Parus atricapillus*). *Journal of Comparative Psychology*, 100: 361–7.

Ratcliffe, L. & Weisman, R. (1987). Phrase order recognition by brown headed cowbirds. *Animal Behaviour*, 35: 1260–2.

Robinson, J. G. (1979). An analysis of the organization of vocal communication in the titi monkey, *Callicebus moloch*. *Zeitschrift fuer Tierpsychologie*, 49: 381–405.

Robinson, J. G. (1984). Syntactic structures in the vocalizations of wedge-capped capuchin monkeys. *Behaviour*, 90: 46–79.

Seyfarth, R. M. & Cheney, D. L. (1986). Vocal development in vervet monkeys. *Animal Behaviour*, 34: 1640–58.

Seyfarth, R. M., Cheney, D. L. & Marler, P. (1980). Vervet monkey alarm calls: semantic communication in a free-ranging primate. *Animal Behaviour*, 28: 1070–94.

Sherman, P. W. (1977). Nepotism and the evolution of alarm calls. *Science*, 197: 1246–53.

Simpson, G. G. (1953). The study of evolution: Methods and present status of theory. In *Behavior and Evolution*, ed. A. Roe & G. G. Simpson, pp. 7–26. New Haven: Yale University Press.

Sinnott, J. M., Beecher, M. D., Moody, D. B. & Stebbins, W. C. (1976). Speech sound discrimination by monkeys and humans. *Journal of the Acoustical Society of America*, 60: 687–95.

Smith, W. J. (1966). Communication and relationships in the genus *Tyrannis*. *Monographs of the Nuttall Ornithological Club*, 6: 1–250.

Snowdon, C. T. (1987). A naturalistic view of categorical perception. In *Categorical Perception*, ed. S. Harnad, pp. 332–54. New York: Cambridge University Press.

Snowdon, C. T. (1988). Communications as social interaction: Its importance in ontogeny and adult behavior. In *Primate Vocal Communication*, ed. D. Todt, P. Goedeking & D. Symmes, pp.108–22. Berlin: Springer Verlag.

Snowdon, C. T. (in press). A vocal taxonomy of the Callitrichids. In *Marmosets and Tamarins: Systematics, Ecology and Behaviour*, ed. A. B. Rylands. Oxford: Oxford University Press.

Snowdon, C. T. & Cleveland, J. (1984). "Conversations" among pygmy marmosets. *American Journal of Primatology*, 7: 15–20.

Snowdon, C. T., Coe, C. L. & Hodun, A. (1985). Population recognition of infant isolation peeps in the squirrel monkey. *Animal Behaviour*, 35: 1146–51.

Snowdon, C. T. & Elowson, A. M. (1992). Ontogeny of primate vocal communication. In *Topics in Primatology, Vol. 1: Human Origins*, ed. T. Nishida, W. C. McGrew, P. Marler, M. Pickford & F. B. M. de Waal, pp. 279–90. Tokyo: University of Tokyo Press.

Snowdon, C. T. & Hodun, A. (1981). Acoustic adaptations in pygmy marmoset contact calls: Locational cues vary with distance between conspecifics. *Behavioral Ecology and Sociobiology*, 9: 295–300.

Snowdon, C. T. & Pola, Y.V. (1978). Interspecific and intraspecific responses to synthesized pygmy marmoset vocalizations. *Animal Behaviour*, 26: 196–206.

Talmage-Riggs, G., Winter, P., Ploog, D. & Mayer, W. (1972). Effect of deafening on the vocal behavior of the squirrel monkey (*Saimiri sciureus*). *Folia Primatologica*, 17: 404–20.

Terrace, H. S., Petito, L. A., Saunders, R. J. & Bever, T. G. (1979). Can an ape create a sentence? *Science*, 206: 891–902.

Waters, R. S. & Wilson, W. A. (1976). Speech perception by rhesus monkeys: The voicing distinction in synthesized labial and velar stop consonants. *Perception and Psychophysics*, 19: 285–9.

Winter, P., Hadley, P., Ploog, D. & Schott, D. (1973). Ontogeny of squirrel monkey calls under normal conditions and under acoustic isolation. *Behaviour*, 47: 230–9.

Part II

Technological skills and associated social behaviors of the non-human primates

Part II Introduction

Generative interplay between technical capacities, social relations, imitation and cognition

KATHLEEN R. GIBSON

The tool-using capacities of modern *Homo sapiens* render our species both the most constructive and the most destructive life form on this planet. What we do with tools not only profoundly alters our own lives, but also those of all other living beings, both animal and vegetable. How we came to be such a constructive and destructive species stands as one, yet unresolved, evolutionary problem. Viewed individually, the diverse human accomplishments seem unimpressive. Termites, honey bees and beavers are accomplished architects. Animals as diverse as digger wasps and Eygyptian vultures use tools. Recently, a bonobo has become a habitual maker and user of stone tools and, in the process, belied the common designation of humans as "the only animals who use a tool to make a tool" (Toth & Schick, 1991). What then does distinguish our species? Chapters in this and other sections address this issue from the diverse perspectives of anthropology, primatology and neurology.

The diminutive South American capuchin monkey (*cebus* spp.), beloved of organ grinders and laboratory primatologists, may well match humans in its destructive tendencies, both in the wild (Terborgh, 1983) and in captivity. Some capuchins are extremely inventive, seemingly able to solve any problem that involves banging, tearing, or probing, with or without tools. Capuchins fall far short of humanity, however, in their constructive (Gibson, 1990) and representational skills (Langer, this volume) and have rarely displayed tool-use in the wild.

The group of *Cebus apella* in the Rome zoo form the basis of a continuing series of studies by Elisabetta Visalberghi. Some of these monkeys have acquired considerable proficiency in the use of hammerstones. In the current study, some of them also mastered the use and modification of sticks to dislodge peanuts from plexiglass tubes. The monkeys, however, showed curious blind spots, often inserting two short sticks at either end of the tube, thereby thwarting their own efforts, a problem Visalberghi attributes to lack of representational capacities (see also Langer, this volume). Nor did a single monkey show any ability to learn from the others. Imitation seemed totally beyond their capacity, and tool-using traditions could not pass from one to the other.

Andrew Meltzoff, a specialist in human infancy, has recently declared that humans deserve the designation *Homo imitans* (Meltzoff, 1988). Visalberghi echoes this theme, noting the exponential effects of combining inventiveness with imitation to allow tool-using and tool-making schemata to spread throughout a population. She also follows Tomasello (1990) in questioning whether any non-human primates possess true imitative capacities. That chimpanzees do imitate has, however, been repeatedly demonstrated in ape-language studies (Terrace, 1979) and is implied by population differences in chimpanzee tool-using traditions in the wild. Boesch corroborates chimpanzee imitational capacities. He notes that mother chimpanzees in the Tai Forest behave as if they expect their young to imitate them. Specifically, one mother chimpanzee was observed demonstrating the proper spatial orientation between a hammer and a nut to her five-year-old offspring. The juvenile then repeatedly attempted to hammer while faithfully maintaining the instrument in the spatial orientation demonstrated by the mother.

The behavior of this mother–young dyad illustrates an important point. Imitation can exist in isolation among species, such as songbirds, that show no attempt to engage in active instruction. Imitation combined with active teaching, however, paves the way for complex cultural traditions. Conse-quently, it is of interest that Boesch interprets the behavior of this mother–offspring pair as an example of true pedagogy. Pedagogy involves joint interactions between tutor and novice such that each observes the other's behavior and the tutor intervenes to correct poor technique on the part of the novice, as did this chimpanzee mother.

It is important to note, however, that this animal's demonstration of an appropriate hammer position is the only example of pedagogy yet described in wild apes. Furthermore, both the imitative and "instructional" behaviors displayed by these animals fell far short of those common to many human endeavors as described by Meltzoff (1988) and Wynn (this volume). The chimpanzees observed by Boesch merely demonstrated and copied the proper positioning of the hammer on the nut. The younger animal did not imitate its mother's actual postures and motor actions. Indeed, the younger chimpanzee, while maintaining the appropriate orientation of the hammer with respect to the nut, repeatedly changed its own body and hand positions, often using extremely poor hammering technique. The mother, however, failed to inter-vene with demonstrations of appropriate postures and muscle movements.

Humans, however, often demonstrate and copy detailed sequences and patterns of motor skills. This can be quite complicated as in the master–apprentice relationships observed among carpenters and tool-makers or in the relationships between instructors and students of ballet, gymnastics and other motor routines. In these instances, the student carefully observes and copies whole patterns of behavior, while the tutor watches and corrects the student's performance, often actively positioning the student's hands, feet, or other body parts. Consequently, as impressive as the Tai and captive chimpanzees are in their imitative and "teaching" skills, humans still seem to far outpace them. Meltzoff's designation of humans as the imitative species holds firm as a matter of degree if not of kind. Humans, however, are also the primary pedagogical species, a point not emphasized by Meltzoff.

Species also differ in the extent to which they integrate their tool-using behaviors with their social behaviors. Capuchin monkey tool-use, as exhibited in the Rome zoo, is entirely a solitary affair. Some reciprocal tool-use, however, has been observed among chimpanzees in laboratory settings (Savage-Rumbaugh & Rumbaugh, 1978). Further, the Tai chimpanzees exhibit communal use of hammering sites, and population-wide interactions between social food-sharing and tool-use.

Thus, Tai chimpanzee subsistence rests upon a strong tool-using and tool-making tradition. The nut-cracking techniques of this population are particularly complex and require approximately 11 years of practice to master. Time spent learning tool-use is time that could otherwise have been spent acquiring foods by non-tool-using methods. Hence, tool-using practice could be viewed as a waste of a young chimpanzee's efforts, especially since chimpanzees can crack the nuts with their own teeth. The pay-off only comes with technical mastery which permits individual animals to process more nuts than they can personally consume. Then, mother chimpanzees can supplement the food supplies of their young and do so until the juveniles are approximately eight years of age, thereby providing their young with leisure time to develop their own tool-using skills. Thus, the advanced tool-using skills of this population and their unusual food-sharing traditions form an interrelated social and technological complex.

Among humans, however, the interdependence between technical and social behavior greatly exceeds that yet demonstrated in chimpanzees. Elsewhere in this volume, Reynolds, Ingold, and Gibson all note the social nature of human tool-making. Gibson points to divisions of labor and the socioeconomic interdependence of tool-producing and tool-using groups. Ingold notes that, among many human populations, production units and kinship units are the same. As a result, people draw on technical skills transmitted through a kinship network. Reynolds postulates a complementation theory of tool-making: i.e. people in all human societies work together in groups of two or more individuals when producing and using tools. Each specializes in a particular task, and each anticipates the action of the other in order to perform the complementary behavior. Human tool-production then is a social activity and does not issue from individuals in isolation. Although the Tai chimpanzees exhibit some social complementarity in their hunting and teaching techniques, no ape population has yet exhibited a human degree of social embeddedness in their tool-making and other constructive activities.

Viewed from some perspectives, the technical skills which chimpanzees and humans apply to their tool-using endeavors display only minimal differences. Thus, as McGrew notes, chimpanzees possess many behaviors, once thought to be the distinctive province of humanity, including the use of tool-kits and tool-sets, and manual complementarity (contra Reynolds, this volume).

Examinations of tool-using contexts, however, indicate substantial species differences. Among chimpanzees, tool-use is primarily restricted to subsistence activities. The main exception is the common agonistic flailing and throwing of objects. In contrast, humans use tools in diverse contexts including the transportation of food and objects, cooking, the building of shelters, and the construction of clothing, jewelry, and art.

Even when only direct food procurement is considered, humans use tools in contexts which have not yet been exhibited by the apes. Thus, wooden spears and digging sticks are structurally simple tools, but humans use them for complex endeavors. Humans, but not apes, engage in long distance tracking of game over periods of hours or even days (Krantz, 1968) and may "call" wild game or birds, hide behind blinds or herd game into corrals. A human gatherer may carefully follow (and talk about) the changing nature of above-ground twigs and vines for months before finally concluding that a tuber is ripe for digging.

Considering the number of distinct components, or what Oswalt (1976) calls "techno-units", employed in the construction of subsistence tools, the average of 1.0 technounits for the tools made by chimpanzees falls only slightly below the corresponding figure of 1.2 technounits reported for the Aboriginal people of Tasmania. This leads McGrew to hypothesize that the differences between chimpanzee and human tool-making techniques fall within the range of cultural variation of a single species. As McGrew recognizes, however, wild chimpanzees have never been observed constructing tools of more than one component, whereas humans in all cultures do (see also Gibson, 1983, Reynolds, 1983). Even in captivity, no chimpanzees have mastered techniques of joining two objects to form a new tool (or other structure) that can be permanently rotated in space regardless of the pull of gravity, except in conditions in which humans have already constructed the joining mechanism (Reynolds, this volume).

On the surface, this may seem a trivial distinction. However, these behaviors demand what have been termed hierarchical mental constructional skills (see Case, 1985; Gibson, 1990; Greenfield, 1991). Hierarchical mental construction involves keeping a number of mental, perceptual, or motor schemes in mind simultaneously, and combining several of these schemes into new wholes which can then be used as subunits of other constructions. Other authors have strongly argued that in human children the ability to engage in complex mental constructional tasks reflects maturing intelligence and that representational capacities, symbolism, mathematical thought and understandings of physical causality are emergent properties reflecting these mental constructional skills (Case, 1985; Langer, this volume). Similarly, it has been argued that apes and humans differ primarily in their mental constructional capacities (Gibson, 1983, 1988, 1990, in press, this volume; Greenfield, in press).

Human mental constructional skills are also evident in the construction of jewelry, art, and other items from two or more components, in the advanced planning of complex tool-making and hunting and gathering endeavors, and in the ability of humans to engage in socially complementary tool-making activities (which require keeping in mind a series of one's own activities as well as those of another person). In addition, humans apply emergent properties of mental construction, such as representation, symbolism, mathematical and physical principles, to their tool-using and tool-making endeavors.

The focus on mental construction can be extended to analyzing species differences in levels of intelligence. The earliest paper to examine tool-use from this perspective was that of Parker & Gibson, (1977). They suggested that species which have a number of tool-using schemata that can be used in varied contexts to meet varied ends possess sensorimotor intelligence as defined by Piaget. This is a level of intelligence characteristic of human children between the ages of 12 and 18 months. It is solely a manipulative intelligence which involves the use of one tool at a time and requires minimal mental construction or representational skills. Subsequent studies have reinforced the view that capuchins possess sensorimotor intelligence at this level. No evidence, however, suggests any greater intelligence in capuchins (commentaries in Chevalier-Skolnikoff, 1989; Gibson, 1990; Langer, this volume).

McGrew's review indicates that chimpanzees also apply sensorimotor intelligence to their tool-using tasks. Since other evidence suggests that chimpanzees have levels of intelligence approximately equal to that of a three- to four-year-old child, one might expect chimpanzee tool-using skills to reflect greater mental constructional capacities than those of cebus. In particular, chimpanzee tool-use should reflect representational capacities. Indeed, the chimpanzee ability to choose appropriately sized and shaped termiting sticks suggests that their representational capacities far exceed those of the capuchin monkeys described by Visalberghi.

In sum, the separate contributions in this volume paint a picture of animal and human tool-using and manufacturing behaviors as being both more similar and yet more divergent than suggested by traditional accounts. Chimpanzees, in particular, seem determined to falsify all sharp behavioral dichotomies devised by human scientists, while still falling far short of human technological achievements. An examination of chimpanzee behavior suggests that in our narrow focus on tool-making, we have missed the more fundamental distinctions between humans and other animals. Humans have a complex behavioral repertoire which incorporates, in an integrative fashion, a variety of discrete behaviors, any one of which may exist in isolation and in more rudimentary form in other animals. In humans, however, the behaviors interact in a generative fashion yielding a cultural product much greater than the sum of its individual parts (see also Visalberghi, this part). This interacting suite of human behaviors includes imitation, teaching, object manipulation and manufacturing skills of all kinds, social behavior, and symbolic communication. All are tied together into a complex web by higher cognitive and neural information processing capacities which permit expanded human mental constructional capacities.

McGrew suggests that chimpanzees can serve as a referential model of the tool-using behaviors of early hominids. His suggestion seems apt, and the population studied by Boesch appears ideal for this purpose. Indeed, on conceptual grounds, Parker & Gibson suggested some years ago (1979) that an ancestral protohominid, possibly of australopithecine grade, would possess the behavioral complex recently described for the Tai chimpanzees. Thus, diverse approaches are now converging on similar evolutionary models. What is now needed is an explanation of how and when, beginning with Tai-chimpanzee-like abilities, the expansion of brain size and cognitive capacity resulted in the evolution of the distinctively human complex of interrelated technical, social, and imitative behavior.

References

Case, R. (1985). *Intellectual Development: Birth to Adulthood.* New York: Academic Press.

Chevalier-Skolnikoff, S. (1989). Spontaneous tool use in *Cebus* compared with other monkeys and apes. *Behavioral & Brain Sciences,* 12, 561–627.

Gibson, K. R. (1983). Comparative neurobehavioral ontogeny: The constructionist perspective in the evolution of language, object manipulation and the brain. In *Glossogenetics: The Origin and Evolution of Language*, ed. E. de Grolier, pp. 52–82. New York & Paris: Harwood Academic Publishers.

Gibson, K. R. (1988). Brain size and the evolution of language. In *The Genesis of Language: A Different Judgement of Evidence*, ed. M. Landsberg, pp. 149–72. Berlin: Mouton de Gruyter.

Gibson, K. R. (1990). Tool use, imitation, and deception in a captive cebus monkey. In *"Language" and Intelligence in Monkeys and Apes: Comparative Developmental Perspectives*, ed. S. T. Parker & K. R. Gibson, pp. 205–18. Cambridge: Cambridge University Press.

Gibson, K. R. (in press). The ontogeny and evolution of the brain, cognition and language. In *Handbook of Symbolic Intelligence*, ed. A. Lock & C. Peters. Oxford: Oxford University Press.

Greenfield, P. M. (1991). Language, tools and brain: The development and evolution of hierarchically organized sequential behavior. *Behavioral & Brain Sciences*, 14, 531–95

Krantz, G. S. (1968). Brain size and hunting ability in earliest man. *Current Anthropology*, 9, 450–1.

Meltzoff, A. (1988). *Homo imitans*. In *Social Learning: Psychological and Biological Perspectives*, ed. T. R. Zentall & B. G. Galef, pp. 319–42. Hillsdale, NJ: Erlbaum.

Oswalt, W. H. (1976). *An Anthropological Analysis of Food-Getting Technology*. New York: John Wiley.

Parker, S. T. & Gibson, K. R. (1977). Object manipulation, tool use and sensorimotor intelligence as feeding adaptations in cebus monkeys and great apes. *Journal of Human Evolution*, 6, 623–41.

Parker, S. T. & Gibson, K. R. (1979). A model of the evolution of language and intelligence in early hominids. *Behavioral & Brain Sciences*, 2, 367–407.

Reynolds, P. C. (1983). Ape constructional ability and the origin of linguistic structure. In *Glossogenetics: The Origin and Evolution of Language*, ed. E. de Grolier, pp. 185–200. New York & Paris: Harwood Academic Publishers.

Savage-Rumbaugh, E. S. & Rumbaugh, D. M. (1978). Linguistically mediated tool-use and exchange by chimpanzees *Pan troglodytes*. *Science*, 201, 641–4.

Terborgh, J. (1983). *Five New World Primates: A Study in Comparative Ecology*. Princeton, NJ: Princeton University Press.

Terrace, H. S. (1979). *Nim, a Chimpanzee Who Learned Sign Language*. New York: Knopf.

Tomasello, M. (1990). Questions regarding imitation, "language", and cultural transmission in apes and monkeys. In *"Language" and Intelligence in Monkeys and Apes: Comparative Developmental Perspectives*, ed. S. T. Parker & K. R. Gibson, pp. 274–311. Cambridge: Cambridge University Press.

Toth, N. & Schick, K. (1991). Early stone technologies and linguistic/cognitive inferences. Paper presented at the American Association for the Advancement of Science meetings, Washington DC, February 17, 1991.

5

Capuchin monkeys: A window into tool use in apes and humans

ELISABETTA VISALBERGHI

The capuchins (*Cebus*, spp.) are arboreal monkeys widely distributed in South America. In captivity, and to a lesser extent in the wild, the tufted capuchin (*Cebus apella*) has been the most studied of the four species of *Cebus*. The data I will refer to, as well as my own experimental work, concern this species. These monkeys grow to about the size of a tomcat, males being larger than females. Their group size varies across and within habitats, ranging from 8 to 15 individuals (Robinson & Janson, 1987; Terborgh, 1983). They are omnivorous; their diet includes fruit, flowers, and nuts, and varies considerably according to what is available. At Iguaçu, in the North of Argentina, the bromeliads (thorny hard plants belonging to the same family as the pineapples) form the bulk of capuchins' diet, whereas at Manu National Park, in Peru, they spend three quarters of their daylight hours searching for animal protein (insects, worms, mollusks, birds, small mammals) (see also Freese & Oppenheimer, 1981; Izawa, 1979). In order to find insects and other tidbits, they frequently search under the bark of trees, poke into holes, and use a variety of quite complex and "rough" techniques to reach food that would otherwise be inaccessible. In contrast with other monkey species, which are very elusive, capuchins are noisy and destructive foragers (Terborgh, 1983). In fact, as they move on, they leave the ground scattered with conspicuous left overs, such as palm leaves 3–4 m long and wasp nests.

The South American primates are the result of 40 million years of independent evolution from Old World ones. Despite this substantial phylogenetic distance, several aspects of their biology and behavior set capuchins close to apes (see also Fragaszy *et al.*, 1990a,b). Neonatal capuchins have large brains relative to their own and their mothers' body weights (Harvey *et al.*, 1987). Martin (1981) demonstrated that if we order the living placental mammals on the basis of the ratio between brain weight and body weight, capuchins hold the fourth position after man. Both motor and cognitive development are slow, being intermediate between monkeys and apes (Fragaszy, 1990), although, in a few characteristics, they are even closer to humans than apes (Antinucci, 1989). Their manipulative skills include a precision grip (Costello & Fragaszy, 1988). Capuchins reach puberty later than other monkeys and have a longer life span (up to 47 years).

Tool use in capuchins. The tube task

The variety of tool-tasks in which captive capuchins are successful is similar to that reported for apes (see Beck, 1980; Passingham, 1982; Gibson, 1990). In fact, no matter what the task is about, and what things are available, capuchins try to fill the gap between themselves and their goal by means of external objects, in a way that is neither rigid nor stereotyped. Stones are employed to crack nuts, paper towels serve to sponge liquids, and sticks can be used as multipurpose tools to rake, to treat wounds, to kill a snake, or to threaten an enemy (Beck, 1980; for an extensive review see Visalberghi, 1990, in press). The question is: are the representational capacities of apes and capuchins further apart than their achievements in tool-use tasks suggest?

To understand better whether their performance hinges upon similar underlying cognitive capacities, we tested capuchins in a task presented at different levels of complexity (Visalberghi & Trinca, 1989). The task consisted of a 30 cm plexiglass tube baited with a peanut in the middle. The reward was obtainable by using a stick; the tool provided for the subject was either effective as it was (simple condition) or required modification (complex conditions). To ensure accurate analysis, the behavior of the monkeys was videotaped. In the simple condition, the monkeys immediately explored the new apparatus, and tried every possible means to reach the reward directly; then, within two hours, three out of six subjects successfully used a stick to push the peanut out of the tube (Visalberghi & Trinca, 1989; Visalberghi, 1990; Visalberghi, unpublished results). The other three capuchins solved the task more slowly (see p. 142). The problem solving styles of the monkeys differed greatly from each other, suggesting a huge behavioral variability.

The subjects were then tested under more complex conditions in which the sticks provided had to be modified. In the first complex condition, the tool was too thick and needed to be broken, or several sticks were held together by rubber tape which needed to be removed (bundle condition); in the second complex condition the stick had blocked ends which needed to be freed (H-stick condition), and in the third complex condition the sticks were too short, requiring two of them to be inserted one behind the other into the tube (short-stick condition). In these complex conditions we wanted to test whether the monkeys could modify and use the stick appropriately. All subjects solved the complex conditions on the first trial within a few minutes. Careful analysis of the videotapes of the ten trials carried out by each subject in each condition showed that their performances were loaded with errors, i.e., attempts to use as tools objects which were not appropriate at all. For example, the monkey inserted in the tube splinters obtained from the bundle which were 1–2 cm long; or after having freed one of the blocked ends of the H-stick, the monkey inserted the other still blocked end into the tube, or even worse, inserted one of the blocks (only a few centimeters long); or after having inserted the first short stick in the tube, the monkey inserted the second at the opposite end of the tube.

A good balance between correct responses and errors may be advantageous, since it allows both new discoveries and appropriate use of previous experience. However, in the case of capuchins' behavior in the tube task it seemed that the monkeys persisted in behaviors which could not possibly be regarded as new and exploratory. To persist in introducing 1–2 cm long splinters or the rubber tape to displace the peanut cannot be considered an attempt to improve the previous, successful and appropriate technique already used to solve the task. A more plausible explanation is that the monkeys did not know why a successful technique was more appropriate than unsuccessful attempts. These kinds of errors continued to be made after a large number of trials; their occurrence suggests that capuchins did not abstract at a representational level the physical characteristics (e.g. length) of the tool required to solve the task (Visalberghi, 1990, in press) and that they did not figure out the rule for avoiding them.

In conclusion, capuchins do not know beforehand how to modify the stick so that when they actually try to insert it, the stick will fit into the tube and push the peanut out. Overall, in the more complex conditions, capuchins' successful performances hinged upon poor understanding; capuchins were successful tool users not because they understood exactly what the task was about, but because they tried every possible way to solve it, using both their own bodies and any external objects they came across. It seems, therefore, that the variety of capuchins' responses and the persistence and vigor of their attempts increase the probability of success. It is as if the capuchins, which are destructive foragers in the wild, in captivity become inventive and persistent tool users. Overall, capuchins successfully solve most of the tasks presented to them. However, despite the fact that they succeed, their performance does not seem to be the result of comprehension of the task requirements (Visalberghi, 1990).

A comparison with chimpanzees and children

In the tool-use literature on both apes and children, there are few systematic data comparable to those obtained in the tube task. Chimpanzees use tools for many different purposes and are able to change the characteristics of a tool to be appropriate to the task (see Goodall, 1986; Tuttle, 1986). However, we lack systematic information concerning how much chimpanzees understand of what they do. For example, is success achieved by modifying step by step the tool until it becomes appropriate for the goal, or do the apes possibly "know" beforehand how to modify the tool and do it accordingly?

Boesch & Boesch (1990, in press) report that wild chimpanzees at Taï National Park on the Ivory Coast are able to choose a tool and modify it appropriately beforehand. The modification is made by chimpanzees on the basis of what task the tool is for. These a priori choices and modifications strongly suggest that chimpanzees can mentally represent what a familiar task requires, without trying out the different possibilities each time.

To assess interspecies differences in representational capacities in tool use, a project to test children and apes in the tube task has started. Visalberghi & Troise (manuscript in preparation) carried out a longitudinal study on eight children tested over an 18-month period (six sessions) in the four conditions of the tube task. In addition, data on the subjects' sensorimotor development were collected (object permanence and means–ends). Children between the ages of 13 and 23 months solved the simple condition. In contrast to capuchins, children who had solved the task in the simplest condition did not become proficient in all of the complex conditions in the same session. At the end of the experiment, most of the children performed without errors in the bundle condition and in the H-stick condition. In the short-stick condition, children still made errors even when 30–32 months old.

Comparable data have also been collected on a total of eleven ape subjects (common chimpanzees, pygmy chimpanzees, and orangs) at the Language Research Center in Atlanta, in a project carried out in collaboration with Sue Savage-Rumbaugh and Dorothy Fragaszy (manuscript in preparation). Results suggest that apes perform much better than capuchins do. Some of the errors which were common in the capuchins' performances were, after extended experience, absent from the performances of the adult apes, and minimal in the performances of the younger subjects. As for children, the short-stick condition was clearly the most demanding for the apes. Therefore, when tested on the same experimental task, apes and young children show a much better understanding of its requirements than the capuchin monkeys do.

Imitation of tool-use skills

Imitation is a way of learning from others which is particularly useful when the observer is not proficient, when opportunities for practice are limited, when costs of errors are high, and when learning by individual experience would be a slow process (see Zentall & Galef, 1988). Restricting evidence for imitation to cases in which *something new is acquired* recognizes the unique function which sets imitation apart from other forms of social learning (Galef, 1988). That is, imitation serves as a means of learning novel behavior directly from others (models). It must be emphasized that the novelty of a behavior cannot be considered as all or nothing: some parts of a behavior may be novel (e.g. its orientation in space), and others may not be. Often, the specific action is already in the animal's repertoire, but the location in which it is performed can render its function completely new (for a definition of imitation and of the logical, methodological steps leading up to it, see Visalberghi & Fragaszy, 1990a).

In the past few years, Dorothy Fragaszy and I have carried out several experiments on tool-use in five different groups of tufted capuchins (*Cebus apella*). In all cases, one or more individuals in a group spontaneously acquired a novel tool-using behavior to obtain desirable food. In the course of these experiments, we became convinced that capuchins did not acquire new tool-using behaviors from watching skillful models, although other processes of social learning, especially local enhancement, were implicated (Fragaszy & Visalberghi, 1990). For example, behaviors such as cracking a nut with a tool did not spread at all (Visalberghi, 1987), or observers who witnessed hundreds of solutions from close up still did not use or even pick up a tool themselves (e.g., in the rod task, Fragaszy & Visalberghi, 1989). Gibson (1990), who carefully studied the behavior of her pet capuchin monkey (probably a *Cebus nigrivittatus*) over a ten-year period, reached similar conclusions. Her monkey's "imitative abilities were quite minimal, seemingly limited to social facilitation" (p. 211).

More systematic data using the tube task were collected to answer the following question: can those capuchins (which were tested in the tube task and did not solve it) learn to solve the task by watching a proficient model repeatedly solving it?

In the tube task, three capuchins out of six spontaneously solved the task within two hours of presentation, whereas the others, a young adult female (Ob 1), a juvenile female (Ob 2), and a juvenile male (Ob 3) did not solve the task, even after repeated sessions. Therefore, these subjects were subsequently presented with the task, which proficient models had solved, during what we called "lessons". Lessons consisted of 57 (for Ob1) and 75 (for Ob2, and Ob3) solutions of the task performed by the model(s) in the presence of the observer(s). The observers' behavior was videotaped and later carefully analyzed. Data were collected to assess (1) whether the observers watched the model(s) solving the tube task or the experimenter baiting it; (2) the effect of watching the models on the observers' behavior toward the tool task.

1.   The percentage time in which Ob1, Ob2, and Ob3 looked closely at the models was 40%, 49%, and 34%, respectively. Likewise, the percentage time spent looking at the baiting was in the range of 40–50%. Instantaneous 5-second sampling was used to score from the videotapes the occurrence of the following "scenes" on the screen: (a) tube with the peanut inside; (b) tube with the peanut and the stick inside; (c) tube with the stick inside. These scenes occurred either without the model (no-M), or with the model close to the apparatus (M). In (b) the presence of the model resulted in tool use; in (c) the presence of the model indicated that a successful use of the stick had just been performed (i.e., solution had just occurred). The analysis of the number of times observers looked at the "scenes" compared with the number expected on the basis of their occurrences, showed that the observers paid significantly more attention to the apparatus when a model was near the apparatus. In addition, the observers did not watch the potentially informative scene (scene b with model, i.e., the solution) more than the other scenes (scene a and scene c with model), from which little could be learnt about the correct problem-solving strategy.

2.   The results of this experiment show that none of the three observers acquired tool-use by watching the models performing the correct action. A comparison of the mean numbers of intervals in which the subjects contacted the tube, manipulated the stick, and contacted the tube by means of the stick in the trials before the lessons (Block I), in the trials interspersed with the lessons (Block II), and in the trials after the lessons (Block III), was performed. The results show that the observers' manipulation of the stick and contact with the tube decreased or remained at comparable levels across blocks. For all three observers the mean number of intervals in which they contacted the tube with the stick increased from Block I to Block II; this trend persisted also from Block II to Block III for Ob3. The lessons led to an increase in stick–tube contacts. However, this behavior was not appropriately directed toward opening the tube, and its spatial orientation did not improve across trials. In conclusion, despite the fact that the observers watched baitings and solutions, the lessons did not allow them to learn how to solve the task. In addition, the data on attention suggest that capuchins did not appropriately select what to look at in order to learn a new behavior.

Capuchins are not, however, the only monkeys which fail to imitate complex new behaviors. In the primatological literature there is a tendency to overestimate the frequency with which novel behaviors disseminate in a group compared to cases in which novel behaviors occur in only one individual (for a discussion of this point see Visalberghi & Fragaszy, 1990a,b). In fact, a critical review of the studies of tool-using in monkeys aimed at identifying possible instances of true imitation revealed little or no evidence of it (Visalberghi & Fragaszy, 1990a; see also Visalberghi 1989).

Why monkeys do not ape

Clearly the imitation of a new tool-using behavior is, at most, a fragile phenomenon in monkeys. Its absence may be due to contextual factors, such as lack of motivation or negative social influences limiting monkeys' behavior (exploitation, limited access to desirable objects; see also Kummer & Goodall, 1985). These variables play an important role in many circumstances (see Fragaszy & Visalberghi, 1990; Visalberghi & Fragaszy, 1990a,b). They cannot, however, account for the results presented above. In the tube task, the monkeys were all motivated to get the reward and to interact with the task. When exposed to the task the behavior of all subjects (non-solvers and solvers) was similar. Finally, during the experiment, competition and social inhibition were never observed. Instead, I would like to explore the hypothesis that imitational learning is prevented by cognitive constraints, which probably hinge upon monkeys' representational capacities (see Mitchell, 1987; Piaget, 1962, Koehler, 1976).

How must an individual learn from a model in order to be able to use a tool? 1. The observer can copy by "rote" the sequence of the motor acts performed by the model, or 2. the observer can learn the "rule" guiding the behavior of the model.

1. Rote copying should be characterized by careful replication, step by step, as we do when instructed in the operation of a new machine, the workings of which we do not understand. Efforts to match the sequence of motor patterns and to monitor the model should be prominent. Monkeys' behavior did not seem to suggest that this careful step by step replication was going on at all. The analysis of the videotapes of the tube experiment did not evidence such copying by the observers. In fact, having watched the model, the observers did not attempt to perform behaviors (or parts of it) similar to the model's ones.

2. Alternatively, the observer might set about understanding the relations among actions and objects. Koehler, for example, wrote that imitation is the ability "to understand and intelligently grasp what the action of the other means" (Koehler, 1976, p. 221). Bruner (1972) argued that visual imagery (rather than motor coding) is implicated in children's (delayed) imitation of other's acts, as the image of the other's acts is used to guide the observer's production of the same acts. If the observer, watching the behavior of a model using a tool, understands the rule, i.e. the cause–effect relationship linking action, objects, and outcome, then its attempts to solve the task with the tool can be effectively

focused on the more salient aspects of the sequence. The results previously reported suggest that capuchins do not observationally learn the rule involved in a tool task.

Many scholars have argued that the intricacies of social behavior have been the major selective force behind the evolution of intelligence in hominids (see, for example, chapters in Byrne & Whiten, 1988). It can be argued that technical intelligence and social intelligence are not independent of one another in the evolution of primates. Data on capuchins suggest that imitation and tool use are different facets of the same capacity: the understanding of how a tool works. An understanding of what makes the tool effective allows errors to be avoided and its functioning to be understood by watching a model. These are two abilities which capuchins do not achieve, despite their success as tool users. Whiten & Byrne (1991) have recently suggested that "imitative copying (at least in the visual mode) like pretence and mindreading, may involve second-order representation, insofar as the acts performed by (and perhaps the intention of) the model have to be translated from what is involved in doing them from the model's point of view (perceived from the imitator's point of view), into the metarepresentation permitting the performance from the imitator's point of view" (pp.278–9). But this translation requirement might not be the crucial factor in accounting for the lack of imitation. In the tube experiment, we often observed the observer side by side with the model sharing the same point of view, but this did not improve the observer's imitative skills. Imitation of novel behavior can occur when the observer knows the goal of the action (in the case of tool use the goal is obviously getting the reward), and understands the strategy necessary for reaching the goal. The strategy can be extracted only from the observation of the cause–effect relationship linking action, objects, and outcome, without translation from the model's to the observer's point of view. In the case of the tube experiment, mindreading does not seem to be necessary, and it can be obviated by the two subjects sharing a similar understanding of the world events, in this case how a tool causes the effect of dislodging a reward from a tube.

At this point, one might explore whether and to what extent a skillful tool-user capuchin understands the effect of the stick on the reward when it inserts and pushes the stick into the tube. An experiment recently carried out by Visalberghi & Limongelli (manuscript in preparation) showed that expert tool-user capuchins do not master the cause–effect relationships involved in tool use. Capuchins achieve success by trial-and-error attempts which are not guided by the knowledge of the conditions that are necessary to achieve solution (e.g., that contact between stick and reward is necessary for pushing the reward out of the tube). Therefore, it seems obvious that if capuchins have a limited understanding of the effects of the tool on the reward when they are actually using the tool, the observation of skillful conspecifics will not allow them to understand how to solve the task and consequently to imitate the model.

Capuchins have been very successfully trained as helpers for quadriplegics (Willard *et al.*, 1982, 1985). In these projects, the monkeys have easily acquired the ability to perform chains of tasks (e.g., remove a plastic bottle from the refrigerator, place it in a slot on a feeding tray, remove the lid and insert a straw). The training strategies were based upon the principles of behavior modification. In the training procedure the desired behavior was split into units which were rewarded. Willard *et al.* (1982) report that part of the training consisted in showing the monkey a unit, giving the command "do this", and rewarding the monkey for behaviors approximating the desired one. Gradually, the monkey produced a behavior resembling more and more closely that shown by the experimenter. From Willard's report it is not possible to assess whether showing the monkey the required behavior was important for its acquisition. However, it is very likely that rewarding the behaviors which increasingly approximate the required behavior can by itself account for its being learnt. Which features differentiate this way of learning behavioral sequences from that of imitating a model performing the whole chain of tasks? Why is the one training technique so effective while the other is not?

Let us examine what happens when an observer sees a model solving a tool task. Whenever the sequence of behaviors leads to success, the reward is most closely associated with the last action preceding it. However, for successful imitation to occur, the repetition of this last action alone is not sufficient. It is necessary to sort out the entire behavioral sequence which leads to success, and replicate it. Therefore, a naive observer would have to have a strategy to get to know the correct behavioral sequence. To figure out the steps of the successful sequence, a cause–effect interpretation which allows for unimportant features to be neglected while keeping the necessary ones is extremely convenient. The identification of the cause–effect relationship needs to be done by analyzing the behavior backward from the effect to the cause: it is necessary to infer the cause from the result. It seems that it is difficult for a monkey to make this kind of inference. For example, vervet monkeys (*Cercopithecus aethiops*) anticipate the presence of a predator on hearing an alarm call, but they do not infer from the tracks left on the ground that a snake has passed there or, from the carcass at a tree fork that a leopard is in the area (Seyfarth & Cheney, 1988).

Training is a completely different affair. In order to teach a monkey a sequence of motor patterns, it is necessary to reward the first action, and not the last. For example, to teach how to solve the tube task, a trainer would start by rewarding the monkey each time it is close to a stick, then when the monkey contacts it, later only when it holds it, and so on. In this case it is not necessary for the subject to show the whole behavioral sequence correctly from the beginning, since gradual approximations are rewarded. In contrast, imitation (and spontaneous trial-and-error behavior too) is rewarded only when the entire sequence is correctly performed.

The relationships among representational abilities in tool use, imitation, understanding of cause–effect relationships and other domains need further investigation. A limited representational capacity in capuchins has also been found in the Piagetian object permanence series (Natale and Antinucci, 1989). In addition, capuchins do not display self-recognition when viewing a mirror image of themselves (Anderson, 1990; Visalberghi *et al.*, 1988). On the other hand, capuchins possess representational capacities in other domains. Robinson (1986) reports that capuchins represent space and locations of objects (fruit trees) in their home ranges. D'Amato & Colombo (1988) and D'Amato *et al.* (1985) report representation of serial order, and transitivity of conditional relations, in capuchins.

One final point warrants further consideration. What are the possible consequences of the fact that tool-use (or other innovations) and imitational learning do not co-occur? Among primates, imitation of novel behaviors seems to be a rarer event than has been assumed in the past. The lack of imitation of new complex behaviors decreases the chances that new skills will catch on in a group or in a population. In short, cultural evolution is perhaps more strictly a human phenomenon than we have been led to suppose, ever since the famous observations of Japanese macaques washing potatoes apparently demolished the Rubicon between human and monkey culture. In fact, passing from monkeys to apes and especially from apes to humans, we see an astronomical increase in both tool-using activities and imitational skills (Meltzoff, 1988a; Meltzoff & Moore, 1983). These two phenomena, proceeding side by side, become mutually reinforcing. Whereas in *Homo sapiens (Homo imitans*, as Meltzoff (1988b) puts it), and to an unknown degree in the ancestral hominids, any tool-using behavior can disseminate quickly by imitation, it is plausible that the imitative spread of innovative behaviors occurs only in circumscribed situations in apes, and may be almost absent in monkeys. The co-occurrence of tool-use and imitation has a snowball effect. The consequences of this amplification process in human evolution have not received adequate consideration, perhaps because of the widespread idea that human and nonhuman primates are all good imitators.

Acknowledgements

I am particularly grateful to Dorothy Fragaszy. Together we discussed and wrote an article which has had a very influential role on this paper.

References

Anderson, J.R. (1990). Use of objects as hammers to open nuts by capuchins (*Cebus apella*). *Folia Primatologica*, 54, 138–45.

Antinucci, F. (ed.) (1989). *Cognitive Structure and Development in Nonhuman Primates*. Hillsdale, NJ: Lawrence Erlbaum.

Beck, B.B. (1980). *Animal Tool Behavior: The Use and Manufacture of Tools by Animals*. New York: Garland STPM Press.

Boesch, C. & Boesch, H (1990). Tool use and tool making in wild chimpanzees. *Folia Primatologica*, 54, 86–99.

Boesch, C. & Boesch, H. (in press). Tool use and tool making in wild chimpanzees. In *Tool Use in Human and Nonhuman Primates*, ed. J. Chavaillon & C. Boesch. Oxford: Oxford University Press.

Bruner, J.S. (1972). Nature and uses of immaturity. *American Psychologist*, 27, 687–708.

Byrne, R. & Whiten, A. (1988). *Machiavellian Intelligence. Social Expertise and the Evolution of Intellect in Monkeys, Apes and Humans*. Oxford: Clarendon Press.

Costello, M.B. & Fragaszy, D.M. (1988). Prehensive grip in capuchins (*Cebus apella*). *American Journal of Primatology*, 15, 235–45.

D'Amato, M. & Colombo, M. (1988). Representation of serial order in monkeys (*Cebus apella*). *Journal of Experimental Psychology*, 14, 131–9.

D'Amato, M., Salmon, D., Loukas, E. & Tomie, A. (1985). Symmetry and transitivity of conditional relations in monkeys (*Cebus apella*) and pigeons (*Columba livia*). *Journal of the Experimental Analysis of Behavior*, 44, 35–47.

Fragaszy, D. M. (1990). Early Behavioral Development in Capuchins (*Cebus*). *Folia Primatologica*, 54, 119–28.

Fragaszy, D. & Visalberghi, E. (1989). Social influences on the acquisition and use of tools in tufted capuchin monkeys (*Cebus apella*). *Journal of Comparative Psychology*, 103, 159–70.

Fragaszy, D.M. & Visalberghi, E. (1990). Social processes affecting the appearance of innovative behaviours in capuchin monkeys. *Folia Primatologica*, 54, 155–65.

Fragaszy, D., Visalberghi, E. & Robinson, J. (1990a). Variability and adaptability in the genus *Cebus*. *Folia Primatologica*, 54, 114–8.

Fragaszy, D., Visalberghi, E. & Robinson, J. (1990b). Adaptation and adaptability of Capuchin monkeys. *Folia Primatologica*, 54, 111–228.

Freese, C.H. & Oppenheimer, C.R. (1981). The capuchin monkeys, genus *Cebus*. In *Ecology and Behavior of Neotropical Primates*, ed. A. F. Coimbra-Filho & R. A. Mittermeier, pp. 331–90. Rio de Janeiro: Academia Brasileira de Ciencias.

Galef, B.G. (1988). Imitation in animals: History, definition, and interpretation of data from the psychological laboratory. In *Social Learning: Psychological and Biological Perspectives*, ed. T. Zentall & B.G. Galef, pp. 3–28. Hillsdale, NJ: Lawrence Earlbaum.

Gibson, K.R. (1990). Tool use, imitation, and deception in a captive cebus monkey. In *"Language" and Intelligence in Monkeys and Apes: Comparative and*

Developmental Perspectives, ed. S.T. Parker & K.R. Gibson, pp. 205–18. Cambridge: Cambridge University Press.

Goodall, J. (1986). *The Chimpanzees of Gombe: Patterns of Behavior*. Cambridge, MA: Harvard University Press.

Harvey, P.H., Martin, R.D. & Clutton-Brock, T.H. (1987). Life Histories in Comparative Perspective. In *Primate Societies*, ed. B.B. Smuts, D.L. Cheney, R.M. Seyfarth, R.W. Wrangham & T.T. Struhsaker, pp. 181–96. Chicago: The University of Chicago Press.

Izawa, K. (1979). Foods and feeding behavior of wild black-capped capuchin *(Cebus apella)*. *Primates*, 20, 57–76.

Koehler, W. (1976). *The Mentality of Apes*. New York: Liveright.

Kummer, H. & Goodall, J. (1985). Conditions of innovative behaviour in primates. *Philosophical Transactions of the Royal Society Series B*, 308, 203–14.

Martin, R.D. (1981). Relative brain size and basal metabolic rate in terrestrial vertebrates. *Nature*, 293, 56–60.

Meltzoff, A.N. (1988a). Infant imitation after a 1–week delay: Long-term memory for novel acts and multiple stimuli. *Developmental Psychology*, 24, 470–6.

Meltzoff, A.N. (1988b). *Homo imitans*. In *Social Learning. Psychological and Biological Perspectives*, ed. T.R. Zentall, & B.G. Galef, pp. 319–42. Hillsdale, NJ: Lawrence Erlbaum.

Meltzoff, A. N. & Moore, K. M. (1983). The origins of imitation in infancy: paradigm, phenomena, and theories. In *Advances in Infancy Research*, vol. 2, ed. L.P. Lipsett, pp. 265–301. Norwood, NJ: Ablex.

Mitchell, R.W. (1987). A comparative-developmental approach to understanding imitation. In *Perspectives in Ethology*, vol. 7, ed. P.P.G. Klopfer & P.H. Bateson, pp. 183–215. New York: Plenum Press.

Natale, F. & Antinucci, F. (1989). Stage 6 object-concept and representation. In *Cognitive Structure and Development of Nonhuman Primates*, ed. F. Antinucci, pp. 97–112. Hillsdale, NJ: Lawrence Erlbaum.

Passingham, R. E. (1982). *The Human Primate*. Oxford: Freeman.

Piaget, J. (1962). *Play, Dreams and Imitation in Childhood*. New York: Norton.

Robinson, J.C. (1986). Seasonal variation in use of time and space by wedge-capped capuchin monkey, *Cebus olivaceus*: implications for foraging theory. *Smithsonian Contributions to Zoology* N. 431.

Robinson, J.C. & Janson, C.H. (1987). Capuchins, squirrel monkeys, and atelines: Socioecological convergence with Old World primates. In *Primate Societies*, ed. B. B. Smuts, D. L. Cheney, R. M. Seyfarth, R. W. Wrangham & T. T. Struhsaker, pp. 69–82. Chicago: The University of Chicago Press.

Seyfarth, R.M. & Cheney, D.L. (1988). Do monkeys understand their relations?. In *Machiavellian Intelligence. Social Expertise and the Evolution of Intellect in Monkeys, Apes, and Humans*, ed. R. Byrne & A. Whiten, pp. 69–84. Oxford: Clarendon Press.

Terborgh, J. (1983). *Five New World Primates*. Princeton: Princeton University Press.

Tuttle, R.H. (1986). *Apes of the World*. Park Ridge NJ: Noyens Publications.

Visalberghi, E. (1987). Acquisition of nut-cracking behaviour by two capuchin monkeys *(Cebus apella)*. *Folia Primatologica*, 49, 168–81.

Visalberghi, E. (1989). Primate tool-use: Parsimonious explanations make better science. Commentary on Chevalier-Skolnikoff article "Spontaneous use of tools and sensorimotor intelligence in *Cebus* compared with other monkeys and apes". *Brain Behavioral Science*, 12, 608–9.

Visalberghi, E. (1990). Tool use in *Cebus*. *Folia Primatologica*, 54, 146–54.

Visalberghi, E. (in press). Tool use in a South American monkey species. An overview of characteristics and limits of tool use in *Cebus apella*. In *Tool Use in Human and Nonhuman Primates*, ed. J. Chavaillon & C. Boesch. Oxford: Oxford University Press.

Visalberghi, E. & Fragaszy, D. (1990a). Do monkeys ape? In *"Language" and Intelligence in Monkeys and Apes: Comparative Developmental Perspectives*, ed. S.T. Parker & K.R. Gibson, pp. 247–73. Cambridge: Cambridge University Press.

Visalberghi, E. & Fragaszy, D. (1990b). Food-washing behaviour in tufted capuchin monkeys *(Cebus apella)* and crabeating macaques. *Animal Behaviour*, 40, 829–36.

Visalberghi, E. & Trinca, L. (1989). Tool use in capuchin monkeys: distinguishing between performing and understanding. *Primates*, 30, 511–21.

Visalberghi, E., Riviello, M.C. & Blasetti, A. (1988). Mirror responses in tufted capuchin monkeys *(Cebus apella)*. *Monitore Zoologico Italiano*, 22, 487–556.

Whiten, A. & Byrne, R.W. (1991). The emergence of metarepresentation in human ontogeny and primate phylogeny. In *Natural Theories of Mind. Evolution, Development and Simulation of Everyday Mindreading*, ed. A. Whiten, pp. 267–81. Oxford: Basil Blackwell.

Willard, M.J., Dana, K., Stark, L., Otwen, J., Zazula, J. & Corcoran, P. (1982). Training a capuchin *(Cebus apella)* to perform as an aide for a quadriplegic. *Primates*, 23, 520–32.

Willard, M.J., Levee, A. & Westbrook, L. (1985). The psychosocial impact of Simian aides on quadriplegics. *Einstein Quarterly Journal of Biology and Medicine*, 3, 104–8.

Zentall, T.R. & Galef, B.G. (1988). *Social Learning. Psychological and Biological Perspectives*. Hillsdale, NJ: Lawrence Erlbaum.

6

The intelligent use of tools: Twenty propositions

WILLIAM C. McGREW

Goodall (1963) first reported tool-use by wild chimpanzees almost 30 years ago, and much has been added to our knowledge since then. The purpose of this chapter is to bring together and to update this accumulated information, going beyond description to interpretation, whenever possible. I shall do so by putting forward, and attempting to justify, 20 propositions concerning the tool use of chimpanzees.

1. Of other primates, only apes are real tool-users

Following Beck (1980), *tool-use* is here defined as "the external deployment of an unattached environmental object to alter more efficiently the form, position, or condition of another object ..." (p.10). *Tool-making* is defined as "any modification of an object by the user or conspecific so that the object serves more effectively as a tool" (p.11).

Reliable reports of natural tool-use by primates other than great apes (Pongidae) extend only to dropping or throwing missiles from above at intruders below. Most other records are anecdotal, few have been replicated, and most pre-date the modern era of scientific field primatology. By comparison with (e.g.) sea otters, monkeys in nature are unremarkable tool-users.

In captivity, the picture is much richer, especially for capuchin monkeys (Visalberghi, 1987). Tool-use abilities shown in the laboratory go beyond what has been seen in nature, although the links to natural origins are clear.

For apes, tool-use is widespread, habitual, varied, and well-documented, both in nature and in captivity. Again and again over the past 25 years new cases of tool-use in the wild have emerged (Boesch & Boesch, 1989) which go beyond what was known from captivity.

Table 6.1. *Use of tools by apes and capuchins in four settings*

	Captive		Free-ranging	
	Spontaneous	Induced	Human-influenced	Natural
Chimpanzee	+ +	+ +	+ +	+ +
Orang-utan	+ +	+ +	+ +	+
Bonobo	+ +	?	+	−
Gorilla	+	+	+	− −
Gibbon	+	?	+	+
Capuchin	+ +	+ +	?	+

Note:
+ + = well-known from several individuals in several populations; + = recorded at
least once somewhere; − − = notably absent from long-term studies of several
populations; − = none seen but data yet sparse; ? = not yet studied.

2. Of apes, only chimpanzees consistently use tools

The technical inclinations of apes (and capuchin monkeys) can be systematically compared by surveying them across four settings (see Table 6.1). In
captivity, *spontaneous* use is unprompted by humans while *induced* use results
from intentional human intervention, often experimental. In nature, *human-
influenced* use is by apes released from captivity or tamed by feeding, while
natural use means by apes living in largely undisturbed conditions (see
McGrew, 1989, for details).

Chimpanzees show frequent and diverse tool-use in all settings. Orang-utans
easily match chimpanzees in all but nature, where tool-use is no more elaborate
than that of monkeys. Bonobos are accomplished spontaneous tool-users in
zoos (Jordan, 1982), but are not yet well-studied elsewhere. Gorillas minimally
use tools in three of the four settings, but the data are notably negative in
natural settings, and their performance is indistinguishable from that of the
lesser apes (Hylobatidae). Captive capuchin monkeys perform better than
most apes but fail in nature, as do all others but chimpanzees.

3. Tool-use by free-ranging chimpanzees is the norm for the species

Wild chimpanzees at various places in Africa have different repertoires of tool-
use. Goodall (1973) produced the first list of tool-use by free-ranging chimpanzees; it compared 10 sites, and all but Gombe's data were based on short-term
studies or single sightings. The most extensive published catalogue is Beck's
(1980) of 20 sites, but it is now outdated.

Table 6.2 lists 34 populations or groups of free-ranging chimpanzees across

Table 6.2. *African study-sites of free-ranging chimpanzees at which tool-use*
has been recorded

Study-site	Country	Subspecies	Major sources (see McGrew, 1992 for details)
Abuko (r)	The Gambia	v	Brewer, 1978; Goodall, 1973
Assirik	Senegal	v	Baldwin, 1979; Bermejo *et al.*, 1989; McBeath & McGrew, 1982; McGrew *et al.*, 1979
Assirik (r)	Senegal	v	Brewer, 1978, 1982
Ayamiken	Equat. Guinea	t	Jones & Sabater Pi, 1969, 1971
Baboon (r)	The Gambia	v	Brewer & McGrew, 1989
Banco	Ivory Coast	v	Hladik & Viroben, 1974
Bassa (r)	Liberia	v	Hannah & McGrew, 1987
Belinga	Gabon	t	McGrew & Rogers, 1983
Bossou	Guinea	v	Albrecht & Dunnett, 1971; Sugiyama, 1981, 1990; Sugiyama & Koman, 1979, 1987
Budongo	Tanzania	s	Sugiyama, 1969
"Cameroon"	Cameroon	t	Merfield & Miller, 1956
Campo	Cameroon	t	Sugiyama, 1985
Cape Palmas	Gabon	t	Savage & Wyman, 1844
Dipikar	Equat. Guinea	t	Jones and Sabater Pi, 1969, 1971
Filabanga	Tanzania	s	Itani & Suzuki, 1967
Gombe	Tanzania	s	Goodall 1964, 1968, 1970, 1973, 1986; McGrew, 1974, 1977, 1979; Teleki, 1979
Ipassa (r)	Gabon	t	Hladik, 1973
Kanka Sili	Guinea	v	Albrecht & Dunnett, 1971
Kanton	Liberia	v	Kortlandt & Holzhaus, 1987
Kasakati	Tanzania	s	Izawa & Itani, 1966; Suzuki, 1966
Kasoje	Tanzania	s	McGrew & Collins, 1985; Nishida 1977, 1980; Nishida & Hiraiwa, 1982; Nishida & Uehara; 1980; Uehara, 1982
Kibale	Uganda	s	Ghiglieri, 1984, 1988
"Liberia"	Liberia	v	Beatty, 1951
Lope	Gabon	t	Tutin & Fernandez, unpubl. data
Mbomo	Congo	t	Fay, unpubl. data
Ndakan	Cent. Afr. Rep.	t	Fay, unpubl. data
Ngoubunga	Cent. Afr. Rep.	t	Fay, unpubl. data
Okorobiko	Equat. Guinea	t	Jones & Sabater Pi, 1969, 1971; Sabater Pi, 1974
Sapo	Liberia	v	Anderson *et al.*, 1983

Table 6.2. *(cont.)*

Study-site	Country	Subspecies	Major sources (see McGrew, 1992 for details)
Tai	Ivory Coast	v	Boesch & Boesch, 1983, 1984a,b, 1989, 1990; Rahm, 1971; Struhsaker & Hunkeler, 1971
Tiwai	Sierra Leone	v	Whitesides, 1985
West Cameroon	Cameroon	t	Struhsaker & Hunkeler, 1971

Notes:
r = released populations, s = *schweinfurthii*, t = *troglodytes*, v = *verus*.

Africa that show some kind of tool-use. The list includes 11 long-term (more than a year) studies of wild chimpanzee populations, only three of which (Bossou, Gombe, Kasoje) are known to be human-influenced through provisioning, plus five others of chimpanzees released into the wild from captivity. Tool-use is well-represented in all three geographical races in eastern, central-western, and far western Africa, and new records are added yearly.

Put another way, *all* long-term studies have yielded evidence of tool-use, with the extent and range of the recorded repertoire being positively correlated with degree of observability. Well-habituated populations are seen to use tools daily.

4. Chimpanzee tool-use is mainly for subsistence

No previous catalogue has sought to distinguish between *habitual* and *occasional* tool-use by chimpanzees. Here, habitual means a pattern shown repeatedly by several members of a group, while occasional refers to rare, idiosyncratic, or unclear records. Also omitted are cases of released chimpanzees, whose behaviour may have been shaped by human companions. (Of course, single cases are instructive in terms of capacity, and anecdotal records often "firm up" to habitual ones.)

Table 6.3 presents a stricter catalogue, limited to habitual tool-use by wild chimpanzees. Only 11 populations showing an overall total of 40 patterns meet the criteria, and these range from one to 11 patterns per site. All of the populations with only one or two patterns are totally or virtually unhabituated to human observations. There is a positive correlation between degree of habituation (or length of study) and number of patterns of habitual tool-use, which suggests that results are incomplete for most populations.

Table 6.3. *Habitual patterns of tool-use of wild chimpanzees*

					Field site			
Pattern	Gombe	Bossou	Kasoje	Tai	Kanka Sili	Assirik	Kanton Sapo Tiwai	Campo Okorobiko
Termite-fish	X		X			X		
Ant-dip	X			X		X		
Honey-dip	X			X				
Leaf-sponge	X	X						
Leaf-napkin	X							
Stick-flail	X	X	X		X			
Stick-club	X	?X	X		X			
Missile-throw	X	X	X		X			
Self-tickle	X							
Play-start	X		X					
Leaf-groom	X		X					
Ant-fish			X					
Leaf-clip		X	X					
Gum-gouge		X						
Nut-hammer		X		X			XXX	
Marrow-pick				X				
Bee-probe				X				
Branch-haul		X						
Termite-dig								XX
Total	11	8	8	5	3	2	1	1

Note:
X = present

No pattern is universal, and the most widespread, hammer-use, is known at only five sites. Weapon-use is probably more common but the data are patchy. Ant-dipping may be the most impressive pan-African pattern, as it occurs from the wet forest site at Bossou to the driest savanna site at Mt. Assirik (McGrew, 1992).

Most (22 of 40) of the habitual tool-use patterns are subsistence activities for acquiring or processing food, especially social insects ($N = 12$) or nuts ($N = 5$). Only one or two others relate to meat, water, or plant foods other than nuts. All are examples of *extractive foraging* (Parker & Gibson, 1977). The weapon-use total ($N = 12$) may be inflated by its being composed of the related patterns of *flail-club-missile*. The five remaining types of non-subsistence tool-use are split between self-directed and apparently ritualised communicatory signals.

5. Some non-subsistence tool-use is mundane but obscure

Three kinds of non-subsistence tool-use by chimpanzees are often seen but remain almost unstudied. Many researchers have seen chimpanzees use probes or prods to investigate the environment. For example, straws may be poked into cracks or sticks against novel objects. There is no comprehensive descriptive account of this for any population, much less a systematic study. There is no firm knowledge in terms of any independent variable like age or sex, or even solid information on the sensory modalities involved, e.g. visual, haptic, olfactory.

Similarly, many observers of chimpanzee youngsters know them to use a variety of objects for both self-stimulation and social interaction (McGrew, 1977). Self-directed or solitary use of stones, sticks, leaves, food-items, etc. may be in self-tickling, masturbation, exploration, and so on. A common social usage is the use of objects to initiate play, especially with peers. Teasing invitations and "catch-me-if-you-can" fleeing are daily occurrences in nursery groups of mothers and young offspring. Again, no systematic comparisons have been drawn, and one wonders if such toys have been dismissed as kids' stuff?

For tool-use in personal hygiene, Goodall's (1986) analysis of leaves being used as napkins is unique. In 90% of cases the substance removed was one of four kinds of bodily fluid: semen, faeces, blood, and urine. Unsuspected sex differences emerged: Males were more fastidious than females; they wiped their penises over 10 times more often than females wiped their vulvas. The efficacy of such "technological grooming" remains untested, and no comparable data have yet been published from any other population.

6. Weapon-use is important but little-known

Goodall (1986) also gave extensive data on weapon-use by the wild chimpanzees at Gombe. Sticks and stones were flailed, clubbed, and thrown at other chimpanzees and at other species, chiefly baboons and humans. Weapon-use at Gombe was mostly shown by males, and there seemed to be differences in targeting between chimpanzees and other species.

Spontaneous weapon-use by chimpanzees has long been known for both captive and wild chimpanzees, but no systematic analyses have been published. Especially striking are graphic descriptions and films of chimpanzees reacting to the sudden presentation of a stuffed leopard (Kortlandt, 1965). Unfortunately, no statistical analyses have been given, so claims of differences between forest-living versus savanna-living chimpanzees in their responses remain as hypotheses yet to be tested.

It seems remarkable that behavioural patterns that have played such an important part in evolutionary reconstructions (Calvin, 1982) are so empirically neglected. Even the simplest experimental studies on the perfectibility of aimed throwing remain undone. Goodall's (1986) findings are welcome, but most come from a provisioning site in an artificial clearing, and comparative analyses await data from the natural setting at Gombe and elsewhere.

7. Chimpanzees use leaves in ways that are certainly ritualised, often communicative, and probably symbolic

In leaf-clipping, the chimpanzee noisily pulls to bits one or more leaves by hand and mouth, leaving only the stripped petiole (Nishida, 1980). The result looks like a fishing tool, but the function is completely different, being most likely a signal. At Kasoje, most cases of leaf-clipping were in courtship, usually from a male to an oestrous female. In other cases, the leaf-clipper was apparently frustrated, usually by lack of access to a tempting incentive such as food possessed by others.

Thousands of kilometres away at Bossou, the same pattern functioned similarly but not identically (Sugiyama, 1981). There, only a few cases were sexual, but almost all others were done in clear frustration or in frustration-related aggression such as when the chimpanzees sought to drive away a persistent human observer. At both sites, the leaf-clippers seemed to be in approach–avoidance conflict, so that the result looked like a ritualised displacement activity. No other population of chimpanzees, wild or captive, has been seen to do leaf-clipping.

Leaf-grooming is more enigmatic. Goodall (1968) first described this calm and deliberate custom at Gombe as occurring when a chimpanzee directed typical grooming motor patterns (handle, peer, mouth, lip-smack) to randomly picked leaves. It was not directly functional, in that the leaves were not cleaned. Wrangham's (unpublished) detailed analysis showed it was always linked to true grooming, usually social but sometimes solitary. Often it served to start or to perk up flagging bouts of grooming with others, but more rarely it occurred when a lone chimpanzee seemed bored. Goodall (1986) likened solitary leaf-grooming to doodling. Leaf-grooming was seen daily at Kasoje (Nishida, 1980), but elsewhere it is unknown.

Both ways of manipulating leaves are ritualised in the sense of being stereotyped and derivative, the behaviour involved having assumed a second-ary significance beyond its original functions in other contexts. Both involve the recruitment of an object into use for communicating esoterically. Both behavioural patterns are directly functionless and completely arbitrary; the groomed or disintegrated leaf is of no interest in its own right, and so seems to function only symbolically, in a non-iconic way.

8. Only chimpanzees have natural tool-kits

The classic cases of tool-use by other species in nature all involve single adaptations. Sea otters smashing molluscs, Galapagos finches probing for larvae, digger wasps tamping nest-hole entrances, etc. are all the *only* habitual pattern shown by those species (Beck 1980). Parker and Gibson (1977) called this "context specific" as opposed to intelligent tool-use, but the point here is that no other species uses tools in more than one context.

Put another way, chimpanzees are the only non-human species in nature to use different tools to solve different problems. They go beyond using the same tool to solve different problems (e.g. a sponge of leaves to swab out a fruit-husk or a cranial cavity) or different tools to solve the same problem (e.g. probes of bark *or* grass *or* vine to fish for termites). Thus, they have a *tool-kit*.

This means that one can study ethnographically the material culture of chimpanzees, and given variance in data-sets across populations, one can subject these to ethnological analyses (McGrew, 1992). If the same data were reported from cross-cultural comparisons of human beings, not an eyebrow would be raised.

9. Only chimpanzees have tool-sets

Chimpanzees in the wild may use different tools either as alternatives in a simple task or in series in a complex task. Two or more tools used sequentially to achieve a single goal comprise a *tool-set*. This entails a true series, in which the unfolding performance shows ordination, i.e. Step 3 must not only follow Steps 1 and 2, but in that order. Finally, a tool-set shows inclusiveness, that is, all steps are necessary for any pay-off to ensue, in a two-step series, Step 1 is of no use unless the concluding Step 2 follows.

Note what this excludes: Use of a facility (see below) does not qualify, as when a chimpanzee perches on a bent-over sapling before starting to dip for driver ants (McGrew, 1974). Even if the still-attached sapling were to count as a tool, it is not necessary for the dipping but merely makes it more comfortable. Use of a bark fishing tool in alternation with a twig does not qualify, as neither in any sense depends on the other. Even the making of flakes as an inadvertent result of cracking open nuts with a hammer-stone (Hannah, unpubl. data) does not qualify unless the flakes are then used further to process the nuts.

Use of tool-sets by captive apes (chimpanzees, Döhl, 1968; orang-utans, Lethmate, 1982) has been elegantly induced experimentally but not seen until recently in free-ranging apes. Brewer and McGrew (1990) described a four-step tool-set used by an 11-year-old female chimpanzee in The Gambia to get honey. She first used a stout stick chisel (with active palmer grip) to break into the outer surface of the arboreal bees' nest, then a finer-pointed chisel (with active and passive palmer grip) to widen the indentation. Then she punctured the wall of the nest with a bodkin (gripped in the teeth as well as hand!), and finally she used a dip-stick probe (with modified pencil grip) to extract the honey. This is only a single case, and so cannot yet be generalised, but it shows a capacity in an ape untutored by humans.

How intelligent is the use of a tool-set? An ambitious explanation might draw an analogy with grammar, in the syntax of performance. A more cautious account might point out that in nest-building, a stereotyped activity shown by all great apes, the same sort of rules apply: Bending over main branches precedes inter-weaving of minor branches, which in turn precedes lining the nest with detached leafy twigs.

10. The same functional taxonomy of tools can be applied to humans and to non-humans

To compare tool-use across species requires, first, a comprehensive but precise, rich yet objective taxonomy that is neither ethno- nor anthropo-centric. The most apt such typology is that of Oswalt (1976). Second, at least two data-sets are needed, collected in similar ways on closely related taxa. These now exist for chimpanzees and humans. Third, an evolutionarily significant focus is needed, a part of daily life that is undeniably subject to natural selection. Food-getting fills this need.

Table 6.4. *Definitions from Oswalt's (1976) taxonomy of elementary technology*

Subsistant	Extrasomatic form that is removed from a natural context or is manufactured and is applied directly to obtain food
Technounit	Integrated, physically distinct, and unique structural configuration that contributes to the form of a finished artifact
Instrument	Hand-manipulated subsistant that customarily is used to impinge on masses incapable of significant motion and is relatively harmless to the user
Weapon	Form that is handled when in use and is designed to kill or maim species capable of motion
Facility	Form that controls the movement of prey or protects it to the user's advantage. *Tended* if physical presence of user is essential for functioning; *untended* if functions in the absence of user
Naturefact	Natural form, used in place or withdrawn from a habitat, that is used without prior modification
Artifact	End product resulting from modification of a physical mass to fulfil a useful purpose
Simple	Retains some physical form before and during use
Complex	Parts change their relationship with one another when form is used

In Oswalt's taxonomy, or *technosystem*, subsistence technology is central, and the basic structure is hierarchical and dichotomous. Any subsistant can be classified, from the simplest naturefact to the most complex instrument or facility (see Table 6.4). The system allows qualitative categorisation of subsistants, and the sum of all subsistants is the food-getting tool-kit for that culture. Quantitative comparison is based on the *technounit*, the building block of the scheme. The number of technounits that make up a finished artifact is a measure of its complexity. Thus, a hafted spear has at least a shaft, point, and binder, giving three technounits, while a sharpened stick used as a spear has only one.

McGrew (1987) sought to examine the narrowest known gap between human and non-human technology, and so chose the subsistence tools of the Aborigines of Tasmania and the chimpanzees of Western Tanzania. Analyses showed the Tasmanians' tool-kit to be surprisingly limited. Most (14 of 18) of their subsistants were artifacts, and the number of technounits per subsistant averaged only 1.2. For the Tanzanian chimpanzees, 18 of their 20 subsistants were artifacts, and the mean number of technounits per subsistent was 1.0.

Other similarities were that neither had any complex forms of any type nor any compound implements. Both used the same types of raw material in the same order of preference: woody vegetation, stone, non-woody vegetation. Both used tools mainly for animal rather than plant prey. Both emphasised tended rather than untended facilities.

However, only humans used subsistants of more than one technounit and untended facilities such as traps. Human subsistants involved manufacturing techniques that are thought (but not proven) to be linked with intelligence: fire, knots, bait. Only humans used secondary tools, that is, tools to make tools, whereas the chimpanzees made all of their tools with their teeth and hands.

Not surprisingly the subsistence ecology of the human society is more complicated than that of the ape. However, the difference is far from wide, and the gap between hominid and pongid technology is bridgeable. In an evolutionary perspective, one can imagine the subsistence technoculture of an intermediate, ancestral hominoid filling the gap. The contrast shown here could easily be cultural, without resort to phylogenetic differences. Given what is known of chimpanzees' ability in captivity (Brink, 1957), they could make and use *all* of the Tasmanians' subsistants.

11. Chimpanzees' tool-making is closer to that of humans than to that of other primates

To infer intelligence in tool-use means attending to operational features, and these are better seen in tool-making than in tool-using.

Beck (1980) recognised four modes of tool manufacture, as defined in Table 6.5a. Detachment is common to many primates; motorically it is no different from plucking a fruit. Subtraction and re-shaping occur only in chimpanzees. Adding or combining in tool-making has never been seen in any free-ranging primates, although orang-utans and chimpanzees will do so in captivity, and in natural nest-building all great apes do so.

Oswalt (1976) also stated four principles of production, as given in Table 6.5b. McGrew (1987) used these principles to compare the Tanzanian chimpanzees with the Tasmanian gatherer–hunters. The humans showed all four, and the apes three: reduction for a fishing probe, replication for a leaf-sponge, linkage between an ant-dipping wand and a sapling facility. Missing was combination, although again nest-building as object manipulation, but not as tool-use, satisfied this condition. No one has yet applied Oswalt's principles to other species, but no other primate seems to go beyond reduction.

Boesch and Boesch (1990) analysed the complexity of the tool-making of wild chimpanzees for comparative purposes. They traced the number of types of modification (up to four but sometimes repeated) in the transformation of raw materials, and listed six types or methods of tool-making, as given in Table 6.5c. The scheme has yet to be applied to any other species, but even orang-utans would likely fail to score on it, at least in nature.

Overall, the technical gap between chimpanzees and other apes is consistently wider than between chimpanzees and humans.

Table 6.5. *Classificatory systems for making of tools*

(a)	*Beck's (1980) four modes of tool manufacture*	
	Detach	Sever fixed attachment between one object and another
	Subtract	Remove object from another unattached object
	Re-shape	Fundamentally re-structure an object's material
	Add/combine	Connect two or more objects
(b)	*Oswalt's (1976) four principles of production*	
	Reduction	Reduce mass of form
	Conjunction	Combine two or more technounits
	Replication	Craft two or more similar structural units to function as one part
	Linkage	Use physically distinct forms in combination
(c)	*Boesch & Boesch's (1989) types of tool-making*	
	Break with hands	
	Cut with teeth	
	Pull (apart) while standing on	
	Hit (and fracture) against hard surface	
	Remove leaves or bark with teeth or hands	
	Sharpen ends with teeth	

12. Only chimpanzees and humans show complementary manual functioning in tool-use

Use of bow and arrow, billiards cue, and knife and fork are all examples of human skilled performance in which the two hands simultaneously but complementarily take part in tool-use. One or more tools may be involved and each hand's grips and motor patterns usually differ. Such patterns appear later in human ontogeny than do one-handed or yoked two-handed acts (Elliott & Connolly, 1974).

In nature, most primate tool-use is one-handed, such as fishing for termites with a probe. Symmetrical and coordinated two-handed tool-use such as hammering a nut is apparently limited to great apes and capuchin monkeys.

Asymmetrical and bimanual tool-use is done only by chimpanzees. In one form, the tool is held in one hand, usually in a precision grip, and the object of action is held in the other, often in a power grip, e.g. using a twig to pick out bone marrow (Boesch & Boesch, 1989). More impressive is what occurs in ant-dipping (McGrew, 1974). One end of the wand is held tightly in a power grip (in either hand or foot!), while the other hand quickly slides the length of the wand in a loose precision grip. The first hand holds the tool steadily vertical while the second collects the ants on the sides of the flexed thumb and forefinger.

Is such skilled performance intelligent? The answer depends on how one defines intelligence. Small-brained creatures still show skilled performance with tools, e.g. the stone-tamping wasps, so the answer may be no. On the other hand, ant-dipping is an ingenious solution to the problem of how to corral, shift, and ingest about 200 biting ants, so the answer may be yes.

13. Laterality in primate tool-use is a dog's breakfast

Laterality of functioning in primates remains a perennial topic of controversy (MacNeilage *et al.* 1987; Fagot & Vauclair, 1991). Every study seems to use different methods and measures, and almost all are limited to handedness (Marchant & McGrew, 1991). Few have studied tool-use and none have taken into account degrees of skilfullness in the task recorded. Most studies have used artificial tasks in impoverished captive settings. We need more evidence from many free-ranging subjects performing a range of skilled tasks in daily life. Sometimes some of these conditions have been met: Schaller (1963) recorded the first-striking hand of eight male gorillas in their chest-beating displays, and all eight started with the right hand.

If laterality of functioning is correlated with asymmetry of structure, and if either or both much pre-date anatomically modern humans (Falk, 1987; Toth, 1985), then can we use data on extinct pongids to elucidate extinct hominids?

14. Across apes, tool-use is unrelated to brain asymmetry

Given the differences in Table 6.1, one can seek correlates with cerebral anatomy. The same highland gorillas that showed laterality in chest-beating showed significant cranial asymmetry (Groves & Humphrey, 1973). Holloway and de la Coste-Lareymondie (1982, Tables 3–7) compared latex endocasts of crania across sizeable numbers of four types of ape, and focused on five measures (of which two were composites). Lowland gorillas showed the most overall asymmetry, while the paramount tool-users, chimpanzees, ranked only third (behind bonobos but ahead of orang-utans).

Lemay (1976) compared fewer numbers of four types of apes for cerebral asymmetry using scaled photographs of preserved brains. She used four measures, only one of which, occipito-petalia, was in common with Holloway and de la Coste-Lareymondie. Lowland gorilla and chimpanzee ranked joint first for greatest asymmetry, with orang-utan third and gibbon fourth. There seem to be no other studies comparing at least four types of ape. What is needed are standardised data on tool-use to correlate with equally standardised data on brain structure for at least six types of ape: chimpanzee, bonobo, orang-utan, gorilla, siamang, and gibbon.

As things stand, one cannot infer tool-use in extinct humans from knowledge of tool-use and cerebral structure in extant apes. Palaeo-anthropologists are entitled to feel frustrated by the data still missing from primatologists.

15. Chimpanzees are the best available model for reconstructing human evolution

Chimpanzees never were, are not now, and probably never will be human beings. The converse is equally true. The phylogenetic lines leading to modern chimpanzees and humans diverged several million years ago, and since then their organic and cultural evolution has proceeded separately. Our pre-modern hominid ancestors are extinct, as are our proto-hominid, hominoid, anthropoid, etc. ancestors. Thus we can never directly know their behaviour and minds, so that their language and intelligence can only be inferred. Thankfully, however, at least some of their artifacts survive. The question is: What did it take to make those tools and how were they used? Re-phrased scientifically, it becomes a question of framing testable hypotheses that will yield useful inferences about the tool-users, based on their artifacts.

One solution is to adopt what Tooby and DeVore (1987) called *referential models*, when one real phenomenon is used as a model for its referent that is less amenable to direct study. Referential modelling has many disadvantages (as detailed by Tooby & DeVore, cf. Stanford & Allen, 1991) but it has one big advantage: It has a strong empirical base. Whatever the costs, the benefit is that we can do science on a present phenomenon whereas we can only guess about an absent one. Chimpanzees provide both tools *and* the acts of their making and use. No other referential model, e.g. social carnivore, baboon, dolphin, other ape, etc. meets this simple condition.

16. The chimpanzee model must be applied selectively and carefully

If chimpanzees are proposed as useful models, the obvious question is "For what?". There seem to be at least four candidates, each corresponding to a different stage in human evolution: Miocene ancestral hominoid, Pliocene proto-hominid, Plio-Pleistocene early hominid, and Pleistocene later hominid. For each of these, use of the chimpanzee model has its advantages and disadvantages.

The safest choice on grounds of homology is a Miocene hominoid descended from some dryopithecine-like ape. Therefore one is seeking the last common ancestor before the split of the pongid and hominid lines in Africa, sometime around 5–7 million years ago (ma). The immediate problem is that there are no known artifacts from that period, and such an African ancestral ape may have been an accomplished tool-user like a chimpanzee or a non-tool-user like a gorilla or something in-between (McGrew, 1989). There are as yet no artifacts to explain, but at least knowledge of tool-use by living chimpanzees can be used as a guide to know what to look for.

The next choice, on some mixture of grounds of homology and analogy, is a Pliocene proto-hominid, an australopithecine of 3–4 ma. By then there had been significant structural change such as the emergence of bipedalism, but confusing taxonomic radiation and no real "explosion" in brainpower. Worst of all, there are still no undeniable artifacts in the palaeo-archaeological record from that time. One can only stretch a point to include the earliest known tools, these being the crude stone artifacts dated at 2.4–2.7 ma from Hadar, Ethiopia (Harris, 1983). Pertinent questions for investigation might be whether or not living apes could and would make such tools, and if so, what for?

The first choice on grounds of clear analogy would be the earliest known Plio-Pleistocene large-brained hominid (*Homo habilis?*) as found at Olduvai Gorge, Tanzania, or at East Turkana, Kenya, some 1.5–2.0 ma. The Oldowan lithic culture differs from anything known for free-ranging apes, but all of the capacities needed to make it are manifested in the non-lithic tools of chimpanzees (Wynn & McGrew, 1989). The challenge is to find *anything* uniquely hominid in the capacities needed to make these artifacts.

Finally, the chimpanzee model might be usefully applied even to a recent (up to 300 000 years ago) hominid such as *Homo erectus* with its Acheulean tools. Some of the features put forward as distinctive for these tools (standardisation, symmetry, measurement, see Wynn, 1988) are arguably present in the material culture of living apes. Here the task for investigation is to sort out those abilities that are shared by ape and human from those found only in either one or the other.

Most attempts so far published (Tanner, 1987) to use chimpanzees as models for human evolution are not precise enough in these terms.

17. Palaeo-anthropologists seeking to make inferences about human evolution inexplicably ignore other primates

The 1980s saw a lively debate about the daily lives of early hominids in Plio-pleistocene times. Essentially it has been the reaction and counter-reaction to the ideas of the late Glynn Isaac (1978) and his students about home-bases, food-sharing, division of labour, pair-bonding, and so on. Much of that framework was based on inferences from artifacts, either stone tools or altered bones, as is the continuing debate, for example, on the relative importance of scavenging.

What is extraordinary to a primatologist is the extent to which participants in the debate ignore other primates in their reconstructions. Instead they concentrate on social carnivores to the total or virtual exclusion of monkeys and apes (Binford 1985; Blumenschine, 1987; Bunn & Kroll, 1986; Shipman, 1986; but cf. Toth, 1985).

The most *extensive* consideration of other primates occurs in Potts' (1988) analyses of material from Olduvai Gorge. There they rate no more than five of 396 pages, far fewer than hyaenas. Only chimpanzees get a look-in; meat-eating baboons (Harding, 1973) are barely mentioned. Again and again, statements crop up like "... hominids and carnivores are clearly the two agents primarily responsible for the bone concentrations at Olduvai" (p.142). It cannot be argued that other primates were absent, for there were more of their fossils present at the seven sites analysed than there were fossils of hominids. It can be argued that no fossil apes were found at Olduvai, but likewise there were no fossil hominids at five of the seven sites either. Thus statements like "... it is now generally believed, though by no means proved, that early *Homo* rather than *Australopithecus* was responsible for the earliest stone tools throughout East Africa" (p.3) are doubly problematical. Both other hominoids and proto-hominids are unjustly ignored.

Wild chimpanzees live in environments virtually indistinguishable from what Olduvai was like in Plio-Pleistocene times (McGrew *et al.*, 1981); they prey regularly on mammals (Riss & Busse, 1977); they scavenge carcasses from other predators (Morris & Goodall, 1977); they use tools to process bones (Boesch & Boesch, 1989); they leave lithic work-sites with characteristically altered tools (Sugiyama, 1981). Most of this had been well-known for years. Everything attributed to hominids at the level of Oldowan culture at Olduvai or Koobi Fora could have been done by pongids.

18. Vocal communication shows no necessary adaptive connection to tool-use

Traditional arguments against the gradual evolution of speech from precursory forms of vocal communication look increasingly threadbare. Research carried out in both field (Cheney & Seyfarth, 1990) and laboratory (Savage-Rumbaugh, 1988) settings show not only greater plasticity and complexity than previously realised, but also functional semanticity. Arguments that vocal communication in non-humans is reactive, stereotyped, and indicative only of motivational states have been dated for some time: Pierce (1985) reviewed 12 studies dating from 1965–1982 that examined the conditionability of vocalisations in non-human primates. Ten of these succeeded on species ranging from ring-tailed lemurs to chimpanzees, suggesting that contractual control is commonplace.

Notably in comparisons across non-human primates, no apparent correlation exists between complexity of communication in any sensory modality and the use of tools. Although chimpanzees always use tools and vervet monkeys never do, there seems to be no real difference in their vocal communication (Cheney & Seyfarth, 1990; Hauser & Wrangham, 1987).

In reconstructing the phylogeny of spoken language during hominisation, there is no reason to link it with tool-use, which likely long antedates it.

19. There is no obvious connection between degree of intelligence and complexity of tool-use

That creatures can be intelligent yet not use tools is clear from the cetaceans (Schusterman *et al.*, 1986). One can hardly imagine a less manipulative organism than a pelagic dolphin. Among primates overall, the correlation is grossly positive (see Meador *et al.*, 1987, for a massive review), but not so across closely related taxa. For apes, none of the standard laboratory measures of intelligence such as patterned string-pulling or transfer index discriminate between tool-using chimpanzees and non-tool-using gorillas (McGrew, 1992). Piagetian measures also fail to do this. Curiously, the only intellectual indicator of tool-use so far found is that of the ability to recognise oneself in a mirror (Gallup, 1987). Chimpanzees and orang-utans do; gorillas and gibbons apparently do not. On the basis of their tool-use one would predict that bonobos *will* be found to recognise their mirror image.

Tool-use is likely to be a by-product of a general ability for problem-solving, and selection pressures favouring better solvers could have rewarded better strategists in either socialising or foraging.

20. The relationship between language and intelligence is one of chicken-and-egg

Even if we knew whether the evolution of greater intelligence were driven more by social or by subsistence pressures, this would not elucidate any links to language, as it functions in both domains. The greatest current challenge to evolutionists is to devise tests that allow language and intelligence to be disentangled in the palaeo-archaeological record.

Acknowledgements

I am most grateful to Kathleen Gibson, Tim Ingold, Jürgen Lethmate, Toshisada Nishida, and Elisabetta Visalberghi for critical comments on the manuscript, though they should not be blamed for any mistakes remaining. Many thanks too to Fay Somerville for patient word processing.

References

Beck, B.B. (1980). *Animal Tool Behavior: The Use and Manufacture of Tools by Animals.* New York: Garland STPM Press.

Binford, L.R. (1985). Human ancestors: changing view of their behavior. *Journal of Anthropology*, 4, 292–327.

Blumenschine, R.J. (1987). Characteristics of an early hominid scavenging niche. *Current Anthropology*, 28, 283–407.

Boesch, C. & Boesch, H. (1989). Hunting behavior of wild chimpanzees in the Tai National Park. *American Journal of Physical Anthropology*, 78, 547–73.

Boesch, C. & Boesch, H. (1990). Tool-use and tool-making in wild chimpanzees. *Folia primatologica*, 54, 86–99.

Brewer, S.M. & McGrew, W.C. (1990). Chimpanzee use of a tool-set to get honey. *Folia primatologica*, 54, 100–4.

Brink, A.S. (1957). The spontaneous fire-controlling reactions of two chimpanzee smoking addicts. *South African Journal of Science*, 53, 241–7.

Bunn, H.T. & Kroll, E.M. (1986). Systematic butchery by Plio/Pleistocene hominids at Olduvai Gorge, Tanzania. *Current Anthropology*, 27, 431–52.

Calvin, W.H. (1982). Did throwing stones shape homid brain evolution? *Ethology and Sociobiology*, 3, 115–24.

Cheney, D.L. & Seyfarth, R.H. (1990). *How Monkeys See the World.* Chicago: University of Chicago Press.

Döhl, J. (1968). Über die Fahigkeit einer Schimpansin, Umwege mit Selbstandigen Zwischenzielen zu uberblicken. *Zeitschrift für Tierpsychologie*, 25, 89–103.

Elliott, J.M. & Connolly, K.J. (1974). Hierarchical structure in skill development. In *The Growth of Competence*, ed. K. Connolly & J. Bruner, pp. 135–68. London: Academic.

Fagot, J. & Vauclair, J. (1991). Manual laterality in non-human primates: A distinction between handedness and manual specialization. *Psychological Bulletin*, 109, 76–89.

Falk, D. (1987). Brain lateralization in primates and its evolution in hominids. *Yearbook of Physical Anthropology*, 30, 107–25.

Gallup, G.G. (1987). Self-awareness. In *Comparative Primate Biology. Volume 2B: Behavior, Cognition and Motivation*, ed. G. Mitchell & J. Erwin, pp. 3–16. New York: Alan R. Liss.

Goodall, J.v.L. (1963). Feeding behaviour of wild chimpanzees. *Symposia of the Zoological Society of London*, 10, 39–48.

Goodall, J.v.L. (1968). The behaviour of free-living chimpanzees in the Gombe Stream Reserve. *Animal Behaviour Monographs*, 1, 161–311.

Goodall, J.v.L. (1973). Cultural elements in a chimpanzee community. In *Precultural Primate Behavior*, ed. E. W. Menzel, pp. 144–84. Basel: S. Karger.

Goodall, J. (1986). *The Chimpanzees of Gombe: Patterns of Behavior.* Cambridge, MA: Harvard University Press.

Groves, C.P. & Humphrey, N.K. (1973). Asymmetry in gorilla skulls: Evidence of lateralised brain function? *Nature*, 244, 53–4.

Harding, R.S.O. (1973). Predation by a troop of olive baboons *(Papio anubis)*. *American Journal of Physical Anthropology*, 38, 587–92.

Harris, J.W.K. (1983). Cultural beginnings: Plio-Pleistocene archaeological occurrences from the Afar, Ethiopia. In *African Archaeological Review*, vol. 1, ed. N. David, pp. 3–31. Cambridge: Cambridge University Press.

Hauser, M.D. & Wrangham, R.W. (1987). Manipulation of food calls in captive chimpanzees. A preliminary report. *Folia primatologica*, 48, 207–10.

Holloway, R.L. & de la Coste-Lareymondie, M.C. (1982). Brain endocast asymmetry in pongids and hominids: Some preliminary findings on the paleontology of cerebral dominance. *American Journal of Physical Anthropology*, 58, 101–10.

Isaac, G. (1978). The food-sharing behavior of protohuman hominids. *Scientific American*, 238(4), 90–108.

Jordan, C. (1982). Object manipulation and tool-use in captive pygmy chimpanzees *(Pan paniscus)*. *Journal of Human Evolution*, 11, 35–9.

Kortlandt, A. (1965). How do chimpanzees use weapons when fighting leopards? *Yearbook of the American Philosophical Society*, 327–32.

Lemay, M. (1976). Morphological cerebral asymmetries of modern man, fossil man, and nonhuman primate. *Annals of the New York Academy of Sciences*, 280, 349–66.

Lethmate, J. (1982). Tool-using skills of orang-utans. *Journal of Human Evolution*, 11, 49–64.

MacNeilage, P.F., Studdert-Kennedy, M.G. & Lindblom, B. (1987). Primate handedness reconsidered. *Behavioral and Brain Sciences*, 10, 247–303.

Marchant, L.F. & McGrew, W.C. (1991). Laterality of function in apes: A meta-analysis of methods. *Journal of Human Evolution*, 21, 425–38.

McGrew, W.C. (1974). Tool use by wild chimpanzees in feeding upon driver ants. *Journal of Human Evolution*, 3, 501–8.

McGrew, W.C. (1977). Socialization and object manipulation of wild chimpanzees. In *Primate Bio-Social Development*, ed. S. Chevalier-Skolnikoff & F.E. Poirier, pp. 261–88. New York: Garland.

McGrew, W.C. (1987). Tools to get food: The subsistants of Tasmanian aborigines and Tanzanian chimpanzees compared. *Journal of Anthropological Research*, 43, 247–58.

McGrew, W.C. (1989). Why is ape tool use so confusing? In *Comparative Socioecology*, ed. V. Standen, & R.A. Foley, pp. 457–72. Oxford: Blackwell.

McGrew, W.C. (1992). *Chimpanzee Material Culture: Implications for Human Evolution*. Cambridge: Cambridge University Press.

McGrew, W.C., Baldwin, P.J. & Tutin, C.E.G. (1981). Chimpanzees in a hot, dry, and open habitat: Mt. Assirik, Senegal, West Africa. *Journal of Human Evolution*, 10, 277–44.

Meador, D.M., Rumbaugh, D.M., Pate, J.L. & Bard, K.A. (1987). Learning, problem solving, cognition, and intelligence. In *Comparative Primate Biology. Vol. 2B: Behavior, Cognition, and Motivation*, ed. G. Mitchell, & J. Erwin, pp. 17–83. New York: Alan R. Liss.

Morris, K. & Goodall, J. (1977). Competition for meat between chimpanzees and baboons in the Gombe National Park. *Folia primatologica*, 28, 109–21.

Nishida, T. (1980). The leaf-clipping display: A newly-discovered expressive gesture in wild chimpanzees. *Journal of Human Evolution*, 9, 117–28.

Oswalt, W.H. (1976). *An Anthropological Analysis of Food-Getting Technology*. New York: John Wiley.

Parker, S.T. & Gibson, K.R. (1977). Object manipulation, tool use and sensorimotor intelligence as feeding adaptations in cebus monkeys and great apes. *Journal of Human Evolution*, 6, 623–41.

Pierce, J.D. (1985). A review of attempts to operantly condition alloprimate vocalization. *Primates*, 26, 202–13.

Potts, R. (1988) *Early Hominid Activities at Olduvai.* New York: Aldine de Gruyter.

Riss, D.C. & Busse, C.D. (1977). Fifty-day observation of a free-ranging adult male chimpanzee. *Folia primatologica*, 28, 283–97.

Savage-Rumbaugh, E.S. (1988). A new look at ape language: Comprehension of vocal speech and syntax. *Current Theories in Research and Motivation*, 35, 201–55.

Schaller, G. (1963). *The Mountain Gorilla.* Chicago: University of Chicago Press.

Schusterman, R.J., Thomas, J.A. & Wood, F.G. (1986). *Dolphin Cognition and Behavior: A Comparative Approach.* Hillsdale: Lawrence Erlbaum Associates.

Shipman, P. (1986). Scavenging or hunting in early hominids: Theoretical framework and tests. *American Anthropologist*, 88, 27–43.

Stanford, C.B. & Allen, J.S. (1991). On strategic story-telling: Current models of human behavioral evolution. *Current Anthropology*, 32, 58–61.

Sugiyama, Y. (1981). Observations of population dynamics and behavior of wild chimpanzees at Bossou, Guinea, in 1979–1980. *Primates*, 22, 435–44.

Tanner, N.M. (1987). The chimpanzee model revisited and the gathering hypothesis. In *The Evolution of Human Behavior*, ed. W. G. Kinzey, pp. 3–27. Albany: State University of New York Press.

Tooby, J. & DeVore, I. (1987). The reconstruction of hominid behavioural evolution through strategic modelling. In *The Evolution of Human Behavior*, ed. W. G. Kinzey, pp. 183–237. Albany: State University of New York Press.

Toth, N. (1985). Archaeological evidence for preferential right-handedness in the Lower and Middle Pleistocene, and its possible implications. *Journal of Human Evolution*, 14, 607–14.

Visalberghi, E. (1987). Acquisition of nut-cracking behaviour by 2 capuchin monkeys (*Cebus apella*). *Folia primatologica*, 49, 168–81.

Wynn, T.G. (1988). Tools and the evolution of human intelligence. In *Machiavellian Intelligence*, ed. R. W. Byrne, & A. Whiten, pp. 271–84. Oxford: Clarendon Press.

Wynn, T.G. & McGrew, W.C. (1989). An ape's view of the Oldowan. *Man*, 24, 383–98.

7

Aspects of transmission of tool-use in wild chimpanzees

CHRISTOPHE BOESCH

The acquisition of a behaviour is accelerated when a model influences the performance of an inexperienced individual. The numerous behavioural differences observed between populations in some species of non-human primates suggest such a mechanism of acceleration, and many of these differences have been attributed to imitation (Beck, 1973; Kawamura, 1959; Kawai, 1965; Nishida, 1987; Sumita, *et al.* 1985). Recently, however, a more critical approach to the phenomenon of imitation has been adopted, and even classical examples, such as potato-washing and wheat-throwing in Japanese macaques, are contested (Visalberghi & Fragaszy, 1990; Galef, 1988 in Whiten, 1989; Tomasello, 1990; Visalberghi, 1988). These critics do not put forward behavioural arguments, but rather claim that the rate of acquisition by other group members was so slow that it could not be accounted for by imitation. This argument is not only based upon the unproven assumption that the motivation to imitate should lead to a quicker dissemination of a given behaviour, but also neglects the limits that social structure imposes on dissemination – for group members are not randomly associated (see Kawai, 1965). Recent observations, however, do seem to support the idea that imitation may be more difficult for primates than previously thought: Cebus monkeys and baboons proved unable to acquire the behaviour of other skilled tool users occupying the same cages, despite prolonged opportunities to observe them (Visalberghi, 1988; Visalberghi & Fragaszy, 1990; Whiten, 1989). Even the imitative abilities of chimpanzees have been questioned, as captive juveniles were shown not to copy precisely a given behaviour but only to learn that a particular tool may be used to obtain a certain food (Tomasello, *et al.* 1987).

In his review, Tomasello (1990) challenges the idea of cultural transmission in wild chimpanzees, mainly on the grounds of his inability to find evidence of true imitation in experiments conducted with captive young chimpanzees. True imitation is the process by which an observer attends to, and attempts to copy, the *behaviour* of another individual rather than trying to reach the *same goal* by whatever means it may find or to use the *same object*, its attention being attracted to that object by the activity of the other individual (Thorpe, 1956; Wood, 1988 in Tomasello, 1990). However, Tomasello's arguments suffer from some logical flaws. First, cultural transmission *must* be studied or demonstrated on sets of behaviours that might have the potentiality to be cultural. The seeming lack of imitation by Yerkes's chimpanzees in sand throwing or the use of reaching sticks (Tomasello, 1990; Tomasello, *et al.* 1987) does not tell us much about hand-grasping, leaf-clipping, nut-cracking or other sets of behaviours showing traditional differences in wild chimpanzees (Boesch & Boesch, 1990; Goodall, 1986; Nishida, 1987; McGrew & Tutin, 1978). Therefore, *cultural transmission should preferably be studied in the field.* Secondly, Tomasello suggests that the appearance of a given behaviour may be put down to ecological conditions at the site where it occurs, and that we cannot assume that the behaviour is conserved when conditions change. If this argument is true for the invention of the behaviour, it does not, however, explain how behavioural differences can persist even where no ecological differences exist to maintain them.

One might expect imitation to work alongside other mechanisms such as trial-and-error learning, social facilitation, stimulus enhancement or maturation processes[1]. Thus, it can be quite tricky to find a behaviour that is attributable only to imitation, or to prove the exact contribution that imitation makes in relation to the other processes. Some authors suggest that only experimental studies could solve this problem (Visalberghi & Fragaszy, 1990), thereby strongly underestimating the negative impact that non-natural social and environmental settings can have on individual histories of learning (see also the argument outlined above on cultural transmission). An alternative view would be to look in the wild for complex novel behaviours that are acquired at an age too early to be mastered cognitively, thus excluding social facilitation, stimulus enhancement and pure maturation processes.

Yet another possibility would be to look for evidence of one animal deliberately performing a piece of behaviour in front of another one, as if in the expectation that the latter could imitate it. In other words, we might *look for evidence of teaching.* It is assumed here that this kind of behaviour would have been eliminated in the course of evolution if that expectation were unfounded. The teaching of one individual by the other is considered to be unique to humanity and to constitute the basis for the existence of human culture (Guilmet, 1977, Passingham, 1982; Premack, in press, Wood *et al.*, 1976). Although teaching in animals is certainly rare, this may be partly because it might be only rarely needed. However, certain contexts may favour teaching, e.g. situations where the model needs to accelerate the acquisition of a behaviour in an inexperienced individual to prevent any damage to its own reproductive success.

I shall present here some evidence for such a transmission mechanism in wild chimpanzees, with regard to the use of tools to open nuts in the tropical rain forest of the Taï National Park, Côte d'Ivoire.

Nut-cracking behaviour

The nut-cracking behaviour is restricted to the chimpanzee populations of the evergreen forest belt in Liberia and Ivory Coast (Boesch & Boesch 1983; 1989, Sugiyama & Koman, 1979). The most commonly cracked nut species, *Coula edulis* and *Panda oleosa*, occur naturally in most Western and Central African forests (Aubreville, 1959). However, no traces of nut cracking have been found in Central African forests where chimpanzees are nevertheless present (Gabon: Tutin & Fernandez 1983; Cameroun: Sugiyama, 1985). This discrepancy between the geographical distribution of the nut-cracking behaviour of the chimpanzee and the nut species has convinced many authors that nut cracking is a traditional behaviour (e.g. Boesch & Boesch, 1990; Goodall, 1986; Nishida & Hiraiwa, 1982).

Typically, the chimpanzees collect nuts on the ground or in the trees, carry them to a root which is used as an anvil, sit and place their provisions on the ground, and start pounding. For this, a nut is placed in a depression of the root (a neat hole of the size of the cracked nuts is obtained after some pounding), and opened by hitting it repeatedly with a hammer tool. Each nut is eaten at once and, having consumed what they have collected, the chimpanzees may search for more nuts and resume cracking at the same anvil or continue at another one, thus moving slowly between nut-yielding trees. On average, an individual cracks nuts with a hammer for 2 hours 15 minutes per day during the 4 months of the *Coula* nut season. Four other nut species are cracked at different periods throughout the year, making the use of hammer tools for processing food a habit.

The availability of hammers, in the form of wooden clubs or stones, is the limiting factor in the forest and they are chosen according to the hardness of the nut species: the harder the nuts, the heavier and the harder the hammer (Boesch & Boesch, 1983). Both the nuts and the hammers are transported. Stone hammers are a rarity in Taï forest and are indispensable for opening the very hard nuts of *Panda oleosa*. Thus, the transport of tools is frequent for this nut species. Our analysis showed that chimpanzees can memorise the locations of up to five stones and select the nearest one to a goal tree (Boesch & Boesch 1984a, average transport distance is above 100 m). This selection is made anew for each different *Panda* tree at which they set out to crack, where no hammer is present. The action is therefore elaborately planned and seems to be more complex than in other chimpanzee tool techniques (Goodall, 1986). Wood for hammers is more abundant than stones and such hammers are transported less and for shorter distances. Many are made, mostly by shortening, from fallen branches found in the vicinity of the tree where a chimpanzee wants to crack nuts – 6.5% of the wooden hammers used were made in this way (Boesch & Boesch, 1990). Techniques for modifying the branch include ones never seen in any other chimpanzee populations, e.g. hitting the branch on a hard surface, or standing on it and pulling upwards, in order to break it (Boesch & Boesch, 1990). Despite the elaborate procedures involved, the nuts cracked with a hammer represent a major food source, providing an adult female with a net benefit of 3800 calories per day during nut seasons (Boesch in prep.).

An analysis of the efficiency of nut cracking (number of hits per nut and number of nuts per minute) showed that it is completely acquired only by the adult individual (Boesch & Boesch, 1984b), and is thus probably the most demanding manipulatory technique yet known to be performed by wild chimpanzees. Such a tool behaviour could only possibly develop in a situation of rich nutritional rewards, including food sharing, in order to compensate for the energy and time invested over the years in acquiring the technique (Boesch & Boesch, 1990). Nut sharing commits Taï chimpanzee mothers for the first eight years of the life of their offspring. It offers a very rich food source for the youngsters, since three- to five-year-old infants receive from their mothers, during the nut season, up to 1000 calories per day in nuts (Boesch, in prep.).

Hence it is precisely in the context of nut-cracking behaviour, which because of its cognitive and manipulatory complexity is acquired much later than any other tool technique (Goodall, 1986; Nishida & Hiraiwa, 1982), that we might expect to find some forms of teaching, in order to ensure that juvenile individuals master it before the mother invests in new offspring.

Mothers' intervention in their infants' nut-cracking attempts

Chimpanzee mothers have been observed to attempt to accelerate the acquisition of this technique in three different ways (Boesch, 1991):

A. Stimulation:

Mothers may stimulate the nut-cracking attempts of their infants in two different ways:

1. *Leaving the hammer on the anvil*: Commonly, when cracking nuts in a group, chimpanzees carry their hammers along when collecting nuts, either to prevent the risk of losing them to others or because they will crack further at another anvil. However, when a three-year-old infant first begins to pound nuts, its mother may leave the hammer more frequently on the anvil during nut collecting (Table 7.1: comparison between columns a and b + c between age classes 2 and 3: $X^2 = 17.29$, df = 1, $p < 0.001$; comparisons of the other adjacent age classes: $p > 0.05$). The infant, remaining near the anvil, may get interested in the hammer and begin to handle it. The mother resumes using her hammer when she comes back from collecting nuts.

2. *Leaving nuts near the anvil*: Infantless chimpanzees were never observed to leave good, whole nuts near the anvil before searching for fresh ones. However, mothers were seen to do so and always left the hammer on the anvil at the same time. This acts as an additional incentive for the infant remaining near the anvil. In fact in all observed cases, the infant would make at least some hits with the hammer after placing a nut on the anvil. Mothers may go a step further by arranging all the necessary elements of nut cracking, so that the infant needs only to perform the actual pounding. Typically, a mother would place one of the nuts she leaves behind in the hole of the root, and the hammer upon it. The infant then needs only to pound the nut with the hammer.

In Table 7.1, the stimulations have been pooled in order to analyse variations in the behaviour of mothers and infants according to the age of the latter. The infants, when 3 years old, start to use the hammer left on the anvil by their mothers more frequently than the younger ones (age classes < 1 to 2 years versus 3 to 6 years; $X^2 = 4.14$, df = 1, $p < 0.05$).

B. Facilitation:

Mothers may facilitate the nut cracking of their infants in two ways:

1. *Providing good hammers*: Nut-cracking infants who are over 4 years old commonly have difficulties in obtaining suitable hammers in a group of chimpanzees, and usually their mothers' tools are better than their own (hammers considered suitable are those with an adequate weight and a regular shape that allow a nut to be opened with fewer than 10 hits). Mothers then start to allow infants to use their own hammers (Table 7.1). This may constitute an important disadvantage for these mothers; for example on 12th February 1989, while cracking in a group moving between fruit-bearing *Coula* trees, Ella let her son, Gérald, take consecutively her four hammers and, as she had to look continually for another appropriate tool, she was able to open only 8 nuts during 40 minutes while her son opened 36. Later, she opened 20 nuts in 9 minutes with her fifth hammer, while her son was working nearby with her handy fourth hammer.

Table 7.1. *Ways that chimpanzee mothers use to influence their offspring's nut-cracking behaviour, according to the age of the infant.*

This table is based on observations of 20 mother–infant pairs. Under "stimulation" are given the numbers of occasions when the mother either (a) took the hammer away from the anvil, (b) left the hammer near the anvil but the infant did not touch it or (c) left the hammer and the infant used it. The figures for "facilitation" refer to the numbers of objects provided by the mother, whereas for active teaching they refer to the number of events of teaching observed.

Age of infant	Observ. time (min)	Stimulation			Facilitation		Active teaching
		a	b	c	Hammer	Nuts	
<1 year	44	5	3	—	—	—	—
1 year	202	22	6	4	—	—	—
2 years	268	16	8	5	—	3	—
3 years	863	23	39	65	8	2	—
4 years	999	15	39	42	25	54	1
5 years	890	7	27	17	38	206	—
6 years	759	7	13	23	48	37	1
8 years	112	—	—	1	153	14	—
Total	4137	95	135	157	272	316	2

Source: Adapted from Boesch 1991.

2.   *Providing intact nuts*: When cracking by themselves, infants have to invest the necessary time to collect the nuts and some mothers may let them take some of their own provisions (Table 7.1). Thus on the same day as the example related above, Gérald took 43 of the 47 nuts he opened from his mother. Similarly, Aurore's daughter took 40 of the 59 nuts that her mother had collected and this resulted in an increase of 20% in the number of nuts eaten per minute by the infant, compared to when she was collecting nuts herself and cracking them with the same tool on the same anvil.

C. Active teaching:

Despite their efforts and the use of convenient tools, infants may face technical difficulties in nut cracking that they are unable to overcome. In such situations, they either try to find another tool or quit and join their mother to beg for nuts. In two cases, the mother, noticing the difficulties of her infant, was seen to make a clear demonstration of how to solve them. We shall describe the two events precisely.

On the 22th February 1987, Salomé was cracking a very hard nut species (*Panda oleosa*) with her son, Sartre. He took 17 of the 18 nuts she opened. Then, taking her stone hammer, Sartre tried to crack some by himself, with Salomé still sitting in front of him. These hard nuts are tricky to open as they consist of three kernels independently embedded in a hard wooden shell, and the partly opened nut has to be positioned precisely each time to gain access to the different kernels without smashing them. After successfully opening a nut, Sartre replaced it haphazardly on the anvil in order to try to gain access to the second kernel. But before he could strike it, Salomé took the piece of nut in her hand, cleaned the anvil, and replaced the piece carefully in the correct position. Then, with Salomé observing him, he successfully opened it and ate the second kernel.

In this example, the mother assisted her son by demonstrating the correct positioning of the nut, although the infant would probably have succeeded in opening it eventually. Salomé interfered before Sartre could make the mistake of hitting the nut badly positioned, thus anticipating the consequence of his actions.

On the 18th February 1987, Ricci's daughter, Nina, tried to open nuts with the only available hammer, which was of an irregular shape. As she struggled unsuccessfully with this tool, alternating her posture (14 times), hammer grip (about 40 times), the position of the nut or the nut itself, Ricci was resting. Eventually, after eight minutes of this struggle, Ricci joined her and Nina immediately gave her the hammer. Then, with Nina sitting in front of her, Ricci, in a very deliberate manner, slowly rotated the hammer into its best position for efficiently pounding the nut. As if to emphasize the meaning of this movement, it took her a full minute to perform this simple rotation. With Nina watching her, she then proceeded to use the hammer to crack ten nuts (of which Nina received six entire kernels and portions of the four others). Then, Ricci left and Nina resumed cracking. Thereafter, she succeeded in opening four nuts in 15 minutes, and as she still had difficulties, she regularly changed her own position (18 times) and the position of the nut, but always held the hammer in the same position as had her mother and never changed her grip nor the position of the hammer. She also whimpered whenever encountering difficulties to attract her mother, but Ricci did not return to her even when she threw a temper tantrum after unsuccessfully attempting for three minutes to open the fifth nut.

In this last example, the mother, seeing the difficulties of her daughter, *corrected an error* in her daughter's behaviour in a very conspicuous way and then proceeded to demonstrate to her how it works with the proper grip. She stopped before having cracked all the nuts so that Nina could immediately try by herself, and her daughter seemingly understood the lesson perfectly, since she *strictly maintained the grip demonstrated to her*, despite increasing difficulties, and attempted to solve these by varying her position and that of the nut.

Discussion

Cultural transmission is possible only if observation and learning from observation go in at least one direction between the novice and the model: imitation goes in one direction, teaching in both (Premack, in press; Tomasello,

1990). In the sets of behaviour reported here, Taï chimpanzee mothers show clearly that they observe their infants' behaviour and intervene in it. In doing so, they demonstrate their ability both to compare their offspring's behaviour to their own conception of how it should be performed, and to anticipate the possible effects of their actions on those of their offspring. These interventions are adjusted to the level of skill reached by the offspring: thus stimulation reaches its maximum at the age of three when the infants start to learn the basic skills of nut cracking, whereas facilitation reaches its maximum at five years (Table 7.1: Spearman correlation coefficient: $rs = -0.48$, $p < 0.05$). Similarly, mothers leave the hammer on the anvil more frequently when their offspring are three years old than when they are only two.

Maternal interventions occur on average once every five minutes (Table 7.1: counting stimulations in columns b and c along with facilitation and active teaching, there were on average 0.215 interventions per minute), but active teaching is rare compared to the two other forms of intervention (0.2% of the interventions (2 teaching instances out of 977 interventions) and with a frequency of 0.0005 per minute (2 instances in 4137 minutes)). To our knowledge, the active teaching episodes reported here in the nut-cracking context are the first to be recorded in a non-human primate population. As suggested earlier, the difficulty of the technique and the importance for the mother that her infant should acquire it before she invests in further offspring may explain its existence in this context (Boesch, 1991). These teaching instances occurred spontaneously in undisturbed mother–infant pairs in their natural habitat. The fact that these conditions are so rarely fulfilled in studies of captive animals may directly explain the failure of such studies to reveal evidence of imitation and teaching: in fact, the motivation to copy or influence behaviour is much greater when the model is a close kin relation than it is with an unrelated or unfamiliar model. An example of how a difference in the status of the model (chimpanzee versus human) can lead to completely opposite results comes from a study in which nut-cracking behaviour was taught to a juvenile female chimpanzee in a zoo; a human demonstrator failed to elicit much interest in 30 hour-long sessions, whereas when exposed to a successful adult female chimpanzee as demonstrator, she tried and succeeded in opening the nut on her own after the very first sessions (Sumita *et al.*, 1985). Experiments under controlled conditions have certain advantages, but the drawbacks should not be obscured as they may make captive conditions inappropriate for studying such processes as learning and the transmission of knowledge (see for example Tomasello, 1990; Visalberghi & Fragaszy, 1990).

In humans, too, true teaching seems to be relatively rare; in a study of children in Nigeria and England (Whiten & Milner, 1984), mothers were seen to teach their children (from six to 14 years old) in only 0.1% of all those interactions having educational effects (two demonstrations out of 1929 educational interactions) and reached a frequency of two instances in 43 800 minutes. It would, however, be premature to conclude that teaching is more frequent in chimpanzee than in human groups, as the contexts of the two studies are rather different: the human study was made of everyday activities with all kinds of natural behaviours (Whiten & Milner, 1984), whereas the chimpanzee study was restricted to the most elaborate technological behaviour known to this species. Nevertheless, the comparison illustrates the fact that events of teaching are rare in both species and that among humans, too, not all types of behaviour are appropriate for observing it.

Human teaching has been shown to fulfil six functions (Wood, *et al.*, 1976): *recruitment*, in which the model attracts the child's attention to a task; *scaffolding*, in which the model simplifies the task by reducing the number of acts required to reach the solution; *direction maintenance*, in which the model keeps the child in pursuit of the solution; *marking critical features*, in which the model accentuates certain features of the task; *frustration control*, in which the model eases the search for the solution; and *demonstration*, in which the model provides the child with information on how to perform a task. In the nut-cracking context, Taï chimpanzees were seen to perform all six functions of tutoring. When stimulating their infants' attempts by leaving the hammer and nuts on the anvil, the mothers appear to fulfil three of these functions: recruitment, direction maintenance and marking critical features. When facilitating the nut cracking of their infants by providing a hammer and collecting nuts for them, they fulfil two other functions: frustration control and scaffolding. Finally in active teaching, they not only control their infants' frustration but, more importantly, they demonstrate the appropriate way of reaching a solution and, in the last example, *corrected an error* in the infant's technique.

Infant chimpanzees in Taï seem to be influenced by the intervention of their mothers. In half of the cases, when the mothers left the hammer on the anvil, the infants used it (see Table 7.1). For an infant, at least three different aspects of nut cracking can be reinforced through observing its mother: it can be attracted to manipulate the same tool, it can attempt by whatever means to reach the same goal or it can directly copy the behaviour of the mother. Of these, only the latter, i.e. imitation, would lead to cultural transmission by making the behaviour stable and similar within a given population (Tomasello, 1990). Nut cracking is strongly influenced by external constraints, such as the physical properties of the materials and the topology of the nut-cracking sites, so that a simple attraction on the part of the infant to use the same tool or to reach the same goal can lead to similar behaviour and a similar result, i.e. nuts are cracked. However, some observations reveal that *imitation of the behaviour of the model* is at work as well. In the second example of teaching presented above, Nina, who was already using the tool and trying, unsuccessfully, to achieve the goal (cracking the nuts), imitated, after her mother's intervention, only a small but important detail in the form of the behaviour of her mother, namely the way she was holding the hammer. In addition, Nina seemed to be aware of the importance of this detail, as in the ensuing difficulties she changed her position many times but never her grip.

The learning of nut cracking is very slow, but some features in this process indicate that imitation of the behaviour of the model takes place at an early stage. Youngsters, under the age of three, still lack the necessary strength to strike a nut successfully and are regularly seen to perform an adequate hitting movement whilst forgetting either to place the nut on an anvil or to use the hammer – directly hitting the nut with the hand or another nut. Older infants were seen to use other types of hammer, such as a piece of hard soil from a termite mound or a hard-shelled fruit. The youngsters were also seen, although more rarely, to forget the nut altogether and to hit the hammer against the anvil. Thus, infants were poorly influenced in their attempts by their mothers' choice of tools, yet it remains possible that they were attempting, albeit inadequately, to reach the same goal as their mothers. Counting against such an interpretation, however, is the fact that they were never seen to use unusual ways of attempting to open a nut such as throwing it against a hard surface or hitting it with a foot or with any other part of the body but the hand (Boesch, in prep). In other words, *infants tried every way possible to open the nut but within the limits of the hitting movements they had seen in the adults.*

Modes of cultural transmission include different types of models and novices: vertical transmission (see Cavalli-Sforza & Feldman, 1981; Hewlett & Cavalli-Sforza, 1986) from parents to infants seems to apply widely to nut cracking among Taï chimpanzees. Besides this mode, horizontal transmission, i.e. between members of the same generation, is the only other mode for which I could collect reliable evidence in nut cracking. Thus Ella, an adult female, has a strong right-hand bias when pounding nuts with a hammer. Two of her three sons are, like her, strongly right handed. The older one has developed the habit of holding the hammer at the right extremity above the ground with his right foot in support of the right hand. He was the only chimpanzee, out of a total of eighty individuals in the study community, to do this, until his youngest brother adopted in a very rigid way exactly the same fashion of holding the hammer.

Turning to the question of the relation between tool-use, tool making and intelligence, Taï chimpanzee mothers demonstrate, in the nut-cracking context, first, a concern for the results of their infants' attempts; secondly, the capacity to compare their offspring's behaviour with their own conception of how it should be performed and, thirdly, attempts to influence such behaviour by anticipitating the possible effects of their acts on those of their infants. Such mental processes are demanding and resemble those needed to make a tool before using it. Indeed, the tool maker needs a precise idea of the shape such an object has to have for it to be considered as an adequate tool, and of the technical steps needed to modify the object so that it conforms to this idea. Only Taï chimpanzees make most of their tools beforehand (in 94% of the cases), whereas Gombe and Mahale chimpanzees reshape them by trial and error during repeated use (Boesch & Boesch, 1990; Nishida & Hiraiwa, 1982). It is possible, therefore, that tool making prior to use and active teaching may be related.

Acknowledgements

Hedwige Boesch and I thank the "Ministère de la Recherches Scientifiques" and the "Ministère des Eaux et Forêts" of the Ivory Coast for accepting this project and the Swiss National Science Foundation, the Leakey Foundation and The Jane Goodall Institute for financing it. The project is part of the UNESCO project Taï-MAB under the supervision of Dr. Henri Dosso. We are grateful to A. Aeschlimann, F. Bourlière and H. Kummer for their constant encouragement and support, and to the director of the Centre Suisse and the staff of the Station IET in Taï for logistic support.

Note

1 Both social facilitation and stimulus enhancement concern acts that are already present in the repertoire of the individual influenced. Only imitation allows an individual to add a novel act to his repertoire (for a review see Visalberghi and Fragaszy 1990). However, a maturation process has a similar effect to imitation by allowing the integration of a novel behaviour in the repertoire of an individual but, contrary to imitation, it should appear at certain specific periods during the development in all individuals. Social facilitation and stimulus enhancement may act along with maturation, so that a novel act may appear in the repertoire of a juvenile as if influenced by a model. Addition of a novel act in the repertoire is thus not enough to prove the existence of true imitation.

References

Aubreville, A. (1959). *La Flore Forestière de la Côte d'Ivoire* 3 vols., Nogent-sur-Marne: C.T.F.T.
Beck, B. (1973). Observational learning of tool use by captive Guinea baboons (*Papio papio*). *American Journal of Physical Anthropology*, 38, 579–82.

Boesch, C. (1991 in press). Teaching among wild chimpanzees. *Animal Behaviour*, 41.

Boesch, C. & Boesch, H. (1983). Optimization of nut-cracking with natural hammers by wild chimpanzees. *Behaviour*, 83, 265–86.

Boesch, C. & Boesch, H. (1984a). Mental map in wild chimpanzees: An analysis of hammer transports for nut cracking. *Primates*, 25, 160–70.

Boesch, C. & Boesch, H. (1984b). Possible causes of sex differences in the use of natural hammers by wild chimpanzees. *Journal of Human Evolution*, 13, 415–40.

Boesch, C. & Boesch, H. (1989). Hunting behavior of wild chimpanzees in the Taï National Park. *American Journal of Physical Anthopology*, 78, 547–73.

Boesch, C. & Boesch, H. (1990). Tool use and tool making in wild chimpanzees. *Folia Primatologica*, 54, 86–99.

Cavalli-Sforza, L.L. & Feldman, M.W. (1981). *Cultural Transmission and Evolution: A Quantitative Approach*. Princeton, NJ: Princeton University Press.

Goodall, J. (1986). *The Chimpanzees of Gombe: Patterns of Behavior*. Cambridge, MA: The Belknap Press of Harvard University Press.

Guilmet, G. (1977). The evolution of tool-using and tool-making behaviour. *Man*, 12, 33–47.

Hewlett, B.S. & Cavalli-Sforza, L.L. (1986). Cultural transmission among Aka Pygmies. *American Anthropologist*, 88, 922–34.

Kawamura, S. (1959). The process of sub-culture propagation among Japanese macaques. *Primates*, 2, 43–60.

Kawai, M. (1965). Newly acquired precultural behaviour of the natural troop of Japanese monkeys on Koshima islet. *Primates*, 6, 1–30.

McGrew, W.C. & C.E.G. Tutin (1978) Evidence for a social custom in wild chimpanzees? *Man*, 13, 234–51.

Nishida, T. (1987). Local traditions and cultural transmission. In *Primate Societies*, ed. S. S. Smuts, D. L. Cheney, R. M. Seyfarth, R. W. Wrangham & T. T. Struhsaker, pp. 462–74. Chicago: University of Chicago Press.

Nishida, T. & Hiraiwa, M. (1982). Natural history of a tool-using behaviour by wild chimpanzees in feeding upon wood-boring ants. *Journal of Human Evolution*, 11, 73–99.

Passingham, R. (1982). *The Human Primate*. London: Freeman.

Premack, D. (in press). The aesthetic basis of pedagogy. In *Festschrift for Jim Jenkins*, ed. R. R. Hoffman & D. Palermo. Hillsdale, NJ: Erlbaum.

Sugiyama, Y. (1985). The brush-stick of chimpanzees found in south-west Cameroon and their cultural characteristics. *Primates*, 26(4), 361–74.

Sugiyama, Y. & Koman, J. (1979). Tool-using and -making behavior in wild chimpanzees at Bossou, Guinea. *Primates*, 20, 513–24.

Sumita, K., Kitahara-Frisch, J. & Norikoshi, K. (1985). The acquisition of stone-tool use in captive chimpanzees. *Primates*, 26, 168–81.

Thorpe, W. (1956). *Learning and Instinct in Animals*. London: Methuen.

Tomasello, M. (1990). Cultural transmission in tool use and communicatory signalling of chimpanzees? In *"Language" and Intelligence in Monkeys and Apes: Comparative Developmental Perspectives*, eds. S. T. Parker, & K. R. Gibson, pp. 274–311. Cambridge: Cambridge University Press.

Tomasello, M., Davis-Dasilva, M., Camak, L. & Bard, K. (1987). Observational learning of tool-use by young chimpanzees. *Journal of Human Evolution*, 2, 175–83.

Tutin, C.E.G. & Fernandez, M. (1983). *Recensement des gorilles et des chimpanzés du Gabon.* Centre International de Recherches Médicales de Franceville, Gabon.

Visalberghi, E. (1988). Responsiveness to objects in two social groups of tufted capucin monkeys (*Cebus apella*). *American Journal of Primatology*, 15, 349–60.

Visalberghi, E. & Fragaszy, D. (1990). Do monkeys ape? In *"Language" and Intelligence in Monkeys and Apes: Comparative Developmental Perspectives*, eds. S. T. Parker, & K. R. Gibson, pp. 247–273. Cambridge: Cambridge University Press.

Whiten, A. (1989). Transmission mechanisms in primate cultural evolution. *Trends in Ecology and Evolution*, 4(3):61–2.

Whiten, A. & Milner, P. (1984). The educational experiences of Nigerian infants. In *Nigerian Children: Development Perspectives*, ed. H. Valerie Curran, pp. 34–73. London: Routledge and Kegan Paul.

Wood, D., Bruner, J.S. & Ross, G. (1976). The role of tutoring in problem-solving. *Journal of Child Psychology and Psychiatry*, 17(2), 89–100.

Part III

Connecting up the brain

Part III Introduction

Overlapping neural control of language, gesture and tool-use

KATHLEEN R. GIBSON

Two classic views of human brain evolution stand in fundamental opposition. Some argue that the human brain contains structures or circuitry not found in the brains of other animals. Not surprisingly, they often assume that major qualitative differences distinguish human language or technology from that of the apes. Others emphasize the massive increase in human brain size and posit that human behavior is an emergent phenomenon which reflects quantitative increases in neural processing capacity. Parallel controversies focus on degrees of neural localization. Does, for instance, each separate behavioral domain have its own discrete neural processing center or module? Alternatively, is the control of seemingly discrete skills widely distributed throughout the nervous system and accompanied by overlapping neural control of diverse behaviors?

The resolution of these controversies requires co-operation between evolutionarily-oriented behavioral scientists and neurobiologists. Only neurobiologists can delineate the exact functions of neural structures and processes. Behavioral scientists, however, can provide knowledge of species-typical behaviors. Such knowledge can dramatically alter the formulation of neural hypotheses and research design. Lieberman (1991), for instance, argues that monkeys and apes lack voluntary control of their vocal behaviors (contra Snowdon, this volume). This leads him to speculate on the presence of an as yet unknown, distinct circuitry in the basal ganglia of the human brain. Others, who recognize the presence of voluntary vocalizations in monkeys, see no need to posit unique human neural circuitry, and assume that quantitative neural differences can account for vocal distinctions (Gibson, 1988).

Similarly, questions concerning whether neural function is localized or distributed may reflect our perceptions of behavior. If vocal language, gesture, tool use and sociality are seen to pertain to distinct behavioral domains, then each is likely to be controlled by separate neural modules. If, on the other hand, these are seen as inseparable aspects of behavior constituting one total domain,

then they may also reflect common underlying mental processes with distributed and overlapping neural control.

In this section, two neurobiologists, Daniel Kempler and William Calvin,
and two anthropologists, Dean Falk and Kathleen Gibson, address neural
function in relationship to language, gesture, tool-use and social behavior.
Kempler focuses on the neural control of gesture and vocalizations in
Alzheimer's patients. The others frame human brain evolution within the
contexts of behavioral evolution.

Both aphasia (language deficit) and apraxia (difficulty in executing purposeful movements of the arm and hand, such as gesture and object manipulation)
may reflect lesions of the left frontal and parietal association areas. In addition,
aphasic patients commonly suffer from apraxia. These findings may imply a
common neural control of language, gesture and object manipulation. Alternatively, aphasic–apraxic patients may possess unusually large lesions extending
over several neural "modules", or an underlying paralysis and motor weakness
which afflicts both oral and manual movements.

Alzheimer's patients are ideal subjects for analyzing relationships between
linguistic and praxis disturbances, because they exhibit both during the early
stages of their disease. In contrast, they develop paralyses only during the final
stages. Kempler finds strong statistical correlations between the loss of verbal
symbolism and of pantomime (gestural symbolism) in these patients, but little
correlation between the loss of either of these skills and syntactical abilities.
This indicates that neurological ties are closer between verbal and gestural
symbolism than they are between semantic and syntactic abilities within either
communicative domain.

Dissociations between disturbances of the gestural and verbal lexicons do
occur, however, and the loss of specific symbols in one domain does not
guarantee their loss in the other. These findings imply that, in adults, manual
and oral symbolism are similarly organized. They may also share some
overlapping neural control. They are not, however, controlled by identical
neuronal circuitry. In Kempler's view, these findings are compatible with a
modular view of language in which some vocal and gestural functions are
independent and others interdependent.

It is important to note, however, that Alzheimer's disease afflicts adults in the
later stages of the life cycle. Consequently, the damaged neural circuits once
possessed full adult functioning. Moreover, prior to the onset of their disease,
most Alzheimer's patients possessed verbal, but not gestural, languages.
Hence, the findings of partial, but not complete, overlap in the circuitry
governing oral and gestural communication in modern human adults may not
tell the entire developmental or evolutionary story. Greenfield (in press) finds

little difference in the neural circuitry of gesture and vocal language in infancy. On this basis, she suggests that modularity *develops* in response to experience. Other findings confirm her view. Neural areas which subserve auditory functions in hearing individuals may exhibit visual functions in congenitally deaf persons who are proficient in sign language (Neville, 1991). On the basis of these data, Greenfield (1991) posits a potentially closer relationship between the neural control of vocal and gestural capacities in early hominid evolution than in modern human adults.

The majority of humans exhibit right-handed dominance for the manipulation of tools and other objects. In most right-handed individuals, both the dominant hand and language are controlled by the left hemisphere. In the left-handed population, language is often, but not always, localized to the right hemisphere. The similar hemispheric control of language and of the primary manipulative hand strongly suggests a commonality of neural control of tool-use and speech. The nature of the relationship appears to reside in the left hemisphere's control of sequential movements (Kimura, 1979). By contrast, the spatial capacities needed for tool-making and other constructions such as drawings lie primarily in the right hemisphere.

The evolution of handedness and lateralization thus represents a special problem in the evolution of tool-use and speech. Over the years, many have speculated that lateralization of function is a unique property of the human brain. Falk's review, however, indicates that neural lateralization is a common mammalian phenomenon which humans have expanded and capitalized upon. In the process, they may also have extended a common mammalian tendency for males to be more highly lateralized than females. Falk suggests that the spatial capacities necessary for human tool-making reflect an evolutionary expansion of earlier male capacities for territorial exploration and its associated cognitive mapping.

Falk's endocast studies indicate that human-like neural lateralization was in place by two million years ago, and that Broca's area had expanded by that time. These findings complement those of Toth and Schick (this volume), that Plio-Pleistocene stone tool-makers were right-handed. They suggest the possibility that rudimentary speech and stone tool-making capacities emerged quite early in the evolution of the genus *Homo*.

Although Plio-Pleistocene representatives of our genus were right-handed and, hence, in all probability exhibited a neural organization similar to our own, their brains were not yet of modern human size. In the ensuing two-million-year evolutionary history of the hominid lineage, the brain more than doubled in size, eventually reaching a gross size approximately four times that of the great apes. Although it is fashionable to denigrate brain size as of little significance, Falk follows other recent scholars (Calvin, 1990; Gibson, 1986, 1988, 1990, Jerison, 1973, 1982; Parker, 1990) in suggesting that major evolutionary increases in the size of the human brain permitted increased intellectual functioning.

The papers by Calvin and Gibson pursue this theme. Calvin notes that language, tool-use, sports, and dance all exhibit complex motor sequencing. Since nerve transmission is too slow for adequate feedback, many motor sequences, such as aimed throwing, speech and hammering must be planned entirely in advance. Calvin argues cogently that advanced planning of motor sequences places a premium on enlargement in the quantity of neural tissue. In particular, accuracy requires the ability simultaneously to keep in mind and evaluate several potential motor sequences. Such planning is so neuron-intensive that a four-fold increase in neural tissue permits only a 26 per cent increase in throwing accuracy.

Calvin further discusses potential neural overlaps between tool-using and linguistic sequencers. He notes that the premotor cortex serves as the primary sequencer for both manual and oral skills (see also Kimura, 1979; Lieberman, 1991). Furthermore, although neurons controlling manual and oral movements are considered to lie in separate portions of the motor cortex, individual neurons may receive information from both the hand and mouth and can send axons to lower motor neurons regulating actions of both anatomical areas. On the basis of these data, Calvin suggests that similar neurons and neuronal circuits can be recruited for quite diverse tasks such as tool-use and speech. In consequence, speech could have emerged as a by-product of selection for tool-using skills, such as aimed throwing.

Gibson presents an information-processing model of brain evolution which focuses on the potential cognitive effects of increased size of the neocortical association areas. According to Gibson, expansion of the association areas yielded increased capacities for hierarchical construction in linguistic, manipulative and social domains. As a result, diverse behavioral domains reflect similar neural processing mechanisms and interact with each other in a generative fashion to yield an integrated human social, linguistic, and technical complex. She also suggests that evolutionary changes have occurred in the neural control of infantile behavior, possibly at subcortical levels. The result of these changes is that human infants possess a repertoire of motor behaviors which channels the development of the neocortex and cognition in linguistic, technical and social directions.

Although Calvin, Falk, and Gibson all concur that the human neural expansion had profound behavioral effects, they differ in their interpretations of the causes of this expansion. Calvin, perhaps because of his heavy focus on aimed throwing, suggests that natural selection could not have acted to produce brain expansion. Rather increased brain size may simply be a by-product of prolonged maturation.

Gibson and Falk, on the other hand, consider that major increases in brain size do reflect the specific operation of selection on the brain *per se*. Increased brain size is potentially subject to intense negative selection because it places enormous metabolic demands on the human organism, can lead to death during childbirth, and can result in permanent retardation of nutritionally deprived fetuses (Armstrong, 1983; Gibson, 1986, 1991; Falk, 1990; Lancaster & Lancaster, 1983; Morgan & Gibson, 1991; Parker, 1990). Consequently, in the views of these authors, the expansion of brain size would have been highly unlikely in the absence of specific selection for increased quantities of neural tissue.

In sum, the chapters in this section suggest substantial overlapping neural control of manual and oral behaviors, but leave open to further research the degree of overlap and its developmental basis. Perhaps, more importantly, they herald a change in our views on the meaning of brain size, and thus open pathways for interrelating cognitive evolution and neural expansion.

References

Armstrong, E. (1983). Relative brain size in mammals. *Science*, 220, 1302–4.

Calvin, W. (1990). *The Ascent of Mind: Ice Age Climates and the Evolution of Intelligence*. New York: Bantam Books.

Falk, D. (1990). Brain evolution in *Homo*: The radiator theory. *Behavioral and Brain Sciences*, 13, 368–81.

Gibson, K. R. (1986). Cognition, brain size, and the extraction of embedded food resources. In *Primate Ontogeny, Cognition and Social Behavior*, ed. J. Else & P. C. Lee, pp. 93–105. Cambridge: Cambridge University Press.

Gibson, K. R. (1988). Brain size and the evolution of language. In *The Genesis of Language: A Different Judgement of Evidence*, ed. M. Landsberg, pp. 149–72. Berlin: Mouton de Gruyter.

Gibson, K. R. (1990). New perspectives on instincts and intelligence: Brain size and the emergence of hierarchical mental constructional skills. In *"Language" and Intelligence in Monkeys and Apes: Comparative Developmental Perspectives*, ed. S. T. Parker & K. R. Gibson, pp. 97–128. Cambridge: Cambridge University Press.

Gibson, K. R. (1991). Myelination and behavioral development: A comparative perspective on questions of neoteny, altriciality and intelligence. In *Brain Maturation and Cognitive Development: Comparative and Cross-Cultural Perspectives*, ed. K. R. Gibson & A. C. Petersen, pp. 29–63. Hawthorne, N. Y.: Aldine de Gruyter.

Greenfield, P. M. (1991). Language, tools and the brain: the development and evolution of hierarchically organized sequential behavior. *Behavioral and Brain Sciences*, 14, 531–95.

192   *Kathleen R. Gibson*

Jerison, H. J. (1973). *Evolution of the Brain and Intelligence.* New York: Academic Press.

Jerison, H. J. (1982). Allometry, brain size, cortical surface, and convolutedness. In *Primate Brain Evolution,* ed. E. Armstrong & D. Falk, pp. 77–84. New York: Plenum.

Kimura, D. (1979). Neuromotor mechanisms in the evolution of human communication. In *Neurobiology of Social Communication in Primates: An Evolutionary Perspective,* ed. H. D. Steklis & M. J. Raleigh, pp. 197–219. New York: Academic Press.

Lancaster, J. & Lancaster, C. (1983). Parental investment and the hominid adaptation. In *How Humans Adapt: A Biocultural Odyssey,* ed. D. Ortner, pp. 33–56. Washington, DC: Smithsonian Institution Press.

Lieberman, P. (1991). *Uniquely Human: The Evolution of Speech, Thought, and Selfless Behavior.* Cambridge, MA: Harvard University Press.

Morgan, B. & Gibson, K. R. (1991). Nutritional and environmental interactions in brain development. In *Brain Maturation and Cognitive Development: Comparative and Cross-Cultural Perspectives,* ed. K. R. Gibson & A. C. Petersen, pp. 91–106. Hawthorne, NY: Aldine de Gruyter.

Neville, H. J. (1991). Neurobiology of cognitive and language processing: effects of early experience. In *Brain Maturation and Cognitive Development: Comparative and Cross-Cultural Perspectives,* ed. K. R. Gibson & A. C. Petersen, pp. 355–80. Hawthorne, NY: Aldine de Gruyter.

Parker, S. T. (1990). Why big brains are so rare: Energy costs of intelligence and brain size in anthropoid primates. In *"Language" and Intelligence in Monkeys and Apes: Comparative Developmental Perspectives,* ed. S. T. Parker & K. R. Gibson, pp. 129–54. Cambridge: Cambridge University Press.

8

Disorders of language and tool use: Neurological and cognitive links

DANIEL KEMPLER

Sometimes when things fall apart, we can see more clearly how they were put together. Therefore, as we consider the relationship between language and tool use, it may be helpful to discuss aphasia and apraxia – disorders of language and purposeful movement. This chapter will briefly describe aphasia and apraxia, review the long-standing debate regarding the relationship between them, and present relevant data from patients with Alzheimer's disease.

Aphasia is language loss typically associated with focal lesions of the left cerebral hemisphere. There have been many different definitions, descriptions and classifications of aphasic symptoms (Darley, 1982; Goodglass & Kaplan, 1983; Jackson, 1878; Lichtheim, 1885; Weisenburg & McBride, 1935), but several points of agreement can be stated:

> Aphasia disrupts the formulation and comprehension of linguistic symbols; it is an impairment of the symbolic system and is not fundamentally a motor disturbance of speech production.
>
> The most common signs of aphasia are anomia (word-finding difficulty) paraphasias (substitutions of one sound or word for another), and comprehension deficits.
>
> Aphasia often affects language functions in all modalities, so aphasic patients frequently suffer deficits in reading (alexia) and writing (agraphia) in addition to problems in formulating and understanding spoken language.
>
> Aphasia is generally associated with lesions to the perisylvian region (distribution of the middle cerebral artery) of the left hemisphere in right-handed individuals.
>
> Aphasic patients differ greatly in the overall severity and the pattern of dysfunction; some patients have very mild word-finding deficits, similar to normal tip-of-the-tongue phenomena, while others are not able to generate or recognize even a single word; some patients have predominant comprehension problems while others have only obvious difficulty with language formulation and production.

Aphasia evaluations are often informal, but several standardized aphasia batteries are available (Goodglass & Kaplan, 1983; Kertesz, 1980; Porch, 1981; Schuell, 1965). A complete aphasia exam will test word-finding, sentence formulation, word and sentence comprehension, repetition, reading and writing. Most aphasia exams also include assessment of gesture, calculation, and occasionally drawing and memory.

Apraxia is a disorder of purposeful movement which cannot be explained by weakness, loss of muscle tone, intellectual deterioration, inability to under-stand instructions or uncooperativeness. Although there are many kinds of apraxias (e.g., constructional and dressing apraxia), this discussion will be limited to limb apraxia – the inability to carry out purposeful movements with the hand and arm[1]. This disorder is of particular interest because it often involves the ability to manipulate objects (or tools), and therefore may reveal something about the neuropsychological foundations of tool use. The most common apraxic symptom is the inability to perform a gesture on command or in imitation. Errors often include hand or arm position misplacement or misorientation, responses which are clumsy, partially correct or substitute a body part for the object (e.g., hammering with a fist rather than pretending to hold a hammer). Although performance generally improves when patients are able to hold an actual object or execute the gesture in context, severely apraxic patients have difficulty executing gestures even in natural, ecologically valid circumstances.

Unlike aphasia, there are no well-standardized apraxia tests. Typical apraxia evaluations require the patient to perform a variety of gesture types, including *meaningless* gestures (e.g., put your thumb on your forehead), *intransitive* gestures which are typically symbolic but do not involve objects (e.g., wave goodbye), and *transitive* gestures in which the patient is asked to pantomime the use of an object (e.g., brush your teeth).

The relationship between aphasia and apraxia

For over a hundred years clinical observation has suggested that aphasia and ideomotor apraxia, more often than not, co-occur (e.g., Duffy & Liles, 1979; Jackson, 1932; Liepmann, 1908). Over the past 25 years many experimental reports investigating gesture in aphasia have appeared in the literature (Cicone, Wapner, Foldi, Zurif & Gardner, 1979; Duffy, Duffy & Pearson, 1975; Duffy, Watt & Duffy, 1981; Gainotti & Lemmo, 1976; Goodglass & Kaplan, 1963; Kertesz & Hooper, 1982; Varney, 1982). Despite different methods of assess-ment and different subject selection criteria, researchers have consistently found significant correlations between gesture and language disturbance. This is particularly true for referential gestures, transitive actions used in the recognition or labelling of common objects (e.g., pantomime of the action of drinking used to indicate a cup). Although we do not yet have a satisfactory explanation for the observed correlations between aphasia and apraxia, many theories have been proposed (e.g., Duffy & Duffy, 1981). These theories can be divided into two broad categories: anatomical and cognitive.

Anatomical explanations

Liepmann, perhaps the first serious researcher in the field of apraxia, proposed that aphasia and ideomotor apraxia are concurrent but independent disorders. He based his claim on the observations of a series of left and right brain-damaged patients. Although he found that aphasia and apraxia often co-occurred, the substantial number of patients who exhibited either aphasia or apraxia without the other convinced him that the two disorders were essentially independent (14 of 20 apraxics were also aphasic; four of 21 nonapraxic patients had aphasia) (Liepmann, 1908). He attributed the correlation of the two disorders to the fact that both language and complex movements are represented in the left cerebral hemisphere of man. This has been upheld by numerous modern studies of both language (e.g., Kertesz, 1979) and movement (Haaland and Yeo, 1989; Kimura and Archibald, 1974). The loci within the left hemisphere which create disturbances of praxis and language are now known to overlap considerably: frontal and parietal lobes are implicated in most cases of apraxia, while frontal, temporal and parietal lobes are most often implicated in aphasia (e.g., Faglioni & Basso, 1985) (See Figure 8.1). In this theory, the correlation between the two disorders is an accident of anatomy: vascular accidents which occur in the distribution of the middle cerebral artery affect areas which are important for two separate functions.

This explanation leaves unaddressed the possible evolutionary reasons for both functions being subserved by this region of the left hemisphere. For instance, it is possible that the left hemisphere may be coincidentally dominant for linguistic and complex motor functions. On the other hand, it has been suggested that the proximity of movement and language cortex is not coincidental at all: both functions involve sequential movement and it is this basic function which is the specialty of the left hemisphere (e.g., Kimura, 1977; Kimura & Archibald, 1974). In either case, language and gesture are viewed as two separate psychological functions subserved by closely related neural tissue. Damage to this area will likely, but not necessarily, disrupt both functions.

Support for this anatomical view has taken several forms. First, Liepmann's findings of double dissociations between aphasia and apraxia have been

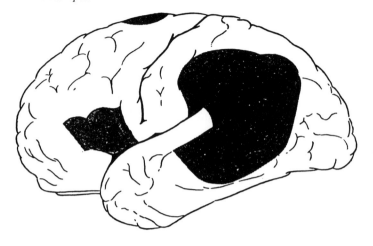

(a)

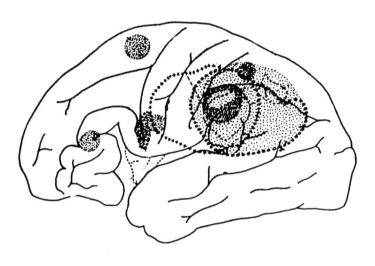

(b)

Fig. 8.1. (a) Typical distribution of lesions which produce aphasia (reprinted from Penfield & Roberts, 1959, "Speech Areas: Evidence from Excision", page 189). Reprinted with permission of the literary executors of the Penfield Papers and Princeton University Press. (b) Typical distribution of lesions which produce apraxia (reprinted from Faglioni & Basso, 1985, p. 22.)

replicated (e.g., Kertesz, 1985). Such double dissociations between cognitive functions are generally taken to indicate that the two functions are neuropsy-chologically independent (e.g., Shallice, 1987). Second, although the correla-tion of apraxia and aphasia is generally high within the aphasic population (range = 0.358–0.76), these figures may overestimate the degree to which aphasia and apraxia co-occur. Generally, subjects are selected on the basis of aphasia, and then evaluated for apraxia. However, there may be many cases of isolated apraxia which would not present to the clinic for evaluation since apraxic disturbances are usually only apparent when subjects are asked to pretend to perform an action, not something people do much in their daily life. Third, there appears to be little correspondence between severity of aphasia and severity of apraxia, again indicating that the two disorders co-vary independently (Goodglass & Kaplan, 1963).

Cognitive explanations

The correlations can also be attributed to a common cognitive basis for language and gesture (e.g., Bay, 1962; Duffy, Duffy & Pearson, 1975; Gainotti & Lemmo, 1976; Goldstein, 1948; Jackson, 1932; Varney, 1982). Several versions of this hypothesis have been proposed. For instance, it is possible that aphasia and apraxia are both secondary to general intellectual impairment associated with brain damage. If brain-injury affects all cognitive operations, it is possible that complex gestural and language systems would both suffer impairment. This theory has some credibility because of the wide range of intellectual deficits which have been found to accompany brain damage in general and aphasia in particular (Hamsher, 1981), but has found little empirical support. For instance, Duffy & Duffy (1981) compared results from several pantomime tests with performance on tests of language and nonverbal problem solving. The results indicated that general intellectual functions contributed relatively little to the performance on the pantomime measures.

A more realistic and popular version of this theory proposes that both aphasia and apraxia are due to a deficit in a central symbolic capacity which is independent of any specific modality. Impairment of this central symbolic process is reflected in simultaneous disruption of verbal and nonverbal behavior. This theory is often attributed to Finkelnburg (Duffy & Liles, 1979), who noted simultaneous deficits in numerous symbolic domains, including language, symbolic gesture, musical notation, and apprehension of money value. Although Finkelnburg championed the notion of a modality independ-ent "asymbolia", it has become apparent that this general position cannot be strongly supported: deficits in one symbolic domain are not always paralleled by impairment in all other symbolic domains (e.g., Glass, Gazzaniga & Premack, 1973; Helm-Estabrooks, Fitzpatrick & Barresi, 1981). This hypothesis has been modified to stress possible local connections between specific symbolic domains. For instance, it appears that the most robust correlations obtain between specific measures of language (e.g., comprehension) and symbolic referential pantomime (Duffy *et al.*, 1975; Duffy & Duffy, 1981; Duffy *et al.*, 1981; Kertesz & Hooper, 1982). Importantly, other aspects of gesture (e.g., intransitive gesture) do not appear to correlate with language impairments (Haaland & Flaherty, 1984).

Unanswered Questions

Although the earliest and most influential theory of apraxia has been Liepmann's anatomical hypothesis (e.g., Goodglass & Kaplan, 1963), the bulk of modern studies have supported the cognitive–symbolic hypothesis (e.g., Duffy & Duffy, 1981). However, two problems in the aphasia literature make it impossible to test several corollaries of the cognitive hypothesis. First, the cognitive hypothesis assumes that ideomotor apraxia is tied to symbolic disturbance of aphasia and is independent from general motor disturbance (e.g., hemiparesis). However, many aphasic patients suffer from both motoric and symbolic deficits making it difficult to determine the contribution of each to the gestural disturbance. In order to establish that the gestural disturbance is independent from the motor deficit, it would be helpful to investigate a population which consistently demonstrates language loss without a predominant movement disorder or hemiparesis.

Second, in the aphasia literature, it has been difficult to associate apraxia with specific language impairments (DeRenzi, Motti & Nichelli, 1980; Duffy & Duffy, 1981; Goodglass & Kaplan, 1963; Seron, van der Kaa, Remitz & van der Linden, 1979). Language is typically described as a componential system, comprised of at least two relatively independent subsystems: the lexicon which carries the burden of symbolic reference; and the computational systems which govern the combinatorial aspects of language (phonology, morphology and syntax). If apraxia and aphasia co-occur because of a common symbolic deficit, we would expect the pantomime deficit to be associated with lexical impairment. On the other hand, if the disorders are secondary to, for instance, a sequencing deficit, we might expect errors to show up as sequential errors in gesture and syntactic errors in language.

In general, there has been a lack of consensus regarding the sort of language impairment which is most associated with apraxia (Duffy & Duffy, 1981; Goodglass & Kaplan, 1963). This is undoubtedly due in part to the lack of agreement on classification and assessment measures for aphasia and the paucity of patients who present with isolated deficits in one aspect of language. Comparison of language and apraxia in a population who present relatively isolated language deficits would therefore be revealing.

Alzheimer's disease

Alzheimer's disease (AD) is a slowly progressive neurological disorder which is accompanied by a wide range of cognitive deficits. However, the degeneration is not global, and these patients present very specific patterns of impairment. Two aspects of the degenerative pattern make investigation of AD ideal to help to resolve some of the unanswered questions of the existing literature on aphasia and apraxia. First, motor and sensory control remain intact until the final stages of the disease (McKhann *et al.*, 1984; Cummings & Benson, 1983), allowing us to assess symbolic gestural performance without contamination from movement disorders or hemiparesis. Second, AD patients, at least through the moderate stages, present with a relatively specific lexical impairment while the disease spares grammatical processes until the late stages (Kempler, 1991; Kempler, Curtiss & Jackson, 1987). It has been shown, for example, that despite Alzheimer patients' extensive naming and word-comprehension impairments (e.g., Bayles & Tomoeda, 1983; Kempler *et al.*, 1987; Martin & Fedio, 1983; Nebes, 1989), they can understand and produce a wide range of diverse and syntactically complex constructions (Kempler *et al.*, 1987; Schwartz, Marin and Saffran, 1979; Smith, 1989). These two aspects of AD will allow us to test two corollaries of the cognitive hypothesis: 1. limb apraxia is independent from general movement disorders , and 2. limb apraxia is associated with specifically referential/lexical impairments in language.

Apraxia in AD

Although apraxia has been known for some time to occur as one of the symptoms of Alzheimer's disease (Denny-Brown, 1958), it has not been one of the hallmark characteristics. Clinical reports suggest that apraxias do not generally occur until the moderate stages of the disease, after language and memory disturbances are firmly established, but before motor disabilities are evident (e.g., Della Sala, Lucchelli & Spinnler, 1987), although it has also been noted to antedate other features of the illness (Cummings & Benson, 1983). Despite a recent upsurge of interest in apraxia in AD (e.g., Foster, Chase, Patronas, Gillespie & Fedio, 1986; Rapcsak, Croswell, & Rubens, 1989; Ska &

Joanette, 1990), no studies have systematically compared the nature and extent of apraxia in AD with language disturbances.

One case study of an Alzheimer patient (Schwartz *et al.*, 1979) did note that although the subject misnamed objects, she was able to "mime the use of depicted objects" (p. 291). This observation of gestural compensation for word-finding difficulty has since been cited as evidence that language deficits in Alzheimer's Disease are specifically semantic, and therefore not symptomatic of general symbolic deficits (Appel, Kertesz & Fisman, 1982; Martin & Fedio, 1983). The inference from this single case study is that Alzheimer's Disease might furnish an example of lexical semantic impairment **without** concomitant impairment in another symbolic domain – pantomime. If true, these results will force us to reject theories which link lexical knowledge to pantomime ability through a common symbolic process.

However, the conclusions based on this study are tenuous for several reasons. First, they are based on observation of a single patient. Second, the observation was anecdotal and was not supported by experimental evidence. Third, there is clinical evidence that gestural disturbances do occur in Alzheimer's Disease (e.g., Della Sala *et al.*, 1987; Rapcsak *et al.*, 1989). The only way to determine whether the gestural difficulties of Alzheimer patients are truly independent of lexical difficulties is to test both comprehension and production of words and gestures in a single group of AD patients. If pantomime and lexical abilities are both dependent on a common symbolic capacity, impairments in the two domains should parallel one another quantitatively (time of onset and severity) and qualitatively (type of errors which occur). These predictions were tested in a previously published study (Kempler, 1988), which will be reviewed here.

Subjects

Eight individuals (five males, three females) meeting the NINCDS–ADRDA criteria for probable Alzheimer's disease and eight age- and gender-matched normal controls were administered a battery of lexical and pantomime tests. All subjects were right-handed Caucasians, schooled in Standard American English. Level of education ranged from eighth grade to college. None of the subjects had any known speech or language pathology prior to the diagnosis of AD. Severity of dementia was moderate to severe as gauged by the Mini Mental State (MMS) Examination (Folstein, Folstein & McHugh, 1975). The mean age of both the patient and control groups was 76.6 years (range: 65–84 years). Subject information is given in Table 8.1.

Table 8.1. *Subject information*

Subject	Gender	Age	MMS*
1	M	72	19
2	M	82	17
3	F	87	15
4	F	82	15
5	F	75	14
6	M	65	14
7	M	76	6
8	M	74	2

Note:
* Mini Mental State score of a possible 30.

Methods

A set of 40 black and white line drawings of common objects was selected from previously published pantomime experiments (Duffy & Duffy, 1984; Duffy & Watkins, 1984). To avoid a familiarity effect two forms of the test were constructed. In Form A, 20 items are presented to be named or pantomimed and a different 20 items are used to test recognition of names and pantomimes. In Form B, the items are reversed. The two forms were alternated with consecutive subjects. The word-frequency of the items in the two forms was equivalent. No norms are available for gestural frequency, but all items were common (e.g., toothbrush, hammer). Lexical production and comprehension were assessed in one session; pantomime comprehension and production were assessed in a separate session.

Lexical procedures: In the naming test, each subject was asked to name 20 line drawings. In the lexical comprehension test, each subject was shown four line drawings and was instructed to point to the picture of the object named by the examiner. The distractor items in the comprehension tests were all semantically related to the target. For example, for the target item "pencil", the three distractors were pictures of a desk, a typewriter and a stack of paper. The subjects were trained on a maximum of five pretest items. The stimuli were repeated as necessary to elicit a meaningful response. Word recognition was assessed prior to naming because it was easier for the subjects and served to orient them to the task.

All comprehension items were scored during testing by the author. In case the subject chose more than one response, or changed his/her mind, the last response was scored. Naming responses were all tape recorded, and transcribed for analysis. For naming, the subjects' *best* (not first or last) response was transcribed and placed into one of eight categories drawn from the literature on naming errors (e.g., visual perception errors, substitution of contrast coordinates, circumlocutions). The responses were categorized by two linguists independently. Overall agreement was 97%.

Pantomime procedures: Pantomime recognition was given before pantomime production so that it could be used to illustrate the task. In the recognition task, the examiner pantomimed the use of an object and the subject selected from four line drawings the appropriate object. In the production task the subject was told to show "how to use" or what they "do with" each of 20 objects. If the subject appeared not to understand, s/he was shown the same line drawing used to elicit the object name and encouraged to pretend to use it. Minimal use was made of the pictures since pilot testing indicated that AD patients tend to become fixated with the picture; when asked to "do" something, they often traced the outline or attempted to lift the object off the page. All pantomime production was video recorded for analysis.

Pantomime scoring was done in much the same way as the lexical scoring. However, pantomime scoring is slightly more complex since no pre-determined target is available: while there is generally only one name for each object, there may be more than one way of depicting how to use it. Therefore, the author determined a prototype gesture for each item by viewing a video tape of the gestures presented by Duffy *et al.* (1975) and the responses given by the normal controls in the present study. Pantomime errors were categorized into one of eight categories drawn from the relevant literature (e.g., body part as object, incomplete). Agreement for two independent raters ranged from 85% to 100% for each subject; mean agreement was 92%. Disagreements occurred only with regard to gestures which were coded as "correct" by one rater and "partially correct" by the other. For instance, on several occasions a gesture was categorized as "excessively vague" by one, and "correct" by the other. There was 100% agreement on incorrect gestures. All disagreements were resolved by viewing and discussing the video tapes.

Results

Lexical comprehension and production

Although both comprehension and production of single words were impaired in the AD population, production was considerably more impaired than comprehension. Twenty one comprehension and 45 naming errors (on 160 trials) were made by the AD group. One naming error was made by the control group.

The naming errors can be characterized several ways. First, 30 (67%) of these errors were semantically related to the target. These 30 are broken down into the five "related" categories below.

1. functional descriptions/circumlocutions (14): e.g., **rowboat:** "for fishing"; **telephone:** "hear things"; **glasses:** "this is for your eyes"
2. part/whole errors (6): e.g., **banana:** "banana peel"; **window:** "house"
3. novel forms (4): e.g., **salt shaker:** "salt holder"; **saw:** "cutter"
4. contrast-coordinates (4): e.g., **car:** "truck"; **sun:** "moon"; **cup:** "pitcher"
5. physical descriptions (2): e.g., **sun:** "it has a glitter to it"

A total of 15 errors (33%) were not semantically related. These errors were of 3 types:

1. visual confusion (8): e.g., **balloon:** "egg"; **cigarette:** "funny stick"
2. Do not know/perseverations (4)
3. uninterpretable (3): e.g., **glass:** "the thing you put in the reflex"; **sink:** "you slide it"

These data indicate that although the patients have difficulty retrieving accurate names, they can often retrieve related semantic information.

Pantomime production and comprehension

The AD patients made 54 pantomime comprehension errors and 72 pantomime production errors. As with the lexical tests, more errors were made in production than in comprehension. Errors in pantomime production were divided into eight error categories. Fifty two of the errors (72%) were related to the target in some identifiable way. These included:

1. Complete pantomimes which contained an error in one component (e.g., handshape or facial expression) ($n = 27$): e.g., **drum:** holding drumsticks, simultaneously humming a tune and swaying back and forth rather than "beating" the drum; **cigarette:** correct hand shape and location, but chewing and/or kissing was substituted for inhaling/exhaling; **salt shaker:** correct picking up motion and shaking motion, but orientation was in error – the patient did not turn the salt shaker over.
2. Incomplete pantomimes (10): e.g., **apple:** chewing motion only, without the appropriate hand holding apple gesture; **door:** knob turning only without subsequent door opening motion.
3. Correct but non-prototypical gestures (4): e.g., **gun:** loading a gun rather than

the typical response of aiming and shooting; **sink:** scrubbing a sink rather than the typical response of turning on and off faucets and washing hands.

4.   Correct but vague (7): e.g., **umbrella:** vague upward motion with the hands without specifically holding the umbrella with either hand.

5.   Body part as object (3): e.g., **apple:** biting into fist rather than pretending to hold apple and bite into imaginary apple; **pen:** using index finger as a pen rather than pretending to hold a pen.

The remainder of the errors (20, or 28% of the errors) were not identifiably related to the target. Twelve responses were "do not know" or refusals. Seven responses were either wrong or uninterpretable (e.g., typing in response to **telephone**, or putting the little finger in the ear in response to **book**). Two responses were perseverations of a previous response.

Comparison of lexical and pantomime abilities: Comprehension across modalities

The most straightforward comparison between gesture and language is in comprehension. The results of lexical and pantomime comprehension tests are presented in Figure 8.2. First, the lower mean number correct for pantomime (13.25 versus 17.5) suggests that this task was harder or less familiar for the subjects than the lexical test.

Second, there appears to be no item-by-item correspondence between pantomime and lexical errors. More than twice as many pantomime items than lexical items were missed. A correlation between test items (e.g., comprehension of the pantomime for **book** and the word "book") yields a phi coefficient of 0.124 (equivalent to a chi square of 2.472, with 1 degree of freedom) which is not significant at the 0.05 level of confidence (Ferguson, 1971). Thus, it appears that the difficulty in the two tasks cannot be directly attributed to difficulty with individual test items.

Third, if performance in the two domains is due to a common symbolic deficit, there should be quantitative parallels between the two. Significant positive Pearson product–moment correlations ($r = 0.78$, $p < 0.01$) support this claim.

Comparison of lexical and pantomime abilities: Production across modalities

While comprehension scores allow a quantitative comparison, an analysis of pantomime and lexical production will allow us to compare both number and type of errors. Quantitative comparison of performance on naming and

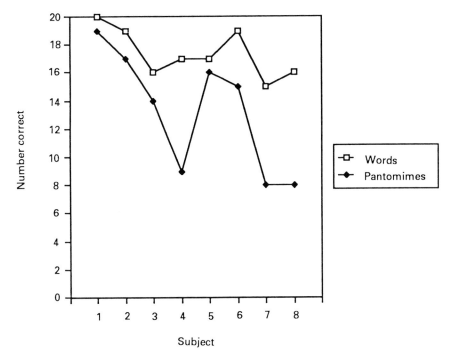

Fig. 8.2. Number of words and pantomimes correctly understood by eight Alzheimer patients, listed in order from mildest (No. 1, mental status = 19 out of 30) to most severe (No. 8, mental status = 2 out of 30). Reprinted with permission from Kempler, 1988.

pantomime production is summarized in Figure 8.3. As with comprehension, a lower mean number correct in the pantomime task than in the naming task (11 versus 14.4) indicates that the pantomime task was more difficult.

Again, we are also interested in whether this correlation is due to difficulty with individual items which cuts across the two modalities, or whether the difficulty is a general one, which is manifested on various (possibly random) items in the two domains. A correlation between the test items revealed a nonsignificant tendency for patients to have difficulty on the same items in both modalities (phi = 0.14, chi square = 3.04, $p < 0.10$).

Quantitative parallels ($r = 0.75$, $p < 0.01$) between the two tasks also support a theory of common origin for the apraxic and aphasic deficits. This calculation, however, does not give credit to partially correct answers which were common in both modalities. In order to compare abilities across tasks more accurately, lexical and pantomime scores were re-calculated for each subject which credit two points for a fully accurate response, one point for a partially accurate response, and no points for no response or an incorrect response. This

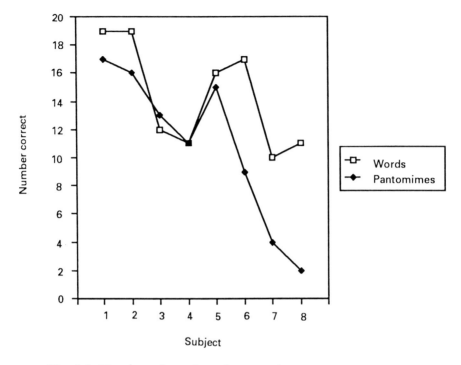

Fig. 8.3. Number of words and pantomimes correctly produced by eight Alzheimer patients. Reprinted with permission from Kempler, 1988.

correlation is slightly higher than the previous one ($r = 0.81$, $p < 0.01$), and suggests that a somewhat finer grained analysis upholds the quantitative parallel between word and pantomime production.

In addition to numerical comparisons, we can make qualitative comparisons between lexical and pantomime errors. Few of the previous investigations of gesture in aphasia discuss the types or distribution of distinct error types, other than to note that apraxic pantomimes include incomplete, awkward, poorly sequenced and wrong movements (Heilman, 1979), and that the most common errors are using a body part as an object (Goodglass and Kaplan, 1963; Rapcsak *et al.*, 1989) and perseverations (Lehmkuhl, Poeck and Willmes, 1983). A more specific analysis is necessary in order to make a meaningful comparison with naming errors.

The comparison between naming and pantomime errors reveals parallels between the two domains. In both modalities, a majority (67% of naming, 72% of pantomime) of errors are related to the target in some identifiable way. There is a significant positive correlation between number of related (or "partially correct") errors in naming and pantomime, of 0.60 ($p < 0.05$) (Figure 8.4).

Since the error categories are not the same for both domains, it is difficult to

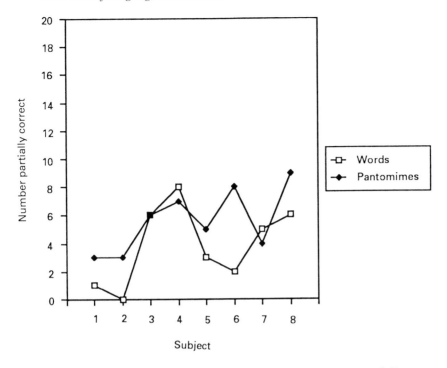

Fig. 8.4. Number of words and pantomimes which were partially correct in production by eight Alzheimer patients.

make detailed comparisons. However, parallels are apparent. For instance, some pantomimes were correct but incomplete (e.g., chewing for **apple**), as were many naming errors (e.g., "hear things" for **telephone**). The non-prototypical pantomimes (e.g., scrubbing a sink rather than turning on faucets) are similar to the novel forms created in naming ("cutter" for **saw**) – atypical responses which suggest that the patients have difficulty accessing prototypical information in both modalities.

Discussion

These data can be used to help evaluate the anatomical and cognitive theories of apraxia and aphasia. In particular, we can discuss corollaries of the cognitive hypothesis: the contribution of motor deficits to apraxia and the relationship between apraxia and lexical as against syntactic deficits.

Liepmann's original formulation proposed that pantomime impairment is due to a movement disorder which affects all purposive movements and is independent of symbolic deficits such as aphasia. One argument against this hypothesis is that the AD population is presumed to have little or no movement disorder at this stage. Specifically, the subjects studied here had no focal neurological/motor signs such as hemiparesis or gait disturbance. Therefore, the parallels between gesture and language cannot be easily attributed to movement disorder *per se*. The data presented here also demonstrate a significant correlation between production (which requires complex purposive movements) and comprehension (which requires only a simple pointing response) of pantomime ($r = 0.7801$; $p < 0.01$). A high correlation between tasks which differ greatly in the complexity of movement can be better explained by a deficit in symbolic function which underlies both simple and complex movements.

Additional results from this study support the notion of a symbolic deficit underlying the pantomime disturbance. Each subject was asked prior to the pantomime expression test to perform ten non-referential movements (e.g., clap, scratch your head). To the extent to which these non-referential movements were easier than pantomime, the pantomime deficit may be attributed to the symbolic disorder, and not to an impairment of all purposive movements. Considerably fewer errors were made on the non-pantomime assessment (10% versus 45%), suggesting that more symbolic actions are more difficult, and that much of the apraxia can be linked to symbolic function independent of movement.

The second question to be addressed by this research is the degree to which gestural disturbance can be related to specific language deficits. If both deficits are related to a common symbolic dysfunction, then we predict that pantomime impairment will parallel lexical but not syntactic abilities. Recall that the AD population has been shown to have relatively preserved phonology, morphology and syntax and significantly impaired lexical abilities (Kempler *et al.*, 1987; Kempler, 1991). The patients studied here have been shown to have intact spontaneous use of a wide range of grammatical constructions and the ability to use syntactic cues in more constrained tasks of writing (Kempler *et al.*, 1987) and these data have been used to demonstrate the dissociation between lexical and syntactic ability in AD. In this context, the parallel impairment in pantomime and lexical ability suggests that the pantomime ability may be more directly parallel to lexical than syntactic ability.

In conclusion, the addition of AD patients to the literature on apraxia and aphasia reveals that limb apraxia can be attributed to a symbolic deficit with little contribution (in this case) from basic motoric deficits. Further, the close quantitative and qualitative parallels between lexical and pantomime deficits support the position that gestural ability is tied to the referential aspects of language, and may be independent of the combinatorial or grammatical aspects. These data can then be used to argue for a common symbolic disturbance which is responsible for the deficits seen across the domains of gesture and language.

The cognitive–symbolic hypothesis can be presented in both strong and weak forms. A strong version might maintain that the mental representations for both names and gestures are the same; a weaker version might claim that the representations overlap, may have similar organizations, and may be accessed in a similar manner, but are essentially independent. In light of the data presented here and elsewhere, it seems unlikely that a strong version of this hypothesis can be supported. The lack of correlation in some individuals and the lack of an item-by-item correspondence between domains both argue against an isomorphism between lexical and gestural representations. A weaker version seems plausible. Specifically, it can be proposed that the two domains are independent, but they are similar in internal organization. The analogous error patterns in AD suggest that words and pantomimes are similar in underlying structure, and that this structure is affected in much the same way and to the same degree in Alzheimer's disease. For example, the structure of lexical and pantomime representations may both consist of complex networks. In each lexical entry there is semantic and phonological information, while each pantomime representation contains hand-shape, hand-movement, and meaning information. Lexical and pantomime entries are embedded in matrices of other similar representations, strongly connected to items close in form and meaning. In Alzheimer's disease, access to semantic targets in both domains is disrupted in a random manner, sometimes allowing access to relatively complete information, at other times allowing access only to partial or nonprototypical information. There does not appear to be a systematic loss of any one type of semantic information (e.g., central or peripheral aspects of semantic categories; name versus function). At the very least, then, we can suggest that the representations for words and pantomimes are simultaneously disrupted, and that the similar internal structures of these representations produce similar error patterns in cases of brain injury.

The present study has more general implications for psycholinguistic theory as well. Current psycholinguistic theories can be broadly divided into those which view language as a system which is intrinsically "of a piece" with nonlinguistic cognition and those which view language as an independent psychological system (see Fodor, 1983; Piatelli-Palmarini, 1980, for discussion). However, aphasia data has suggested that syntactic aspects of language may well be dissociable from both lexical semantics and aspects of general cognition (Schwartz *et al.*, 1979; Kempler, 1984; Kempler *et al.*, 1987; Caramazza & Zurif, 1976). The data presented here suggest, in addition, that while lexical semantics may be dissociable from the computational aspects of language function, it is not independent of nonlinguistic cognitive functions, particularly symbolic functions such as those reflected in pantomime. An independence of syntactic function alongside an observably close tie between semantics and nonlinguistic cognition has been proposed for the acquisition of language as well (e.g., Curtiss, Kempler & Yamada, 1981). The data presented here are compatible with a modular view of language in which some components may be functionally independent while others are relatively more interdependent with nonlinguistic aspects of cognition.

A point about further investigations can be made. If, as proposed, lexical and pantomime abilities are compromised due to impairment in underlying symbolic function, then reflections of this deficit might be seen in other symbolic activities as well. One such area to investigate is representational drawing. Since drawing differs from both language and pantomime it is difficult to apply the same standards of analysis. However, if symbolic functioning is generally disrupted, we would predict that as the pantomime and lexical deficits worsen, we will see an equivalent deterioration in drawing ability. Pilot data suggest that representational drawings deteriorate as well. Figure 8.5 shows examples of daisies drawn by normal and AD subjects. As the disease progresses, the figures become less complete, and less daisylike, although they generally retain enough information to be identified as a flower. Eventually, similar criteria may be applied to drawing as we have for pantomime and language. We can then compare deficits across domains in terms of severity. Additionally, we should be able to see if there is improvement in copying a drawing which may reduce the amount of symbolic capacity necessary. We can also systematically investigate the impairment in production and apprehension of various other symbols (e.g., logos, numbers) as has been attempted with aphasics and alexics (Bay, 1962; Gardner, 1974).

In closing this discussion, we must consider a very important caveat. It is possible that all symbolic activities (like all mental activities) may not be equally tied to one another in normal function and in cases of deterioration. Lexical semantic ability appears to be relatively more tied to symbolic gesture than either of them are to syntactic abilities. We have demonstrated a substantial link between pantomime and word knowledge but such a close tie 1. may not be absolute, and 2. may not exist between other symbolic domains. There is evidence that gestural and language impairments do occur independently of one another (DeRenzi *et al.*, 1980; Heilman, Coyle, Gonyea & Geschwind, 1973; Poeck & Kerschensteiner, 1971). This dissociation, if upheld with methods comparable to those used here, must be taken to indicate that the two domains, despite possible neurological and functional similarities, are not merely mapped one onto the other. With regard to the second point, deficits in

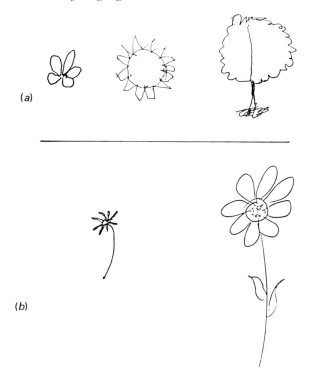

(a)

(b)

Fig. 8.5. (a) Drawings of a daisy by Alzheimer patients. (b) Drawings of a daisy by healthy elderly controls.

one symbolic domain may not be paralleled in another: it has been shown that aphasic patients with a deficit in symbolic use of language are capable of learning to use other symbol systems (Glass *et al.*, 1973; Helm-Estabrooks *et al.*, 1981). In addition, many category specific asymbolias (e.g., alexia, agraphia, acalculia) have been described (Hecaen and Albert, 1978), demonstrating that having symbolic content does not necessarily imply that two domains are closely tied functionally or anatomically. The present study has not shown that gesture and language are inseparable, but rather that in at least one clinical population, gesture and one aspect of language – the lexicon – are impaired in a parallel manner.

Summary and conclusions

The population of Alzheimer's patients presents a unique opportunity to study the relationship between language and manual gesture. Since in this population, the lexical aspects of language deteriorate disproportionately compared to other aspects of the linguistic system (phonology, morphology and syntax remain relatively intact to the final stages), we can explore the relationship between the specifically symbolic (as opposed to combinatorial) aspects of language and nonlinguistic abilities. In addition, since the motor system remains intact until the final stages, early and moderate AD patients provide an ideal opportunity to compare the dissolution of symbolic gesture and lexical knowledge. By contrast, typical aphasic patients who have been studied in relation to the issue of "asymbolia" are 1. often hemiparetic and 2. exhibit a range of speech and language problems including agrammatism, anomia, etc.

The present study found, in the absence of motor involvement and with morpho-syntactic ability intact, a significant correlation between comprehension and production of words and pantomimes. The parallel is both quantitative (as the number of errors in one domain increases, so do errors in the other) and qualitative (the errors made in the two domains appear similar in type). The types of errors seen indicate that the meaning networks surrounding the target items can be accessed, but that precise information within a network is often difficult to locate in both language and gesture. These findings are taken to support a theory which links language and gesture through shared symbolic capacity.

Acknowledgements

I am grateful to Catherine Jackson for help in coding and scoring the patient responses.

Note

1 Discussions of limb apraxia traditionally include two major subtypes: ideational apraxia and ideomotor. In ideational apraxia, there is thought to be inadequate formulation of the motor program, and particular difficulty in sequencing a series of movements such as folding a letter, placing it in an envelope, sealing it, and then putting a stamp on it. Ideomotor apraxia is thought to result when an intact motor program is disconnected from the kinesthetic-innervatory motor engram, which is necessary for the motor program to be carried out, resulting in inappropriate hand shapes and movements during voluntary limb movements. However, there is genuine dispute over whether these actually constitute independent syndromes, and the distinction will not be important for this discussion.

References

Appel, J., Kertesz, A. & Fisman, M. (1982). A study of language functioning in Alzheimer patients. *Brain and Language*, 17, 73–91.

Bay, E. (1962). Aphasia and non-verbal disorders of language. *Brain*, 85, 411–26.

Bayles, K. & Tomoeda, C. (1983). Confrontation naming impairment in dementia. *Brain and Language*, 19, 98–114.

Caramazza, A. & Zurif, E. (1976). Dissociation of algorithmic and heuristic processes in language comprehension: Evidence from aphasia. *Brain and Language*, 3, 572–82.

Cicone, M., Wapner, W., Foldi, N., Zurif, E. & Gardner. H. (1979). The relationship between gesture and language in aphasic communication. *Brain and Language*, 8, 324–49.

Cummings, J. & Benson, D.F. (1983). *Dementia: A clinical approach*. Boston: Butterworths.

Curtiss, S., Kempler, D. & Yamada, J. (1981). The relationship between language and cognition in development: Theoretical framework and research design. *UCLA Working Papers in Cognitive Linguistics*, 3, 1–61.

Darley, F. (1982). *Aphasia*. Philadelphia: W.B. Saunders.

Della Sala, S., Lucchelli, F. & Spinnler, H. (1987). Ideomotor apraxia in patients with dementia of Alzheimer type. *Journal of Neurology*, 234, 91–3.

Denny-Brown, D. (1958). The nature of apraxia. *Journal of Nervous and Mental Disease*, 126, 9–33.

DeRenzi, E., Motti, F. & Nichelli, P. (1980). Imitating gestures: A quantitative approach to ideomotor apraxia. *Archives of Neurology*, 37, 6–10.

Duffy, R.J. & Duffy, J.R. (1981). Three studies of deficits in pantomimic expression and pantomimic recognition in aphasia. *Journal of Speech and Hearing Research*, 46, 70–84.

Duffy, R.J. & Duffy, J.R. (1984). *New England Pantomime Tests*. Tigard, Oregon: CC. Publications.

Duffy, R.J., Duffy, J.R. & Pearson, K.L. (1975). Pantomime recognition in aphasics. *Journal of Speech and Hearing Research*, 18, 115–32.

Duffy, R.J. and Liles, B.Z. (1979). A translation of Finkelnburg's (1870) lecture on aphasia as "asymbolia" with commentary. *Journal of Speech and Hearing Disorders*, 44, 156–68.

Duffy, J.R., Watt, J. & Duffy, R.J. (1981). Path analysis: A strategy for investigating multivariate causal relationships in communication disorders. *Journal of Speech and Hearing Research*, 24, 474–90.

Duffy, J.R. & Watkins, L.B. (1984). The effect of response choice relatedness on pantomimic and verbal recognition ability in aphasic patients. *Brain and Language*, 21, 291–306.

Faglioni, P. & Basso, A. (1985). Historical perspectives on neuroanatomical correlates of limb apraxia. In *Neuropsychological Studies of Apraxia and Related Disorders,* ed. E. A. Roy, pp. 3–44. The Netherlands: Elsevier Science Publishers.

Ferguson, G.A. (1971). *Statistical Analysis in Psychology and Education*. New York: McGraw-Hill.

Fodor, J. (1983). *The Modularity of Mind, an Essay on Faculty Psychology*. Cambridge, MA: The MIT Press.

Folstein, M.F., Folstein, S.E. & McHugh, P.R. (1975). Mini-Mental State: A practical method for grading the cognitive state of patients for the clinician. *Journal of Psychiatric Research*, 12, 189–98.

Foster, N.L., Chase, T.N., Patronas, N.J., Gillespie, R.N. & Fedio, P. (1986). Cerebral mapping of apraxia in Alzheimer's Disease by positron emission tomography. *Annals of Neurology*, 19, 139–43.

Gainotti, G. & Lemmo, M. (1976). Comprehension of symbolic gestures in aphasia. *Brain and Language*, 3, 451–60.

Gardner, H. (1974). The naming and recognition of written symbols in aphasic and alexic patients. *Journal of Communication Disorders*, 7, 141–53.

Glass, A.V., Gazzaniga, M.S. & Premack, D. (1973). Artificial language training in global aphasics. *Neuropsychologia*, 11, 95–103.

Goldstein, K. (1948). *Language and Language Disturbances*. New York: Grune and Stratton.

Goodglass, H. & Kaplan, E. (1963). Disturbance of gesture and pantomime in aphasia. *Brain*, 86, 703–20.

Goodglass, H. & Kaplan, E. (1983). *The Assessment of Aphasia and Related Disorders*. Philadelphia: Lea and Febiger.

Haaland, K.Y. & Flaherty, D. (1984). The different types of limb apraxia errors made by patients with left vs. right hemisphere damage. *Brain and Cognition*, 3, 370–84.

Haaland, K.Y. & Yeo, R.A. (1989). Neuropsychological and neuroanatomic aspects of complex motor control. In *Neuropsychological Function and Brain Imaging*, ed. E. D. Bigler, R. A. Yeo & E. Turkheimer, pp. 219–44. New York: Plenum Press.

Hamsher, K. (1981). Intelligence in aphasia. In *Acquired Aphasia*, ed. M. T. Sarno, pp. 327–60. New York: Academic Press.

Hecaen, H. & Albert, M. L. (1978). *Human Neuropsychology*. New York: John Wiley and Sons.

Heilman, K. (1979). The neuropsychological basis of skilled movement in man. In *Handbook of Behavioral Neurobiology, Vol.2: Neuropsychology*, ed. M. Gazzaniga, pp. 447–61. New York: Plenum Press.

Heilman, K.M., Coyle, J.M., Gonyea, E.F. & Geschwind, N. (1973). Apraxia and agraphia in a left-hander. *Brain*, 96, 21–8.

Heilman, K. and Valenstein, E. (eds.) (1979). *Clinical Neuropsychology*. New York: Oxford University Press.

Helm-Estabrooks, N.A., Fitzpatrick, P.M. & Barresi, B. (1981). Visual action therapy for global aphasia. *Journal of Speech and Hearing Disorders*, 47, 385–89.

Jackson, J.H. (1878). On affectations of speech from disease of the brain. *Brain*, 1, 304–30.

Jackson, J.H. (1932). *Selected Writings, 2*, ed. J. Taylor. London: Hodder and Stoughton.

Kempler, D. (1984). Syntactic and Symbolic Abilities in Alzheimer's Disease. Unpublished Ph.D. dissertation, UCLA.

Kempler, D. (1988). Lexical and pantomime abilities in Alzheimer's disease. *Aphasiology*, 2, 147–59.

Kempler, D. (1991). Language changes in dementia of the Alzheimer type. In *Dementia and Communication: Research and Clinical Implications*, ed. R. Lubinski, pp. 98–114. Toronto: Decker Publishing.

Kempler, D., Curtiss, S. & Jackson, C. (1987). Syntactic preservation in Alzheimer's disease. *Journal of Speech and Hearing Research*, 30, 343–50.

Kertesz, A. (1979). *Aphasia and Associated Disorders: Taxonomy, Localization and Recovery*. New York: Grune and Stratton.

Kertesz, A. (1980). *The Western Aphasia Battery*. London, Ontario: University of Western Ontario.

Kertesz, A. (1985). Apraxia and aphasia: Anatomical and clinical relationship. In *Neuropsychological Studies of Apraxia and Related Disorders*, ed. E. A. Roy, pp. 163–78. Amsterdam: Elsevier/North Holland.

Kertesz, A. & Hooper, P. (1982). The extent and variety of apraxia in aphasia. *Neuropsychologia*, 20, 275–86.

Kimura, D. (1977). Acquisition of a motor skill after left-hemisphere damage. *Brain*, 100, 527–42.

Kimura, D. & Archibald, Y. (1974). Motor functions of the left hemisphere. *Brain*, 97, 337–50.

Lehmkuhl, G., Poeck, K. & Willmes, K. (1983). Ideomotor apraxia and aphasia: an examination of types and manifestations of apraxic symptoms. *Neuropsychologia*, 21, 199–212.

Lichtheim, L. (1885). [On Aphasia] *Brain*, 7, 433–485. (Originally published in *Deutsches Archiv fuer Klinische Medizin*, 1885, 36, 204–68).

Liepmann, H. (1908). *Drei Aufsatze aus dem Apraxiegebeit*. Berlin: Karger.

Martin, A. and Fedio, P. (1983). Word production and comprehension in Alzheimer's disease: The breakdown of semantic knowledge. *Brain and Language*, 19, 124–41.

McKhann, G., Drachman, D., Folstein, M., Katzman, R., Price, D. & Stadlan, E. (1984). Clinical diagnosis of Alzheimer's disease: Report of the NINCDS–ADRDA Work Group under the auspices of the Department of Health and Human Services Task Force on Alzheimer's Disease. *Neurology*, 34, 939–44.

Nebes, R.D. (1989). Semantic memory in Alzheimer's disease. *Psychological Bulletin*, 106, 377–94.

Penfield, W. & Roberts, L. (1959). *Speech and Brain Mechanisms*. Princeton: Princeton University Press.

Piatelli-Palmarini, M. (ed.) (1980). *Language and Learning: The Debate Between Jean Piaget and Noam Chomsky*. Cambridge, MA: Harvard University Press.

Poeck, K. & Kerschensteiner, M. (1971). Ideomotor apraxia following right-sided cerebral lesion in a left-handed subject. *Neuropsychologia*, 9, 359–61.

Porch, B. (1981). *Porch Index of Communicative Abilities*. Palo Alto: Consulting Psychologists Press.

Rapcsak, S.Z., Croswell, S.C. & Rubens, A.B. (1989). Apraxia in Alzheimer's disease. *Neurology*, 39, 664–68.

Schuell, H.M. (1965). *The Minnesota Test for Differential Diagnosis of Aphasia*. Minneapolis: University of Minnesota.

Schwartz, M., Marin, O. & Saffran, E. (1979). Dissociations of language function in dementia: A case study. *Brain and Language*, 7, 277–306.

Seron, X., van der Kaa, M.A., Remitz, A. & van der Linden, M. (1979). Pantomime interpretation in aphasia. *Neuropsychologia*, 17, 661–8.

Shallice, T. (1987). Impairments of semantic processing: Multiple dissociations. In *The Cognitive Neuropsychology of Language*, ed. M. Coltheart, G. Sartori, & R. Job, pp. 111–27. London: Lawrence Erlbaum Associates.

Ska, B. & Joanette, Y. (1990). *Pantomimes in dementia of the Alzheimer's type*. Poster presented at the Canadian Psychological Association, Ottawa, May 31–June 2.

Smith, S. (1989). *Syntactic Comprehension in Alzheimer's Disease*. Paper presented at the Academy of Aphasia. October, 1989. Santa Fe.

Varney, N.R. (1982). Pantomime recognition defect in aphasia: Implications for the concept of asymbolia. *Brain and Language*, 15, 32–9.

Weisenburg, T. & McBride, K.E. (1935). *Aphasia*. New York: Commonwealth Fund (Reprinted in 1964, New York: Hafner).

9

Sex differences in visuospatial skills: Implications for hominid evolution

DEAN FALK

Over a century ago, Paul Broca concluded from clinical observations that a specific area of the left frontal lobe (Broca's area) was involved in speech. Language thus became the first behavior for which an asymmetrical cortical substrate was discovered. Because speech was found only in humans, the incorrect assumption was made that brain lateralization was also unique to humans. That assumption held sway until quite recently. However, it is now known that a variety of animals including rodents, birds and monkeys are characterized by right/left differences of the cerebral cortex (Glick, 1985). Brain lateralization, it seems, is phylogenetically old.

Humans are animals in whom brain lateralization has been carried to an extreme (Falk, 1987a). Language is found in 100% of the normal population, and unidirectional (right) handedness occurs in approximately 93% of people. Both behaviors are highly dependent upon the left hemisphere of the brain, and neither typifies any other population of primates. Human left hemispheres are also specialists in analytical, time-sequencing tasks. More global, holistic abilities are associated with the right hemispheres of humans, on the other hand. These include visuospatial imagery, musical abilities, and emotional processes.

The above should not be taken to imply that the brains of all humans are organized in precisely the same way. Nervous systems are highly "plastic" and a good deal of variation exists in the "wiring" from brain to brain (Ojemann, 1983). Further, McGlone (1980) has convincingly argued that males and females differ in their patterns of brain lateralization. Such differences in cortical asymmetry are apparently associated with different performances on behavioral tests. In general, female populations excel at language skills, rightward bias for fine motor skills, and emotional decoding and expression. Populations of males perform better at visuospatial skills, mathematics, and musical composition. However, it should be noted that there is wide overlap in the distributions of performances of males and females, and that differences in mean performances are often subtle but statistically significant (Falk, 1987a).

Sex differences in brain lateralization may have important implications for the topic under discussion in this volume. For historical reasons mentioned above, language (and therefore the left hemisphere) has been a primary focus of investigations of brain evolution and cortical lateralization. Tool production, on the other hand, requires visuospatial skills that are dominated by the right hemisphere (Wynn, 1989). Below, I attempt to outline what is known about (a) sex differences in human visuospatial skills, (b) the evolution of brain lateralization in mammals, (c) brain lateralization in primates, and (d) tool production and "intelligence" in hominids.

Visuospatial skills and sex differences

Of all the specific cognitive skills, spatial ability shows the most consistent sex difference, with males scoring higher than females (Plomin, DeFries & McClearn, 1990). In a study of tactile perception of ambiguous figures (Lenhart & Schwartz, 1983), males performed much better with right and especially with left hands (right hemisphere) than did females, but *only* under conditions where participants were instructed to rely exclusively on mental imagery. When participants were instructed to rely on verbal strategies, females outperformed males but the differences were not as dramatic as for imagery strategies. In control groups given no specific coding instructions, males performed as if they had been given verbal rather than imagery coding instructions even though males have "privileged access to right-hemisphere imagery codes".

In a study by Guay & McDaniel (1978), the sexes were not separated by a visual task that involved flat two-dimensional figures, but males performed significantly better on four tasks that were based on three-dimensional figures, such as those shown in Figures 9.1–9.4. It is interesting that females in the study reported that they enjoyed spatial thinking to a greater extent than reported by males, and Guay and McDaniel concluded that cultural variables may account for much of the discrepancy in spatial aptitude between the sexes.

Results of another (earlier) study did not support a pure environmental explanation for sex differences in visuospatial ability (Bock & Kolakowski, 1973). Tests for visuospatial ability were administered to parents and offspring in 167 families. Analysis confirmed that males perform better than females on spatial visualization tests, and revealed a pattern of correlations that is consistent with a recessive X-linked gene of intermediate frequency. The authors therefore concluded that visuospatial ability is determined by a gene

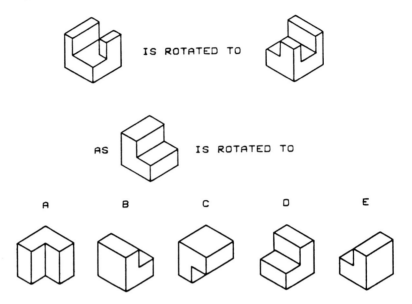

Fig. 9.1. Mental rotation test. Study how the object in the top line is rotated and select the object from the bottom line that looks as the figure in the second line would appear if it were rotated in exactly the same way. Example provided by Roland Guay.

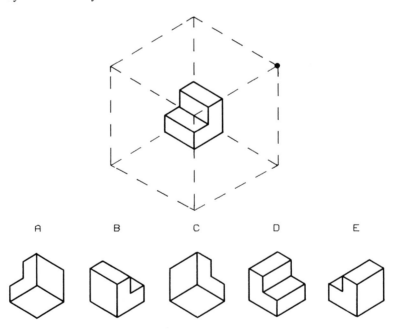

Fig. 9.2. The example shows an object positioned in the middle of a glass box, and drawings A–E show the same object as seen from different views. Select the view that would be seen if the black dot were located between the object and the viewer's eyes. Courtesy of Roland Guay.

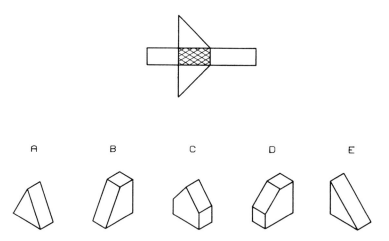

Fig. 9.3. The upper diagram shows a three-dimensional figure that has been flattened; the shaded portion indicates the bottom surface. Select the appropriate figure that represents what the upper figure would look like when it is folded into a three-dimensional object. Provided by Roland Guay.

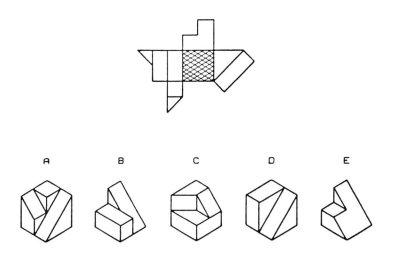

Fig. 9.4. The instructions for this problem are the same as for Figure 3. While Figures 1–3 provide very simple examples of mental imagery tests, this example (courtesy of Roland Guay) provides a more complicated task. As a population, males perform better than females on the kinds of visual imagery tests represented in Figures 1–4. See text for discussion. Solutions for the tests are the following: Fig. 1 = D, 2 = E, 3 = E, 4 = B.

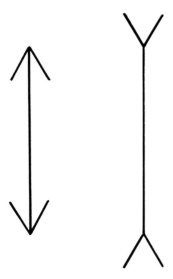

Fig. 9.5. The Mueller–Lyer figure for which subjects are required to judge the relative lengths of the vertical shafts. The vertical shaft on the right is generally seen as longer (from Deregowski, 1989, with permission).

for spatial ability that is sex-linked. They also concluded (for other reasons) that the gene for spatial ability may be testosterone-limited in its expression. However, since the Bock & Kolakowski study, data from larger studies have failed to confirm the hypothesis of sex-linkage for spatial ability (e.g., DeFries *et al.*, 1979).

Research on susceptibility to optical illusions provides another area in which sex differences in spatial processing could be explored. Unfortunately, although there is a large literature on cross-cultural responses to visual illusions, few of these studies address the question of sex differences (Deregowski, 1989). More work has been done on the Mueller–Lyer illusion (Figure 9.5) than on all other optical illusions combined (Coren & Girgus, 1978). This illusion occurs in animals ranging from fish to monkeys (*ibid*). As one might predict, when the Mueller–Lyer illusion was presented first to the right and then to the left hemispheres of the brains of male and female college students (i.e., through opposite visual fields), the right hemisphere showed a larger effect than the left (Clem & Pollack, 1975). However, preliminary data analyses failed to reveal reliable sex differences in susceptibility to the Mueller–Lyer illusion.

In light of the widely documented differences between men and women in visuospatial ability, it might seem surprising that the sexes do not differ in responses to the Mueller–Lyer optical illusion. As noted above, Guay & McDaniel (1978) found that a two-dimensional task did not separate the sexes,

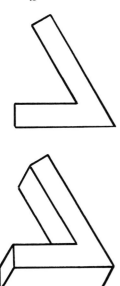

Fig. 9.6. Two of the drawings to which subjects set the arms of wooden callipers used in the Kwengo Callipers Test (Deregowski and Bentley, 1986). A subject who sees the bottom figure as depicting a 3D object is likely to see the angle between the represented arms as larger than that of the top figure (from Deregowski, 1989, with permission).

whereas four three-dimensional visual tasks did. Thus, the Mueller–Lyer illusion may fail to separate the sexes because it is based on figures that are essentially two-dimensional. It would be interesting to discover whether susceptibility to three-dimensional illusions, such as that shown in Figure 9.6, separates the sexes.

As is clear from the foregoing, efforts to sort out the respective influences on visuospatial abilities due to nature and nurture have not been successful. Environmental explanations for differential cross-cultural responses to optical illusions include both "carpentered world" and "ecological" hypotheses (Segall, Campbell & Herskovits, 1966). According to the former, people who live in carpentered environments that contain many solid right angles are more likely to misperceive ambiguous angles as right angles and will therefore be more susceptible to certain optical illusions than individuals from non-carpentered environments. On the other hand, the ecological hypothesis suggests that inhabitants of open terrain (e.g., desert) will be more susceptible than people who live in closed environments to optical illusions in which ambiguous linear stimuli are perceived as extending away from the viewer. Other explanations for susceptibility to optical illusions focus on physiological mechanisms of vision, such as the amount of pigment in various parts of the eye (Coren, 1989), or genetic explanations (Coren & Porac, 1977).

Similarly (and as noted above), both environmental and genetic explanations have been offered to explain sex differences in visuospatial abilities. Geschwind & Galaburda (1987) come close to merging nature and nurture with their interesting suggestion that different chemical environments (e.g., amounts of testosterone) during fetal development of males and females are responsible for (later) sex differences in visuospatial skills. In sum, it seems that visuospatial abilities are complex and probably based on a combination of innate and environmental factors. Two things are for certain: First, visual stimuli are compelling (Deregowski, 1989) – e.g., try *not* to respond to the conflicting stimuli presented in Figure 9.7! Second, for whatever reasons, populations of males outperform populations of females on three-dimensional visuospatial tasks.

The evolution of brain lateralization in mammals

As noted above, lateralized nervous systems are found in a wide range of mammals ranging from rodents to birds to primates (Glick, 1985). A review of experimental studies for various mammals (based on postural/motor, neurochemical, lesion, and neuroanatomical measures) shows that sex is a *major* determinant for patterns of behavioral and brain asymmetries (Robinson *et al.*, 1985). However, patterns of sex differences in asymmetries are complex and the direction of sex differences must be considered on species-by-species and region-by-region bases.

Of the measures surveyed by Robinson *et al.*, postural/motor asymmetries are perhaps most relevant to our discussions of visuospatial abilities (above) and tool production (below). It is interesting that sex differences in postural/ motor asymmetries (e.g., preferred direction of circling in rodents) are "the rule rather than the exception" (*ibid*, p. 195), and are surprisingly consistent across the literature. In all cases of postural/motor asymmetries showing sex differences, females are more strongly lateralized than males. Human females also appear to be more lateralized for postural/motor (but not for other) asymmetries than males (see Porac & Coren, 1981 for review).

Over ten years ago, Webster (1977) diagrammed how an asymmetrical nervous system would permit an organism to label left–right information, and he therefore suggested that the evolution of brain lateralization in mammals was related to analysis and memory of spatial position. Webster also speculated that development of an asymmetrical brain came about in conjunction

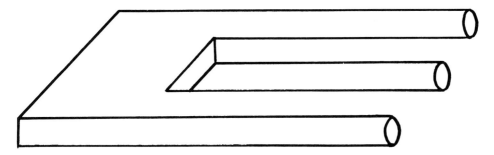

Fig. 9.7. The two-pronged trident, an "impossible" figure (from Deregowski, 1989, with permission).

with selection for behaviors related to reproduction and territoriality (1977:217–218):

> The adaptive survival aspects of territorial behavior, especially its reproductive aspects, provide the mechanism of selection favoring the evolution of asymmetrical brains. If successful mating takes place in a territory, if that is predicated on a successful definition and defense of that territory, and if such definition and defense are facilitated by the existence of a state of brain asymmetry, then there will be selection for whatever neural systems are lateralized to produce that asymmetrical state.

Webster points out that territories are presumably associated with cognitive spatial maps, and that there is also a widespread association of vocalization with territoriality (e.g., howler monkeys). These ideas are interesting and warrant consideration.

Did sex differences in visuospatial ability evolve in certain species because they contributed differentially to their reproductive success? Gaulin & Fitzgerald (1989) tested this idea for two species of congeneric voles (rodents) whose mating systems differ. Results of their radiotelemetric studies show that male *polygynous* meadow voles expand their ranges during the breeding season, whereas females do not. This suggests that range expansion is a male reproductive tactic. In contrast, neither male nor female *monogamous* prairie voles expand their ranges during breeding season. The investigators subsequently trapped voles of both species and tested them for visuospatial ability in laboratory mazes. In keeping with the above, only meadow voles showed consistent male superiority on the spatial tasks.

Because of their habitats, some arboreal species rely more on auditory than on visual senses in their reproductive strategies. For example, Oscine birds such as canaries rely on songs to announce their presence to neighbors, and to claim breeding territories (Nottebohm, 1989). More important for our discussion, *male* songbirds sing to attract mates. Further, birdsong is neurologically lateralized to left forebrain and hypoglossal structures (Nottebohm, 1977), and these structures are three to four times larger in adult male canaries that sing complex songs than in adult females that sing simpler songs (Nottebohm, 1989). Apparently, birdsong contributes to reproductive success, and it is reasonable to conclude that sex differences in this behavior, as well as the neurological asymmetries underlying birdsong, were selected for in this context.

Brain lateralization in primates

Unlike the bird and rodent examples presented above, primates rely heavily on *both* vocal and visual communication systems, although the complexity (subtlety) of primate visual and auditory signals seems to increase between arboreal and terrestrial habitats (Marler, 1973). Further, both vocal communication ("language") and visuospatial systems are generally lateralized, in humans, to left and right hemispheres respectively. Taking a cue from other mammals, could sex differences in visuospatial and vocal communication skills have (a) existed and (b) enhanced reproductive success in the primate ancestors of early hominids? These are interesting questions, but there is little firm evidence with which to address them.

One may, however, engage in "educated" speculation. Although very little work has been done on sex differences in brains of primates, a number of recent investigations nevertheless suggest that the substrate for human brain lateralization was present during early anthropoid evolution (Falk *et al.*, 1990). For example, hemispheric specializations similar to those that characterize *Homo sapiens* appear to be present in macaque monkeys (*Macaca*) who are left-hemisphere dominant for processing species-specific vocalizations (Heffner & Heffner, 1984; Petersen *et al.*, 1978) and right-hemisphere dominant for discriminating faces (Hamilton & Vermeire, 1988; Ifune, Vermeire & Hamilton, 1984). Macaque and human populations are both characterized by a significant directional asymmetry in the shape of the frontal lobes (known as "petalia"), whereby the right frontal lobe protrudes further than the left (Falk *et al.*, 1990). In humans, this trait has been shown to have a significant relationship with right-handedness (Galaburda *et al.*, 1978; LeMay, 1977). Significant heritabilities have been found for frontal petalias of rhesus monkeys (Cheverud *et al.*, 1990) and these are consistent with a hypothetical genetic component for cortical lateralization (Geschwind & Galaburda, 1987).

A survey of handedness studies across the primate order also suggests that early anthropoids were neurologically lateralized (MacNeilage, Studdert-

Kennedy & Koob, 1987). In general, nonhuman primates tend to be left-handed when reaching for food and right-handed for manipulative tasks. The distribution of results across extant primates led MacNeilage *et al.* to develop an evolutionary model based on structural and functional adaptations to feeding. According to this model, the initial primate adaptation (indicated by studies on prosimians) involved right-hemisphere specialization for visually guided reaching with the left hand. Results of the survey for "higher" (anthropoid) primates, on the other hand, suggest that the left hemisphere subsequently became specialized for right-hand manipulation and bimanual coordination. This model therefore gives primacy to specialization of the right hemisphere for visuospatial functions, during early primate evolution (however, it should be noted that vocal communication systems were not under consideration).

From the above, we may conclude that anthropoid primates are *very* lateralized in their brain organization, and that this trait has enjoyed a long evolutionary history. Of all the anthropoids, of course, *Homo sapiens* has the most lateralized brains. Since sex is a major determinant for patterns of behavioral and brain asymmetries in mammals including humans (see above), we may speculate that sex probably interacted with brain lateralization in ancestral primates. However, more evolutionary studies on brain lateralization *that attend to sex* are needed to "flesh out" this speculation. Meanwhile, we are left wondering if early anthropoid males were specialized for right-hemisphere visuospatial tasks and, if so, how would sex differences in brain organization have related to reproductive strategies?

Tool production and "intelligence" in hominids

The neurological substrate for brain lateralization that was probably brought to earliest hominid evolution has been described above. In short, right hemispheres were already specialized for visuospatial tasks, and left hemispheres were dominant for vocal communication systems, when hominids first arrived on the scene, some five to six million years ago. It is also reasonable to speculate that there may have been sex differences in patterns of asymmetry in the earliest hominids.

During the first several million years of hominid evolution, things were slow in the "tools and intelligence" department. Brain size did not increase appreciably. In fact, it never got above 600 cm³ in australopithecines (Falk, 1987b). Nor was there much in the way of stone tools. However, australopithecines changed physiologically during their evolution, as bipedalism continued to develop (Falk, 1990a,b). In particular, a "radiator" network of cranial veins began to develop in gracile, but not in robust, australopithecines. This network is well developed in extant humans, in whom it functions to cool heat-sensitive brains under conditions of hyperthermia. (It does this by delivering cool surface blood from the head into the braincase.) Presumably, the "radiator" network of veins that began to develop in conjunction with bipedalism in gracile australopithecines also helped regulate brain temperature. In the long run, a thermal constraint that had previously kept brain size in check was released by the "radiator". This proved to be extremely advantageous for the descendants of gracile australopithecines, i.e., members of the genus *Homo*.

The fossil and archaeological record for *Homo* picks up around two million years ago in East Africa. And what a record it is! Brain size "took off" and subsequently doubled from approximately 700 cm^3 to 1400 cm^3 (the "radiator" continued to develop right along with brain size, Falk, 1990a). Recorded tool production also accelerated in *Homo*, spanning from initial clunky stone tools to contemporary computer, space, and biological engineering technology. Elsewhere, I have argued (Falk, 1990a), and defended (Falk, 1990b), the assertion that the evolution of brain size was the result of selection for general intelligence during the course of hominid evolution.

The fossil and archaeological record suggests that early *Homo* had brains that were lateralized along the lines of our own and that, subsequently, cortical asymmetry developed even further. An endocast from a nearly two-million-year-old *Homo habilis* skull (KNM-ER 1470) reveals a sulcal pattern in the left frontal lobe that looks similar to that which is associated with Broca's speech area in people (Tobias, 1981; Falk, 1983). In keeping with this, early *H. habilis* presumably had at least the beginnings of speech. (For reasons that are beyond the scope of this chapter, I think that "language" subsequently underwent a long evolution in *Homo*; Falk, 1991.) Toth (1985) suggests that populations of early *Homo* were also right-handed, another trait that depends on the left frontal lobe. Finally, there is evidence that the right hemisphere became more specialized for visuospatial processes in conjunction with tool production during the evolution of *Homo* (Wynn, 1989).

Unfortunately, attention to differences of sex is missing from most hominid studies that pertain to brain lateralization. For example, since living females excel over males in language skills (see Falk, 1987a for review), it would be interesting to explore the relationship between sex and the evolution of this "tool" of the left hemisphere. Similarly, it would be useful to know the sex distribution of early stone tool knappers. Were they predominantly male? As spatial competence increased during the evolution of tool production (Wynn, 1989), were sex differences in patterns of behavioral and brain asymmetries also evolving? If so, what were their implications for reproductive strategies?

I have an image of one of our distant mammalian ancestors. The ancestor is male and it is mating season. He's good at getting around in space, and temporarily extends his daily range in search of mates. Because of a good "cognitive map", he is able to return home later and to repeat the process during the next mating season. During his life, he fathers many offspring.

Thanks to Wynn (1989), I also have an image of one of the mammal's distant, future descendants. It is a knapper making a bifacial stone tool, some 300 000 years ago. As the tool takes shape, the knapper uses notions of perspective, the control of spatial quantity, and an understanding of composition. Overall shape is very important to the knapper. When it is finished, the tool has a regular cross section – it is pleasing. And all because the knapper's distant ancestor was good at getting around, and "delivered" the genetic bases for skilled visuospatial abilities to his offspring. The knapper is satisfied with the way the tool has turned out. *She packs up and goes home.*[1]

Notes

1 Two things to remember: 1. Because of the way genes are transmitted from one generation to the next, characteristics that are selected for in one sex are extremely likely to affect the other, and 2. there is a *huge* amount of "overlap" in the abilities of males and females.

References

Bock, R. D. & Kolakowski, D. (1973). Further evidence of sex-linked major-gene influence on human spatial visualizing ability. *American Journal of Human Genetics*, 25, 1–14.

Cheverud, J., Falk, D., Hildebolt, C., Moore, A. J., Helmkamp, R. C. & Vannier, M. (1990). Heritability and association of cortical petalias in rhesus macaques *(Macaca mulatta)*. *Brain, Behavior & Evolution*, 35, 368–72.

Clem, R. K. & Pollack, R. H. (1975). Illusion magnitude as a function of visual field exposure. *Perception and Psychophysics*, 17, 450–54.

Coren, S. (1989). Cross-cultural studies of visual illusions: The physiological confound. *Behavioral and Brain Sciences* (commentary), 12, 76–7.

Coren, S. & Girgus, J. S. (1978). *Seeing is Deceiving: The Psychology of Visual Illusions*. New Jersey: Lawrence Erlbaum Associates.

Coren, S. & Porac, C. (1977). Family correlations in visual-geometric illusions. *Behavior Genetics*, 7, 50–1.

DeFries, J. C., Johnson, R. C., Kuse, A. R., McClearn, G. C., Polovina, J., Vandenberg, S. G. & Wilson, J. R. (1979). Familial resemblance for specific cognitive abilities. *Behavior Genetics*, 9, 23–43.

Deregowski, J. B. (1989). Real space and represented space: Cross-cultural perspectives. *Behavioral and Brain Sciences*, 12, 51–119.

Deregowski, J. B. & Bentley, A. M. (1986). Perception of pictorial space by Bushman. *International Journal of Psychology*, 21, 743–52.

Falk, D. (1983). Cerebral cortices of East African early hominids. *Science*, 221, 1072–74.

Falk, D. (1987a). Brain lateralization in primates and its evolution in hominids. *Yearbook of Physical Anthropology*, 30, 107–25.

Falk, D. (1987b). Hominid paleoneurology. *Annual Review of Anthropology*, 16, 13–30.

Falk, D. (1990a). Brain evolution in *Homo*: The "radiator" theory. *Behavioral and Brain Sciences*, 13, 333–81.

Falk, D. (1990b). Evolution of a venous "radiator" for cooling the cortex: "Prime releaser" of brain evolution in *Homo*, response to open peer commentary. *Behavioral and Brain Sciences*, 13, 368–81.

Falk, D. (1991). Implications of the evolution of writing for the origin of language: Can a paleoneurologist find happiness in the Neolithic? In *Language Origins: A Multidisciplinary Approach*, ed. J. Wind, P. Lieberman & B. Chiarelli, pp. 245–51. Dordrecht, The Netherlands: Kluwer Academic Publishers.

Falk, D., Hildebolt, C., Cheverud, J., Vannier, M., Helmkamp, R. C. & Konigsberg, L. (1990). Cortical asymmetries in frontal lobes of rhesus monkeys *(Macaca mulatta)*. *Brain Research*, 512, 40–5.

Galaburda, A. M., LeMay, M., Kemper, T. L. & Geschwind, N. (1978). Right–left asymmetries in the brain. *Science*, 199, 852–56.

Gaulin, S. J. C. & Fitzgerald, R. W. (1989). Sexual selection for spatial-learning ability. *Animal Behavior* 37, 322–31.

Geschwind, N. & Galaburda, A. M. (1987). *Cerebral Lateralization: Biological Mechanisms, Associations, and Pathology*. Cambridge, MA: The MIT Press.

Glick, S. D. (1985). *Cerebral Lateralization in Nonhuman Species*. New York: Academic Press.

Guay, R. & McDaniel, E. (1978). *Correlates of performance on spatial aptitude tests*. Final report grant No. DAHC 19-77-G-0019. Alexandria, VA: US Army Research Institute for the Behavioral and Social Sciences.

Hamilton, C. R. & Vermeire, B. A. (1988). Complementary hemispheric specialization in monkeys. *Science*, 242, 1691–94.

Heffner, H. E. & Heffner, R. S. (1984). Temporal lobe lesions and perception of species-specific vocalizations by macaques. *Science*, 226, 75–6.

Ifune, C. K., Vermeire, B. A. & Hamilton, C. R. (1984). Hemispheric differences in split-brain monkeys viewing and responding to videotape recordings. *Behavioral and Neural Biology*, 1, 231–35.

LeMay, M. (1977). Asymmetries of the skull and handedness. *Journal of Neurological Sciences*, 32, 243–53.

Lenhart, R. E. & Schwartz, S. M. (1983). Tactile perception and the right hemisphere: A masculine superiority for imagery coding. *Brain and Cognition*, 2, 224–32.

MacNeilage, P. F., Studdert-Kennedy, M. G. & Koob, G. F. (1987). Primate handedness reconsidered. *Behavioral and Brain Sciences*, 10, 247–303.

Marler, P. (1973). A comparison of vocalizations of red-tailed monkeys and blue monkeys, *Cercopithecus ascanius* and *C. mitis*, in Uganda. *Zeitschrift fur Tierpsycholgie* 33, 223–47.

McGlone, J. (1980). Sex differences in human brain asymmetry: A critical survey. *Behavioral and Brain Sciences*, 3, 215–63.

Nottebohm, F. (1977). Asymmetries in neural control of vocalization in the canary. In *Lateralization in the Nervous System*, ed. S. Harnad, R. Doty, L. Goldstein, J. Jaynes & G. Krauthamer, pp. 23–44. New York: Academic Press.

Nottebohm, F. (1989). From bird song to neurogenesis. *Scientific American*, February, 1989, 74–9.

Ojemann, G. A. (1983). Brain organization for language from the perspective of electrical stimulation mapping. *Behavioral and Brain Sciences*, 6, 189–230.

Petersen, R. R., Beecher, M. D., Zoloth, S. R., Moody, D. B. & Stebbins, W. C. (1978). Neural lateralization of species-specific vocalizations by Japanese macaques *(Macaca fuscata)*. *Science*, 202, 324–7.

Plomin, R., DeFries, J. C. & McClearn, G. E. (1990). *Behavioral Genetics, A Primer*, 2nd edn. New York: W. H. Freeman and Company.

Porac, C. & Coren, S. (1981). *Lateralized Preferences and Human Behavior*. New York: Springer-Verlag.

Robinson, T. E., Becker, J. B., Camp, D. M. & Mansour, A. (1985). Variation in the pattern of behavioral and brain asymmetries due to sex differences. In *Cerebral Lateralization in Nonhuman Species*, ed. S. D. Glick, pp. 185–231. New York: Academic Press.

Segall, M. H., Campbell, D. T. & Herskovits, J. M. (1966). *Influence of Culture on Visual Perception*. Bobbs-Merill.

Tobias, P. V. (1981). The emergence of man in Africa and beyond. *Philosophical Transactions of the Royal Society of London*, Ser. B, 292, 43–56.

Toth, N. (1985). Archaeological evidence for preferential right-handedness in the Lower and Middle Pleistocene, and its possible implications. *Journal of Human Evolution*, 14, 607–14.

Webster, W. G. (1977). Territoriality and the evolution of brain asymmetry. *Annals of the New York Academy of Science*, 299, 213–21.

Wynn, T. (1989). *The Evolution of Spatial Competence*. Urbana: University of Illinois Press.

10

The unitary hypothesis: A common neural circuitry for novel manipulations, language, plan-ahead, and throwing?

WILLIAM H. CALVIN

In considering transitions of organs, it is so important to bear in mind the probability of conversion from one function to another ...
Charles Darwin, *On the Origin of Species*, 1859

That evolution, over all-but-infinite time, could change one physical organ into another, a leg into a wing, a swim bladder into a lung, even a nerve net into a brain with billions of neurons, seems remarkable, indeed, but natural enough. That evolution, over a period of a few million years, should have turned physical matter into what has seemed to many, in the most literal sense of the term, to be some kind of metaphysical entity is altogether another matter.
Derek Bickerton, *Language and Species*, 1990

Introduction

It is traditional to talk about language origins in terms of adaptations for verbal communication, to talk about toolmaking in terms of natural selection shaping up the hand and the motor cortex, to talk about the evolution of intelligence in terms of how useful versatile problem-solving and look-ahead would have been to hominids. But *might natural selection for one of these improvements serve to haul along the others* as well? Indeed, might some fourth function be the seed from which the others grew? What are the chances of coevolving talk, technique, and thought?

Getting "something for nothing" is, I know, profoundly anti-Calvinist. And the Puritan ethic seems to require us to look for a function's antecedents in their usefulness to that very function. But, as the epigram shows, new uses for old things is not anti-Darwinian. Conversions of function can be profound, providing an enormous boost toward a new ability. Capabilities occasionally arrive unheralded by gradual predecessors. In the familiar case of the origin of bird flight, one observes that it takes a lot of wing feathers to fly even a little. Presumably natural selection for thermal insulation shaped forelimb feathers up to the threshold for flight. Natural selection for a better airfoil shaped feathers thereafter, but the switch-over from the insulation function to the flying function was presumably a surprise, leaving the protobirds to explore their newfound gliding abilities rather as we might try to figure out a holiday gift that arrived without an instruction manual.

Inventions are the novelties in evolution, though from most of the literature you would think that shaping-up streamlining was what it was all about. But adaptations are only improvements on an existing function; what we're talking about here is the invention itself before streamlining, and that's often a matter of a Darwinian conversion of function. Nature does take leaps, and such conversions of function are even faster than those anatomical leaps envisaged by proponents of punctuated equilibria and hopeful monsters.

Why might we expect one of these hominid improvements (tools, language, intelligence – and, I might add, accurate throwing and making music) to serve as a preadaptation for the others? First the general reason, then some specific ones.

In general, the brain is better at "new uses for old things" than any other organ of the body. Sometimes two digestive enzymes, each having evolved separately for a different food, can act in combination to digest a third foodstuff. But a brain can easily combine sensory schemas and movement programs in new ways, since it tends to use a common currency that allows disparate things to be compared. From whatever source, an excitatory or inhibitory input is first converted into positive or negative millivolts; nerve cells then add and subtract in this substitute value system. For one input to substitute for another, it is only necessary that they produce similar voltage changes in the relevant nerve cells. One can add apples and oranges to get so many pieces of fruit.

More specifically, the following functions are all examples of *stringing things together*, that specialty of the left brain:

> For toolmaking and tool use, one usually needs to make a novel *sequence* of movements.
>
> To go beyond the categories of ape language (chimpanzees have about 36 standard vocalizations, with 36 associated standard meanings) and say something novel, our ancestors seem to have hit upon the trick of assigning meaning to a *sequence* of individually-meaningless vocalizations.
>
> To create a novel plan of action, one needs to *spin a scenario* connecting the past with the future. One has then to contemplate it before acting (since most

novel courses of action are either nonsensical or dangerous), perhaps going through a number of rounds of variation and selection to improve its prospects.

This "get set" phase is particularly important for versatile throwing; one has to produce trains of nerve impulses going to dozens of muscles, timed like the finale of a fireworks display sequence.

And music is a time series much like language; instead of phonemes with various frequency components, we have chords. Instead of stringing phonemes together to make words and sentences, we string chords together to make phrases and melodies, build an emotionally-satisfying structure in time and sound.

But why might all these functions overlap in the brain? Why might some neural machinery be used interchangeably for one, then the other?

"Borrowing" neural machinery

Thanks to our map-making tendencies, we usually assume that the neural machinery for moving the hand should have little to do with that for moving the mouth: at least on the motor strip, they're in "different compartments" (though, I might note, neighboring ones). But it is premotor and prefrontal cortex that has the reputation for stringing things together, getting them ready to finally execute, not motor cortex. It isn't even clear that such sequences feed exclusively through motor strip, since premotor cortex has as many direct connections to spinal cord as does motor strip itself.

Furthermore, both inputs and outputs tend to be broadly wired in cerebral cortex: the functional maps may suggest segregation, but the anatomy is one big smear (another example of Alfred Korzybski's "The map is not the territory"). For example, in the somatosensory path from a fingertip to the cortex's sensory strip, a given thalamocortical neuron responding *only* to that small patch of skin will nonetheless send axon branches to nearly all of the cortical map for the hand, not just the map of that particular finger (presumably the synaptic strengths are stronger there than elsewhere). And a cell near the thumb–face boundary line of sensory strip seems to receive inputs from both; in some weeks the face connections are stronger, in other weeks the hand connections dominate that cell's functioning (references in Calvin, 1989).

A cell in motor cortex, seemingly specializing in a hand movement, will nonetheless send axon branches to numerous other segments of the spinal cord; perhaps overcoming the clumsiness of youth involves tuning up the synaptic strengths to emphasize one destination over the others. So to suggest that the parts of cerebral cortex which optimize hand movements might also aid mouth movements is not such a radical proposal. Indeed, Charles Darwin made such neurological observations in his 1872 book, *The Expression of the Emotions in Man and Animals*:

> [Persons] cutting anything with a pair of scissors may be seen to move their jaws simultaneously with the blades of the scissors. Children learning to write often twist about their tongues as their fingers move, in a ridiculous fashion.

It looks as if that broad smear of wiring may sometimes cause simultaneous activation of rather different muscle groups. The tendency of right-handed gestures to accompany left-brain-generated speech (Poizner *et al.*, 1987) is only the most commonly noticed manifestation of this higher-order control of sequence.

Cortical specializations for sequence *per se*

A hundred years after Darwin, the neuropsychologists also began to emphasize sequencing. A. R. Luria noted (a modern review is Stuss & Benson, 1986; pp. 225–226) how patients with prefrontal damage had difficulty in planning a sequence of movements. For example, a patient is in bed with his arms under the covers. He is asked to raise his arm. He doesn't seem able to do so. But if you ask him to remove his arm from under the covers, he can do that. If you then ask him to raise it up and down in the air, he does it all correctly and smoothly. There's no paralysis (which would suggest damage to motor strip or below), and no difficulty with executing a fluent sequence (which would suggest premotor cortex damage) – just a difficulty in *planning* the sequence, getting stuck on the condition of working around the obstacle of the confining bedcovers.

Further posteriorly, there was demonstrated an overlap between language disorders and sequential hand movements. In aphasic patients without any paralysis, Doreen Kimura (1979) and her coworkers showed that these stroke patients had trouble stringing together a sequence of hand movements, even though they could carry out each individual motion with ease. To use an everyday example (rather than the novel manual–brachial sequences they tested), the patient might be able to turn a key in a lock, turn the doorknob, and push open the door – but not all in one facile sequence. It was as if the stroke had damaged some sequencing machinery that was often used for language but also for planning novel hand–arm sequences.

And then the neurosurgeon George Ojemann began repeating Wilder Penfield's classic brain stimulation studies in awake epileptic patients, using much more sophisticated tests of language and related sensory- and movement-sequencing functions (Ojemann, 1983; Ojemann & Creutzfeldt, 1987; Calvin & Ojemann, 1980). "Language cortex" includes such left-frontal-lobe regions as premotor and prefrontal, and extends posteriorly to include the parietal and temporal lobe regions surrounding the Sylvian fissure. While highly variable from patient to patient, the peri-Sylvian "core" of the cortex from which function could be disrupted was always related to sequencing. Tests of nonlanguage auditory sequences were disrupted from the same regions as tests of nonlanguage oral–facial movement sequences, suggesting that such cortex subserves both receptive and expressive sequencing. This "sequencing core" is typically surrounded by a peripheral region where stimulation disrupts short-term verbal memory or some other aspect of language.

So it would appear that the brain has some regions which are particularly specialized for generating and analyzing sequences, and that they may be used by multiple sensory and motor systems.

Why a versatile sequencer?

The left brain has some specializations for sequencing that go back at least as far as the monkeys – we don't know how it evolved, but the left hemisphere seems to be better at deciphering a novel string of sounds than is the right hemisphere (reviewed by Falk, 1987). But there isn't much evidence that animal calls are sequenced for an additional meaning, except for the escalating loudness of monkey cries possibly indicating predator proximity.

On the motor side of things, it is easier to see why some neural sequencing machinery might be needed. Normally, it is not essential to make a detailed movement plan before starting to move. That is because goal-and-feedback works pretty well, even for most novel movements, as the brain gets "progress reports" along the way. But there are some movements which are too rapid for feedback to guide them: they are over-and-done-with by the time that the first feedback arrives. Compared to most electronic and hydraulic systems, neural feedback is quite slow: human reaction times require hundreds of milliseconds. One of the fastest loops is from arm sensors to spinal cord and back out to arm muscles: it takes 110 milliseconds for feedback corrections to be made to an arm movement (Cordo, 1987).

But dart throwing doesn't take much longer. Thus feedback from joints and muscles is wasted – you might use it to help you plan for the next time, but your arm is an unguided missile shortly after the throw has begun. You must plan perfectly as you "get set" to throw, create a chain of muscle commands, all ready to be executed in exactly the right order. The same is true for hammering, which both baboons and chimpanzees use effectively to crack open shells. You need something like a serial buffer memory in which to load up all the muscle commands in the right order and with the right timing relative to one another – and then you pump them out blindly, without waiting for any feedback.

Recall player pianos, where you punch an instruction roll that remembers which keys should be hit, and when. It may not look like a sheet of music (it is 88 channels high, one for each key) but it functions like one. The number of muscles that you need to control for throwing or hammering is similar, once you start counting up all those finger flexors and shoulder muscles.

To reprogram a player piano roll is a nuisance – to make corrections requires tape and a punch at a minimum – and to slow down an arpeggio in the middle of a long piece makes one wish for a recording medium that could be locally stretched a little. Since the brain is always doing little variations on a theme in improving its hammering or throwing repertoire, it has probably found a less awkward way of "recording motor tapes". Especially for throwing, where the distance to the target and the desired speed tend to vary widely with different targets – and as a hunter, one is most often in situations never encountered before.

More than one planning track for sequences?

As you get set to throw (at least in a situation less standardized than a basketball free-throw), the brain has to discard a lot of possibilities, narrow down the search to a few good candidates, then select one to be let loose. So at least two serial buffers are required, one to hold the current candidate train while it is graded by memories of past successes and failures, and another buffer to hold onto the best-so-far.

But the more planning tracks, the sooner a better candidate sequence might emerge: think of dozens or even hundreds of planning tracks in simultaneous operation. It would be like a railroad marshalling yard, many candidate trains vying for a chance to be let loose on the main track. You might keep the muscles inhibited (much as in dreaming sleep) until you have found a good-enough choice, then you could "play that tape", acting out that movement scenario (I suspect, however, that some reproduction steps intervene). Until then, you presumably shuffle the trains-in-formation after each round, occasionally substituting something linked in memory.

One obvious way for this to operate would be for the higher-rated trains to "reproduce themselves" (with, of course, copying errors on occasion) in the tracks of the trains rated poorly by the memorized environment. In this manner, the "best sequence" would also be the most common sequence (and attaining a certain percentage of the total tracks might represent the threshold for a sequence to be "let loose" on the main track of action).

In Fig. 10.1, I show words being sequenced, to use a serial-order example that is more easily communicated than muscle command sequences. I show

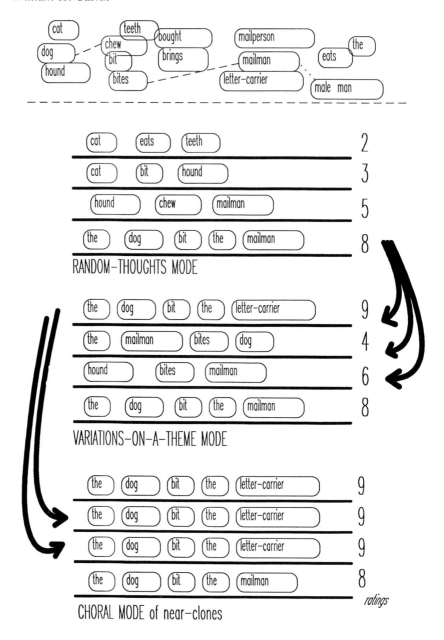

Fig. 10.1. The basic design and operation of a Darwin Machine:
• String together schemata rather than DNA (modular movements, phonemes, words, etc.); here words are shown for ease of illustration, the ones at the top being ones currently in short-term memory and thus more likely to be utilized.
• But there are many parallel "tracks" for such "trains", like a railroad marshalling yard (here only four planning tracks are shown);

only four of the many planning tracks, each randomly assembling a train of words, grading it against memories of how similar sequences have performed in the past, keeping the best one or two sequences, and trying again. Such rounds of shaping up, as Dawkins (1986; ch. 3) showed for shaping up random letter sequences to a Shakespearian model, can quickly converge on a good-enough sequence.

The brain as a Darwin Machine

This should be beginning to sound familiar. What other process has lots of serially-coded individuals, the sequence judged by the environment, and some individuals correspondingly more successful at reproducing themselves (albeit

> • new trains are random permutations and mutations of old trains (the middle panel is the successor of the top panel);
> • grading (numbers at right) is by serial–sequential memories of past successes and failures of strings, using dimensionless ratings, e.g., the economists' Subjective Expected Utility scores.
> • The better trains reproduce with variation into the tracks occupied by low-ranked trains, as shown by the arrows connecting the panels.
> • There are many rounds of shaping up, just like there are many generations of animals, antibodies. This is the Darwinian Two-Step: new injections of randomness, then another shaping-up, back and forth.
> What should this shaping up look like after a few rounds? (Two rounds are shown.)
> • Is there just one "best train" let loose on the "main track" to muscles by a comparator-and-decision-maker?
> • Or might half the trains come to be identical clones? (Perhaps mass action initiates the decision to move?)
> • Population thinking requires that we ask how the variability in the whole population changes, not just concentrate on "the best."
> • Other Darwinian processes suggest near-duplication of the most successful.
> •• Speciation involves the favored-by-environment type reproducing far better;
> •• Immune response involves selective reproduction of variants that better fit the invaders.
> • Thus near-clones might well come to occupy half the movement-planning tracks, as a string becomes well-formed.
> • The initial rounds would have enough diversity between the trains to be called *Random-thoughts mode*.
> • Intermediate rounds might be called the *Variations-on-a-theme mode*, which (if the grading scheme is maintained, i.e., attention doesn't shift) evolves into a population of near-clones.
> • This *Choral mode* might be what was under natural selection during hominid evolution, simply because of its usefulness in reducing timing jitter during throwing. But the secondary uses of the Darwin Machine might largely involve the first two modes of operation.

with a little shuffling of the code)? Actually, there are now two good examples: 1. DNA strings of units called *genes* being shaped up by an environment of prey–predators–pathogens, and 2. amino acid strings called *antibodies*, shaped up during the immune response by the differential reproduction associated with finding a foreign protein to destroy. Shuffling the code a little seems to be far more important than inducing errors, e.g., the *permutations* of crossing-over during meiosis is the common source of new variants, not *mutations per se*.

As I have recently argued (Calvin, 1987, 1989), having many neural sequencers from which to select gives rise to a very interesting property: brain processes can, in effect, simulate the process used by Darwinian evolution. They can shape up new serially-coded throws, thoughts, plans or sentences rather than new species or antibodies, using innocuous remembered environments rather than real-time noxious ones, operating in *seconds* rather than the *days* of the immune response or the *millennia* of species evolution. With this "Darwin Machine", the brain selects sequences of schemas. Here are some possible examples:

> Any of the manual-brachial ballistic movements (*throwing, hammering, clubbing*) might benefit in situations where a standardized movement could not be used. Achieving the "well-grooved" movement of the free-throw or dart throw might also occur through the Darwin Machine's process.
>
> *Kicking* and *dancing* might also benefit, to the extent that the neural sequencer makes good connections to output pathways for the lower body.
>
> *Language* would benefit, producing sequences of words to make a sentence suitable to the occasion and obeying the local rules of grammar as well.
>
> *Plan-ahead* in general might benefit, to the extent that more general schemas could be sequenced instead of movement commands.
>
> *Insight*, to the extent that it depends on testing possible sequences of action by simulating them first inside the head, ought to benefit from such machinery and so allow a hominid to "do the right thing" the first time without a lot of *overt* trial-and-error.
>
> *Musical abilities* could similarly benefit from a more general-purpose sequencer (stringing together chords to make melodies) without any natural selection for music *per se*.
>
> Serial-order *recreations* have surely been augmented by secondary use of a sequencer, since it seems so unlikely that chess and contract bridge have been exposed to natural selection themselves. It is hard to identify games that do not have a strong element of either serial-order, ballistic movement, plan-ahead, or fancy finger sequences.

Language implications of a neural sequencer

One can imagine a manual–brachial sequencer being adapted to simple kinds of language. A verb is usually a stand-in for a movement. And the targets of a ballistic movement are examples of nouns: "Throw at that rabbit!" or "Hit that nut." This seems just like the predicate of a sentence: the verb and its object. Spontaneous sign-language constructions in apes are often of this type, as are the earliest utterances of human children.

All animals exhibit associative learning, but such sequencing machinery might greatly augment associative abilities, quite without making any initial use of the exact ordering. All humans go through a stage of protolanguage (Bickerton, 1990) before the age of two, where the mere association of words serves to carry a simple message (and many a traveller reverts to protolanguage when the foreign-language phrase book fails). Were chimpanzees able to use combinations of their 36 standard vocalizations to convey nonstandard messages, they could acquire some of the power of pidgin languages of today.

Using a noun as the subject of a sentence-like construction seems a bit more novel, with less precedent in movement planning (where the subject is most often implied: oneself!). There is nothing particularly verbal about such sentences: "You go there" can be communicated with gestures, even eye movements. Having two nouns in the same sentence may create some ambiguity as to who is the *actor* and who is the *acted-upon* (see Rumbaugh and Savage-Rumbaugh, this volume). If the nouns are of very different classes (a person and a place), there may be no ambiguity. Yet a construction such as "John called Julie", where both nouns are capable of playing either role, starts requiring syntax, such as the English convention that the word order in a simple declarative sentence is actor–action–object. "Julie called John" means something different than "John called Julie". The mere addition of word order conventions would have enormously expanded the number of possible messages that could be constructed.

Bickerton (1990) cautions that much of modern language requires more of a structural analysis than these examples of sequencing provide. Beyond the second year of life, most normal humans acquire an ability to construct sentences with lots of grammatical words (*below, before, because* ...) and embedded phrases (*He ate it with a spoon*). Embedded phrases tend to be nested like Russian dolls rather than sequenced like a string of beads, requiring what linguists call a phrase structure analysis. And argument structure serves to identify the thematic roles associated with a verb, as when "give" tends to require a "who", "what", and "to whom". The verb "sleep" merely sends one in search of a "who" in the sentence.

To accomplish such modeling of the possible message, in a manner far more complex than mere word order allows, suggests considerably more neural machinery for sequencing than two tracks. Alternative interpretations of ambiguous sentences would need to be maintained simultaneously for comparison against memories, and better models evolved. It is this level of language which needs a Darwin Machine in the brain, not simple association or word ordering.

And so sequencing machinery could play a triple role: 1. even a single sequencer augments the *association* of various schemas, 2. it can also allow some conventions about *word order* to be used to greatly expand the possible messages that can be constructed, and 3. combinations of sequencers can permit an evolution of mental models for "Who did what to whom?" that use episodic memories of similar situations and sequences. Phrase structure can be mimicked in a railroad-yard-like branching of sequencer tracks; argument structure (and the other six theorems of modern grammar) may simply be part of the judgements that shape the evolving model. Any remaining ambiguity in the message (as in Bickerton's example of *The cow with the crumpled horn that Farmer Giles likes*) can often be resolved from memories (perhaps Farmer Giles is known to collect horns, suggesting that it is the horn that he likes, rather than the cow).

Besides the issues associated with innovative secondary uses, we have the evolutionary issue of which function came first. Or at least, which was fastest: Fast tracks in evolution tend to preempt slow tracks. Even if the advantages of plan-ahead intelligence or embedded phrases might suffice to evolve the more elaborate neural sequencing machinery in the long run, might nut-cracking hammering have done it faster?

While we might agree that music ought to have so little feedback on natural selection as to be near the bottom of the list, what should head the list will surely be the subject of an interesting debate. Linguists will be able to envisage the advantages of improvements in representations and communications, and will suggest that language improvements ought to be a fast track. Behaviorists will probably favor problem-solving versatility as a fast evolutionary path. Since I'm a neurophysiologist originally concentrating on the neurons that make decisions about movements, my contribution to the debate will be an evolutionary scenario for why accurate throwing might be the fastest track of all.

Accurate throwing as a growth industry

Many evolutionary improvements are "one-shot deals" that cannot be repeated for further improvement; in a sense, they are like college courses that cannot be repeated for additional credit. Yet some, like the Music Department's performance courses, can be taken again for credit, and so generate a significant growth curve. For example: To an aquatic mammal with sufficient fat, less hair seems to be better. And even less, better still. And so on until the growth curve reaches a plateau (one can only become so naked).

What is the hominid growth curve for hammering nuts? Given how skillful the chimpanzees studied by Boesch & Boesch (1981) seem to be (at least, at positioning the targets; the hammering sequence may not be as carefully varied), that curve might have been close to having reached a plateau even in our common ancestor. The profitability of nut-cracking was probably not redoubled very many times during hominid evolution, if our common ancestor was chimpanzee-like in this aspect of behavior.

This is in contrast to throwing's growth curve. Chimpanzees mostly engage in threat throwing, trying to intimidate leopards, snakes, and fellow chimpanzees; it seems to be an adjunct to flailing about with a branch, rocks being secondarily used. The great growth potential comes with converting the threat throw into a predatory throw (see Plooij, 1978, for one possible scenario). Action at a distance is an important invention for hunters, as it reduces the chances of injury from hooves, horns, and teeth – but this is a good example of an improvement that cannot be repeated for additional advantage.

What can be improved is the distance of the throw: twice as far is always better, no matter how many times the distance redoubles. And because throwing twice as far usually means throwing twice as fast, the kinetic energy (or "stopping power" in military terms) quadruples as distance doubles.

Accuracy can also be improved time and again for additional advantage, as one comes to be able to hit smaller and smaller targets at the same distance. The supply of small birds and mammals is much more reliable than that of large mammals. Attaining sufficient accuracy to strike the head is an important improvement when dealing with medium-sized mammals.

Adding some technology to the rock throw further increases the productivity of predatory throwing. The line of improvement from spear to bow-and-arrow to crossbow is familiar to us, though archaeology suggests that these were the products of the last ice age and the present interglaciation; the line of improvement from the common ancestor through *Homo erectus* to early *Homo sapiens* has to be based largely on theory at this point.

A throwing-based evolutionary scenario

In this exercise (which I hope that those concerned with hominid evolution based on tools, language and intelligence will consider emulating, so as to facilitate comparisons), I will start with the skills that I will ascribe to our common ancestor with the chimpanzees and bonobos. Thus I will assume that the basic ballistic skills were present (and in the case of hammering, even well-developed). For throwing, we do not know how accurate chimpanzees really are; the anecdotes only record their occasional (perhaps haphazard) hits. I think we can assume that if chimpanzees were accurate enough to bring down game with any regularity, we would have heard about it (given how much they enjoy fresh kills, they would probably be the terror of Africa).

In evolution, changes in behavior tend to precede adaptations in anatomy, so what kind of behavioral improvement could a chimpanzee-like ape make that would be rewarded by a big improvement in diet? Or in defense? If we assume that they initially lacked accuracy in throwing, this implies a big target or some special circumstance. And the waterhole provides both.

Transition from chimplike behaviors

I suggest that hominids practiced the familiar carnivore strategy: lying in wait at the waterhole for herds to come to drink. Since chimpanzees like to flail large tree branches around and then fling them, let us assume that one threw such a branch into the midst of a herd that was busy drinking. Even before the branch landed, the herd would startle from the sudden movement. By the time that the branch landed somewhere in their midst, the herd would be wheeling around and beginning to stampede.

Even if the branch didn't knock an animal down, some animal might well become entangled with the branch and trip. Ordinarily, such an animal would quickly get back on its feet and successfully run away from the hunters. But in the midst of a stampede, it would likely be knocked down again by other herd animals in their panic. Some such animals could be seriously injured; even if the hominid had merely intended to chase away competitors for scarce water, there might soon have been meat to scavenge.

Once the hominids got the idea of flinging branches to gain meat, the yield would have increased: the animal need not be fatally injured by the branch or stampeding herd. It need only be delayed enough in regaining its feet so that the hunters could apprehend it. Certainly, chimpanzees have no trouble in dispatching small pigs or monkeys, once they manage to lay hands upon them; one imagines the hominids starting with some herd animal smaller than zebra (Thomson's gazelles, for example) and tackling the larger ones after their finishing-them-off technique improved.

I suggest that the waterhole fling technique survives to the present day in the form of throwing sticks like the knobkerrie and boomerang, used to trip up a running animal; its virtue as a candidate for the earliest technique lies in its minimal requirements for accuracy and its continuity with known abilities and preferences of the chimpanzees. Flinging branches into herds at a waterhole would have made a good transition onto the bottom of the ladder of predatory throwing; the herd not only makes a large target (it doesn't matter which one is hit) but the herd "amplifies" the effect of the projectile by delay or injury of a downed animal. And the technique would be well-rewarded, what with a new-found ability to eat meat every day; one can imagine a "new niche" level of boom times resulting for the hominids that invented this technique.

Transition to rock throwing

Lacking a handy branch (and the technique might well exhaust the supply near a waterhole), chimpanzees tend to throw a big rock. Thrown into a herd, it might well land between animals (and so be less effective than a large branch). But herds that are feeling threatened by predators tend to cluster together tightly, and this is especially notable upon visits to the waterhole. Even if you miss, it's easier next time!

A large rock landing on the back or side of a small grazing animal might well injure it sufficiently; however, all the rock really has to do is knock the animal off its feet and the stampeding herd will serve to delay its escape. For larger herd animals, however, one can imagine a rock as being frequently ineffective, except on the rare occasions when an animal is hit on the head. What might improve the technique?

Enter the handaxe

A flat rock (a chunk of slate, for example) may spin in flight like a frisbee. If thrown in the plane of spin, air resistance is minimized and distance increased. If it hits an animal this way, it may also do more damage, as all of the impact is concentrated in a narrow edge; the sharper the edge, the more damage to the skin.

By itself, this does not seem very promising. I mention it because it seems to describe the enigmatic Acheulean handaxe in several respects: nice spin symmetry, and sharpened edges all around. Those sharpened edges are the prime reason that "handaxe" is surely a misnomer: pounding with one would cause it to bite the hand that held it. While damaged versions of the classic shape might well serve a chopping function, surely the symmetry, the edging, and the point are indicative of some function that we have failed to comprehend. I suggest that classic handaxes may be better appreciated in the context of the waterhole ambush.

Following the suggestion of Jeffreys (1965) that handaxes might have been thrown with spin, e.g., into a flock of birds, O'Brien (1981, 1984) undertook to test the aerodynamic properties of a fiberglass replica of one of the very large Acheulean handaxes from Olorgesaile. Even when thrown by experienced athletes, the handaxe did exactly what frisbees do when thrown by beginners: it turned from horizontal to vertical orientation in mid-flight and then landed on its edge. I recently repeated her experiments with five heavily-weathered handaxes from the Sahara, with much the same results. Unlike frisbees that roll along the ground for some distance, a spinning handaxe will soon impale its point in the ground and come to a sudden halt.

While of no significance when throwing into a flock of birds, I can see distinct advantages of such properties when lobbing a handaxe into a herd visiting a waterhole. An ordinary rock, lobbed into a tightly-packed herd, is most often going to hit an animal on its back or side. Generally the rock will then bounce off – but if it were to somehow impale itself, it would transfer all of its momentum to the animal, rather than only a fraction.

Consider throwing a handaxe into the herd. It lands on edge, doing considerable damage to some animal's skin (whether or not the edge actually penetrates it). The handaxe spins forward and the point tends to snag the skin (catching a roll of skin pushed forward by the impact, even if it fails to penetrate). Then the handaxe drops to the ground, its forward momentum exhausted since it was effectively transferred to the animal's back or side. This is far more likely to knock the animal off its feet than is the impact of an ordinary rock that bounces away.

Neurological knock-down

While the foregoing might suffice, I have discovered (Calvin, 1990) some additional neurological reasons why even a small handaxe (and most will fit comfortably into the hand) might bring down a large animal: the skin damage tends to countermand the reflex commands that ordinarily would prevent toppling after impact.

As background, note that all mammals have righting reflexes; a blow pushing the animal to its right will tend to generate extension commands to the right legs. If the response is quick enough, the animal's slow rightward movement from the momentum transfer will be countered and the animal will manage to keep its center of gravity to the left of its right legs. Once the center of gravity passes the supports, the animal will collapse (the prime virtue of higher velocities and bigger rocks is that the rightward drift becomes so rapid that instability is reached before the reflex loop can operate).

But there are other reflexes which may overwhelm righting reflexes, e.g., the withdrawal reflexes which tend to create movement away from a site of injury, as when one lifts the leg after stepping on something sharp. Withdrawal reflexes depend on the site of injury; for a four-legged animal, injury to its back (which would most often come from an overhanging tree branch or rock outcrop) causes both hindlegs to flex, dropping the back down to disengage it from the overhang. Pain is the effective stimulus for a withdrawal reflex, and the handaxe's sharpened edge would surely cause much more pain than the featureless rock. Still, there are many examples (see Calvin, 1983a) of people not noticing skin cuts if busy with something else.

It is manipulation of the edges of an incision which cause the most requests from patients for boosters in the local anesthetic. A handaxe that injured the skin, and then yanked hard on the damaged area (as would occur when the handaxe's point snagged the skin), seems likely to cause a massive flexion reflex of the hindlimbs. This alone might cause the animal to collapse, even if the sideward momentum transfer was minimal because it was a small handaxe.

So the point need not penetrate, nor need the edges incise, the skin for such a use to be effective. Nor need the handaxe be heavy, or have a lot of horizontal velocity, as pushing the animal sideways is no longer the only way to make it collapse. (Needless to say, the hunter need not understand neurological knock-down mechanisms in order to have noticed that even a juvenile's small handaxe was effective, so long as it had a sharp edge and point.) Still, this knock-down might merely result in a limping animal running away from the pursuing hunters – were it not for the herd stampeding and delaying the downed animal. The context of the herd is still essential (which also means that the technique would be ineffective against fellow hominids – unless they clustered together for protection), at least until the side-of-the-barn accuracy improves.

I suggest that the handaxe's characteristic features (flattened spin symmetry, all edges sharpened, and something of a point) make it an excellent projectile for waterhole predation, one that might have greatly increased the yield. That handaxes (Howell, 1961, Isaac, 1977, Kleindienst & Keller, 1976) are regularly found along watercourses and lake margins (as if lost in the mud like a golf ball; often they are standing on edge in situ) supports this explanation. While handaxes also make good fleshers (and broken ones may suffice as choppers), none of the other uses proposed for handaxes can explain all of the aforementioned characteristic features or the archaeological peculiarities. I would suggest that the Acheulean cleavers, and the handaxes without all-around edging, may be damaged versions of the classic design; if they were thought to be less effective in downing herd animals, they might well have been "retired" and converted to less demanding secondary uses such as chopping or fleshing. The handaxes found in lake beds are probably the ones that the hunter lost in the mud (Indian arrowheads in North America, as well as golf balls, similarly accumulate at waterholes). Thus the theoretical prediction: waterhole hand-axes ought to be more completely edged, by comparison to the ones found elsewhere.

Improvements beyond the handaxe lob

The waterhole lob ought to work well on the savannah, with its herds and permanent waterholes that must be visited during the dry season, predators or not. While flinging branches into such herds suggests a natural transition from chimpanzee-like behaviors, and while the handaxe might have been the "high technology" of this waterhole ambush technique, we must ask where the growth curve goes from there. There are some important inventions, such as persuading animals to injure themselves by stampeding over a cliff, but their growth curves are minor.

Longer and longer lobs would be useful as herds became wary, but the real growth curve lies with accuracy: being able to hit even a small herd. Or being able to prey on animals that do not cluster together so conveniently and compound the injury done to the downed animal. Or being able to prey on the smaller mammals and birds which are far more plentiful (and which, when their numbers are depleted by heavy predation, can regain population numbers more quickly because of their shorter gestation times and juvenile periods).

The only problem is that accuracy is expensive, whether trying to hit a smaller target from the same distance or trying to hit the same target from twice as far. This is because of one of the fundamental differences between the electrical operation of our excitable cells (both nerve and muscle) and those of modern electronics; because of the way that ion channels through the cell membrane function, the cells that control our movements cannot time events with any great accuracy. They are jittery, irreducibly noisy.

Throwing has a crucial timing step that is not present in hammering or clubbing or kicking: the projectile must be released at just the right instant (Fig. 10.2). Too early, and the launch angle will be too high, the projectile overshooting the target. Too late, and the projectile hits the ground in front of the target. We can talk of the "launch window" as that span of release times wherein the projectile will hit the target somewhere between its top and bottom. For a throw at a rabbit-sized target from a four meter distance (about the length of an automobile), this launch window is 11 ms wide. So at the end of a throw which started several hundred ms earlier, one must time the relaxation of the grip to stay within that 11 ms. The typical spinal motorneuron has an intrinsic timing jitter of at least that much (Calvin & Stevens, 1968; Calvin, 1983b, 1986) when the cell is generating action potentials at 200 ms intervals.

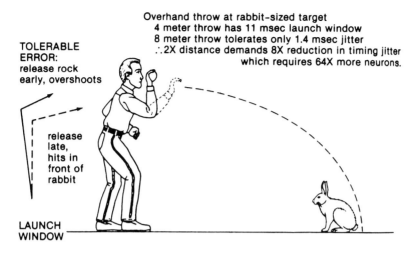

Fig. 10.2. The launch window for a simple overhand throw at a rabbit-sized target (10 cm high and 20 cm deep). Figure reproduced from Calvin (1989); calculations from Calvin (1983b).

But most people can, with practice, hit a rabbit-sized target at such a comfortable distance on perhaps half of the tries, so let us assume that the timing mechanisms are sufficient. Move the target out to twice the distance, eight meters, and practice enough to achieve hits on half the tries. The reason that this is so hard is that the launch window shrinks by a factor of eight to only 1.4 ms. The solid angle subtended by the target fell by a factor of four; furthermore, throwing twice as far with a reasonably flat trajectory means throwing twice as fast, so the time scale is halved. An electronic instrument would have no trouble in keeping its timing jitter to within a 1.4 ms window over several hundred milliseconds. But neurons would probably have to be redesigned from scratch.

The law of large numbers

Nature seldom redesigns, but it does one thing very well: it duplicates cells endlessly. And it seems that the way around cellular jitter is simply to assign a great many neurons to the same task, timing the launch, and then to average their recommendations. Need to halve the jitter? Just use four times as many cells as originally sufficed. To double the throwing distance and maintain hit rate, one needs an eight-fold reduction in jitter: that merely requires 64 times as many timing cells as originally sufficed. Triple the distance? Only 729 times as many cells.

Now becoming good enough to re-attain your hit rate at twice the distance is surely not a matter of growing a lot of new cells for the movement control circuits. It seems more likely that they are borrowed as one gets set to throw, the experts (probably the premotor cortex and the cerebellum) recruiting some temporary help from other brain regions, much as the choir in the Hallelujah Chorus temporarily borrows the audience.

Might, however, our four-fold increase in brain size have something to do with this "law of large numbers", our development-controlling genes instigating the growth of more nerve cells as a result of throwing accuracy coming under natural selection for feeding families during hominid evolution? Alas, a four-fold increase in neuron numbers will only "buy" a 26 per cent increment in throwing distance – and the first 10 per cent increase in neuron numbers (roughly equivalent to what the natural variation in the human population might presently be) will only buy a 2 per cent longer throw. And so I doubt that it was a simple matter of bigger-brains-throw-better; I suspect that recruiting the helpers is easier with a more juvenilized brain (Calvin, 1990), and that the bigger brain is simply a side-effect of neoteny. In other words, selection was not for size per se.

Concluding remarks

Many things have contributed to hominid evolution, and humans would not be what they are today without effects from many of them. The inventions that would seem to place the greatest demands on brain reorganization, however, are accurate throwing and structured language.

Predatory throwing seems possible from such transitional stages as envisaged in my waterhole lobbing proposal; accurate throwing is not a single invention but a long climb, fortunately with a growth curve that seems unlimited. Eating meat regularly would have greatly expanded the hominid niche, allowing hominids to expand well outside the savannah, even into the temperate zone (where the major constraint is that the gathering is so poor in the winter months that either hoarding or meat-eating seems essential).

Throwing seems to be a fast track for hominid evolution; whether it is the fastest track to a versatile sequencer and its Darwin Machine secondary uses can be better judged when a similar analysis is available for language, intelligence, toolmaking, and other candidates. What are their transitions from apelike behaviors, their growth curves, their niche-expanding properties, and their secondary uses?

Acknowledgements

I thank Gareth Anderson for assistance in the handaxe experiments; Derek Bickerton, Kathleen Gibson, and Katherine Graubard for comments on the manuscript; revision of the manuscript was greatly aided by conversations with Christopher Boehm, Christophe Boesch, David Brin, Iain Davidson, Daniel C. Dennett, Dean Falk, Susan Goldin-Meadow, Doug Hofstadter, Nick Humphrey, Ray Jackendoff, Adam Kendon, Marcel Kinsbourne, Philip Lieberman, Andy Lock, Peter MacNeilage, William McGrew, George Ojemann, Martin Pickford, Duane Rumbaugh, Vince Sarich, Sue Savage-Rumbaugh, Mark Sullivan, Nick Toth, Peter Warshall, Pete Wheeler, and Thomas Wynn.

References

Bickerton, D. (1990). *Language and Species*. Chicago: University of Chicago Press.
Boesch, C. & Boesch, H. (1981). Sex differences in the use of natural hammers by wild chimpanzees: a preliminary report. *Journal of Human Evolution*, 10, 585–93.
Calvin, W. H. (1983a). *The Throwing Madonna: Essays on the Brain*. New York: McGraw-Hill.
Calvin, W. H. (1983b). A stone's throw and its launch window: timing precision and its implications for language and hominid brains. *Journal of Theoretical Biology*, 104, 121–35.
Calvin, W. H. (1986). *The River that Flows Uphill: A Journey from the Big Bang to the Big Brain*. New York: Macmillan.
Calvin, W. H. (1987). The brain as a Darwin Machine. *Nature*, 330, 33–4.
Calvin, W. H. (1989). *The Cerebral Symphony: Seashore Reflections on the Structure of Consciousness*. New York: Bantam Books.
Calvin, W. H. (1990). *The Ascent of Mind: Ice Age Climates and the Evolution of Intelligence*. New York: Bantam Books.
Calvin, W. H. & Ojemann, G. A. (1980). *Inside the Brain: Mapping the Cortex, Exploring the Neuron*. New York: New American Library.
Calvin, W. H. & Stevens, C. F. (1968). Synaptic noise and other sources of randomness in motoneuron interspike intervals. *Journal of Neurophysiology*, 31, 574–87.
Cordo, P. (1987). Mechanisms controlling accurate changes in elbow torque in humans. *Journal of Neuroscience*, 7, 432–42.
Darwin, C. (1872). *The Expression of the Emotions in Man and Animals*. London: J. Murray.
Dawkins, R. (1986). *The Blind Watchmaker*. London: Longman.
Falk, D. (1987). Brain lateralization and its evolution in hominids. *Yearbook of Physical Anthropology*, 30, 107–25.
Howell, F. C. (1961). Isimila: A Paleolithic site in Africa. *Scientific American*, 205, 118–29.
Isaac, G. L. (1977). *Olorgesailie*. Chicago: University of Chicago Press.
Jeffreys, M. D. W. (1965). The hand bolt. *Man*, 65, 154.
Kimura, D. (1979). Neuromotor mechanisms in the evolution of human communication. In: *Neurobiology of Social Communication in Primates*, ed. H. D. Steklis & M. J. Raleigh, pp. 197–219. New York: Academic Press.
Kleindienst, M. R. & Keller, C. M. (1976). Towards a functional analysis of handaxes and cleavers: The evidence from East Africa. *Man*, 11, 176–87.
O'Brien, E. M. (1981). The projectile capabilities of an Acheulian handaxe from Olorgesailie. *Current Anthropology*, 22, 76–9.
O'Brien, E. M. (1984). What was the Acheulean hand ax? *Natural History*, 93, 20–3.

Ojemann, G. A. (1983). Brain organization for language from the perspective of electrical stimulation mapping. *Behavioral and Brain Sciences*, 6, 189–230.

Ojemann, G. A., & Creutzfeldt, O. D. (1987). Language in humans and animals: contribution of brain stimulation and recording. In *Handbook of Physiology. Section 1: The Nervous System, Volume 5 part 2, The Higher Functions of the Brain*, ed. V. B. Mountcastle, F. Plum, & S. R. Geiger. Washington DC: American Physiological Society.

Plooij, F. X. (1978). Tool-use during chimpanzee's bushpig hunt. *Carnivore*, 1, 103–6.

Poizner, H., Klima, E. S. & Bellugi, U. (1987). *What the Hands Reveal About the Brain*. Cambridge: MIT Press.

Stuss, D. T. & Benson, D. F. (1986). *The Frontal Lobes*. New York: Raven Press.

11

Tool use, language and social behavior in relationship to information processing capacities

KATHLEEN R. GIBSON

Anthropologists once thought that humans were the only animals who could symbolize and the only ones who could make tools. We now know that wild chimpanzees make simple tools and that, in captivity, all species of great apes can learn rudimentary symbolic communication. Still, the linguistic and technological achievements of apes fall far short of human achievements.

This chapter suggests that at least three factors underlie differences in achievement levels between humans and apes. First, humans possess a greater degree of brain-size-mediated information processing capacity and apply this capacity to their tool use, language and social behavior. Second, human tool use, language and social behavior are mutually interdependent. Third, tool use and language are so strongly canalized in the human species that they emerge in most mentally normal children by one and a half to two years of age. In contrast, apes exhibit minimal, if any, genetic canalization in these behavioral domains.

The recognition of these ape–human distinctions has important implications for the evolution of human intellectual skills. For instance, since human social intelligence, tool use and language all depend on quantitative increases in brain size and in its associated information processing capacities, none could have suddenly emerged full-blown like Minerva from the head of Zeus. Rather, like brain size, each of these intellectual capacities must have evolved gradually. Further, since these capacities are interdependent, none could have reached its modern human level of complexity in isolation. Rather, the attainment of fully human capacities in each of these domains depended on the ability to integrate achievements in all three.

Differences between apes and humans in language, object manipulation and intelligence

According to modern investigators, human intelligence is inherently constructional in nature (Case, 1985; Dennett, 1988; Langer, 1980, 1986; Piaget, 1954). In other words, humans break perceptions, motor actions and concepts into small component parts and then combine and recombine these parts into higher order constructs. These new constructs can, in turn, function as subunits of still higher mental syntheses. Mental constructional capacities increase during human growth and maturation with the result that tool use, language and social behavior become increasingly complex and contain increasing numbers of embedded concepts.

As a result of the ability to embed mental constructs within still higher order constructs, human intelligence has a layered or hierarchical pattern. This is of interest because others have suggested that human tool use and language exhibit greater degrees of construction and hierarchization than those of the apes (Gibson, 1983, 1988, 1990, in press; Greenfield, 1991; Reynolds, 1983).

In Case's view, the increasing power of human mental constructional skills during maturation reflects increased information processing capacities. That is, the greater the amount of information that can be processed and held in mind simultaneously, the more complex the potential mental constructions (Case, 1985). This chapter suggests that differences between ape and human behaviors may also reflect quantitative differences in information processing capacity.

Language in relationship to information processing capacity

All modern human languages are constructional in nature in that they all consist of numerous words or symbolic gestures which are combined and recombined in varied patterns to form more complex syntactic and semantic utterances. Thus, words or gestures may be combined into phrases, phrases into sentences and sentences into oral "paragraphs", stories or other larger units. These constructions are layered or hierarchical in nature in that components of sentences or "stories" form subunits of larger wholes.

The longer the linguistic communication, the greater the information processing capacity required as more words must be held in mind to produce and interpret the communication. Longer communications also demand greater syntactic capacity; they may be more "information-rich", and they may demand more internalized knowledge on the part of both speaker and listener.

Thus, the linguistic utterances of very young children are limited in both length and information content (Piaget, 1955). Children whose "sentences" are limited to one to two words may request food, toys or other objects, attention and help. They may also spontaneously note or name objects in their environment and ask or answer questions of who, what or where (Brown, 1980). The information content of these communications, however, is "sparse" and limited to actions experienced by both listener and speaker and to objects known to both. Usually, only one object or action is requested at a time.

As linguistic lexicon and sentence length increase, so too does information content (Piaget, 1955). By four to five years, children may request explanations about causality, with the proverbial, "why" questions. They may also describe their own actions verbally, give others brief instructions in sentence format, or describe objects with a series of words. Even at this stage, however, children have difficulty making themselves understood unless the actions, objects and events are known to both speaker and hearer. In fact, they must often resort to pointing and gesture to clarify their linguistic meanings (Karmiloff-Smith, 1979; Piaget, 1955).

Not until the elementary school years of seven to nine can children fully describe events to listeners unfamiliar with them by incorporating large amounts of information in appropriately structured series of sentences. It is also at this time that children become capable of debating and absorbing factual knowledge transmitted by formal education or other non-experiential means. Such abilities depend on sophisticated grammatical understanding and clear comprehension of the plurifunctional meanings of certain "marker" words, such as "the" (Piaget, 1955; Karmiloff-Smith, 1979). In addition, these abilities demand sufficient information processing capacities to be able to master large amounts of factual knowledge.

Apes can master a number of symbolic gestures or other visual symbols and they can combine and recombine these into short syntactic utterances (Gardner, *et al.*, 1989; Miles, 1983; Greenfield and Savage-Rumbaugh, 1990). From the standpoint of both syntax and content, however, ape symbolic communications display information processing capacities and mental constructional skills similar to those displayed by very small children, of three years of age or less. Thus, ape syntactic constructions are usually limited to short "sentences" composed of two or three separate gestures. Further, apes do not appear to construct meaningful sentence sequences and the content of ape communications is usually "information-sparse". Thus, apes usually focus on requests or note objects in the immediate environment. To my knowledge, no verifiable reports exist of apes giving appropriately sequenced descriptions of events or discussing, in a comprehensible fashion, objects and events not part of the common experience of both ape "speaker" and human "listener".

Information processing capacity in relationship to tool use, construction, classificatory skills and mathematics

Humans have a strong propensity to manipulate objects with respect to other objects (object–object manipulation, as defined by Parker and Gibson, 1977). They do so in at least three contexts – the formation of temporary object groupings, tool use, and object construction. Like language, each of these behaviors occurs in information-sparse and information-rich contexts.

Children's constructions of temporary object groupings serve to promote classificatory and logico-mathematical skills (Langer, 1980; 1986; Inhelder and Piaget, 1958). Thus, children may compose sets of like objects (as, for instance, placing blue objects in one group, red in another). They may also change the size of sets by adding and subtracting objects. These actions vary in constructional complexity and in information processing capacity required from single groupings of two objects, to multiple groupings of many objects. Multiple groupings can result in classificatory sorting of objects. Alternatively, object compositions can be seriated into groups of progressively larger numbers of objects. Older children construct larger groups and more groups at once, than infants. Older children also "play" at counting the numbers of objects in groups, at rearranging object groupings and then recounting or at adding and subtracting objects and then recounting.

Tool use and object construction focus on causal rather than mathematical relationships between objects (Langer, 1986). Just as humans possess a number of words which they combine into higher order linguistic constructs, they also possess a number of object–object manipulation schemes which they combine and recombine in complex hierarchical patterns. Thus, most humans possess a basic tool-using and constructional repertoire which consists of hammering, using leverage (as in digging), rubbing, scraping, probing, stacking, containing, cutting, tying, adhering with glue or other sticky substances, mixing, weaving, throwing and heating. These schemes can be used individually in information-sparse contexts such as hammering a nut or as parts of complex information-rich object constructions as in building a skyscraper or a car. These latter constructions are hierarchical in the sense that subcomponents of the building or machine will be constructed first and joined together with others to make one larger whole.

Small children just learning to use tools tend to use one of these schemes at a time, i.e., hammering. As children grow and become versed in the technology of a particular culture or trade, they begin to combine schemes into lengthier and more information-rich tool-using sequences. Thus, small children in almost any culture may bang objects against each other, dig in the dirt, place objects in containers, or throw objects. Adults, however, combine varied schemes into purposeful tool-using sequences. For instance, a hunter may first construct a spear from several subcomponents (stick, stone, adhesive material and poison), then throw it at prey, then cut the prey into pieces, then cook it. A gatherer may first fashion a kaross, a pot and a stone knife, use the knife to sharpen a digging stick, then use the digging stick to dig tubers from the ground, the kaross to transport them and the pot to cook them.

These latter activities require sufficient information processing capacities to construct, in advance, a logical and effective sequence of tool-using procedures. Other typical hunter–gatherer activities, such as making clothing, constructing tepees or thatched-roofed huts and making baskets may place even greater demands on information processing capacities by requiring a larger knowledge base and the ability to quantify materials and measurements.

Apes exhibit some spontaneous object–object manipulations. As in language, however, ape manipulative behaviors exhibit less overall information processing and mental constructional capacity than human behaviors. Thus, although apes may group branches as in building a nest, they rarely, if ever, spontaneously group large numbers of objects into multiple sets of logical classes or in accord with mathematical principles (Gibson, personal observations). Similarly, apes, like humans, possess a number of tool-using and constructional schemes. Many chimpanzees can hammer, probe, rub, throw and stack (Goodall, 1986). Their tool-using and constructional lexicon, however, is limited in comparison to the human. Apes do not use adhesive materials, tie knots, weave or use heating procedures (Parker and Gibson, 1979; Reynolds, 1983); hence, they do not create permanent junctions between two or more separate objects. In general, the tool-using schemes lacking in apes are those which develop relatively late in human children and which demand keeping several spatial relationships or objects in mind simultaneously.

In addition, apes rarely exhibit advanced planning of a series of tool-using schemes. Thus, chimpanzees may probe for termites, stack branches to make nests, hammer nuts or throw branches, but only one report exists of a single wild chimpanzee using several tools in succession to meet a single end (see McGrew, this volume). Even in this instance all of the tools were probes, as opposed to the common human tendency of using several different types of tools in succession. Nor do wild apes use materials such as bone or skin which must first be prepared by a series of sequential tool-using procedures.

Information processing capacity in relationship to social intelligence

Like tool use and language, social intelligence varies in accordance with the amount of information processing capacity demanded. Thus, studies of the development of empathy, cooperation and morality indicate that as children age social behaviors and social judgements incorporate increasing amounts of information (Case, 1985; Kohlberg, 1984). Older children even take mathematical concepts into account when making social judgements as, for instance, when trying to determine which child is happiest with his birthday presents (Case, 1985).

Finally, in order to make effective judgements about others' behavior humans must have a theory of mind or of "intentionality". That is, they must have theories about what others are thinking. Dennett (1988) has suggested that human intentional systems exhibit varying levels of embeddedness. A first order intentional system might consist of "x believes that p". A second order system would be "x wants y to believe that x is . . .". A still higher order system is illustrated by Dennett's sentence ". . . I suspect that you wonder whether I realize how hard it is for you to be sure that you understand whether I mean to be saying that you can recognize that I can believe you to want me to explain that most of us can keep track of only about five or six orders, under the best of circumstances" (Dennett, 1988, pp. 185–186).

Although comparative studies of social intelligence remain in their infancy, Whiten and Byrne (1988) argue that chimpanzees may comprehend three orders of intentionality, while Premack (1988) has concluded that chimpanzees have a theory of mind approximately equivalent to that of a three- or four-year-old human child. Thus, humans also appear to apply greater information processing capacity than apes in their social analyses and judgements.

Summary

Apes possess social intelligence as well as tool-using and linguistic capabilities. No sharp qualitative dichotomies separate humans from apes in any of these domains. Humans, however, bring greater informational processing capacity to bear in each domain and, hence, produce more complex and information-rich linguistic, social and technological behaviors.

The interdependence of human tool use, language, social behavior and mathematical concepts

In humans, tool use, social behavior, language and mathematical thought are mutually interdependent and reinforcing. These interdependent relationships are a fundamental aspect of the human "cognitive niche".

Human technology, for instance, is inextricably linked with human social structure. The human life-style is not only tool-based, but many tools and constructions are used and reused over long periods of time. This long-term use of the same tools and dwellings is a distinguishing feature of the human adaptation. It is also an essential one because if humans had to manufacture new dwellings, pots, baskets, clothing, etc., on a daily basis, they would lack time for foraging activities. The reuse of dwellings and tools, however, entails a kind of sociality quite different from that of the apes, in which the objects become foci of long-term social relationships indicated by such notions as "property" and "household".

In addition, human tool use is cooperative and social in nature. Not only do humans cooperate in activities such as building shelters and hunting with tools, all cultures also possess social networks characterized by divisions of labor and redistributions of the products of tool-using endeavors. Even when humans appear to be engaging in solitary hunting, gathering or tool-making tasks, they are usually acting as part of a complex social network (Ingold, 1988; Reynolds, this volume). Thus, the solitary gatherer or hunter takes food home to be shared with family or other members of the band. The more a society depends on complex machinery and architectural constructions, the more it depends on social divisions of labor and redistributive processes. It would, for instance, be impossible for one person to manufacture his own three or four bedroom house, computer, television, automobile, clothing, etc. and still find time to grow or hunt all of his own food. Thus, modern technology both demands social divisions of labor and is fostered by it. For these reasons, arguments about which is more important, social intelligence or technological intelligence (e.g., Byrne and Whiten, 1988), may be fruitless. In humans, the two depend upon each other.

Social divisions of labor place major demands on language, social intelligence and quantitative thought. Linguistically articulated and transmitted systems of social conventions are critical for the continuing functionality of hunter–gatherer systems, while larger societies promulgate complex written laws (Fortes, 1983; Ingold, in press; Parker, 1985). Quantitative skills lie at the basis of food-sharing and redistributive processes. Thus, in all societies, those who "gather" and prepare food for groups of people must estimate food quantities if people are not to go hungry and/or to waste food. Large societies also possess formal systems of trade and barter, often using "money" or other systems of exchange that depend on formal mathematical skills.

Language and technology are also interdependent. All societies possess oral traditions pertaining to subsistence methods which include, if not tool-making techniques, at least discussions of appropriate times and places to use tools (appropriate time to hunt, harvest, etc.). Modern societies use tools such as pens and manufactured materials such as paper to record and promulgate laws, trade, history, scientific and technological traditions, myth and literature.

In non-human primate groups, the interdependence of social, communicative and tool-using endeavors is less extensive than among humans. Thus, although groups of monkeys and apes may mob predators and throw branches in their direction (Goodall, 1986), ape tool use is primarily an individual effort. Thus, apes who use tools in cracking nuts or in probing for termites generally do so alone and rarely share the products of their tool-using efforts. Similarly, each builds its own nest. This lack of cooperative tool use does not reflect a lack of sociality. Quite the contrary, most apes are very social creatures. Chimpanzees also cooperate in hunting small prey, and they may share meat (Boesch and Boesch, 1989; Teleki, 1973).

Rather, the minimally social nature of ape tool use may reflect two factors. First, apes lack the necessary information processing capacity to engage in constructional activities that place a premium on divisions of labor, such as building large dwellings, or in those activities that make food sharing practicable, such as making baskets and karosses. Second, the same lack of information processing capacity that renders ape tool use and speech less complex than our own may make it difficult for them to process large amounts of technical and social information at once.

Whatever the cause, however, the minimal integration of technology with communication and social behavior in wild apes must have retarding effects on their "cultural" behavior. Although apes do share environmental information, their methods do not permit the "information-rich" communications of modern humans. This is probably one reason why wild apes appear to fall short of their seeming technological and intellectual potential. Their cultural information processing capacities lag further behind our own, than do their individual information processing capacities.

Brain size in relationship to information processing capacity

At 1200 to 1300 grams, the average human brain is approximately three to four times the size of the average ape brain. When allometric relationships between brain size and body size are considered, the human brain emerges as relatively very large indeed (Jerison, 1973). Some areas of the human brain such as the neocortex, the cerebellum and certain thalamic and limbic nuclei have increased in size more than others. Proportionately, the greatest increase occurred in the neocortical association regions (Passingham, 1975).

Since deficiencies in language, object manipulation and social behavior may follow lesions in cortical association regions, these regions have long been considered to mediate higher intellectual functions. For instance, patients with lesions in the left parietal association cortex may experience difficulty in naming objects, in understanding and producing grammatical constructs, in dressing themselves, in using objects and manual gestures, in map reading and in simple mathematical calculations (Geschwind, 1965). Patients with lesions in the left temporal association areas may have problems recounting events or stories in proper sequence; those with lesions in the right temporal cortex, with recognizing faces (Milner, 1967). Patients with frontal lobe lesions may exhibit some of these same problems, plus difficulties in planning sequential actions, lack of self discipline and inabilities to inhibit socially undesirable behaviors (Luria, 1966).

According to Luria (1966), these neurological deficits indicate that the association areas have synthetic functions. In other words, they take separate sensory perceptions or motor acts and integrate them into higher order simultaneous or sequential constructs. For instance, in order to comprehend a painting a patient must perceive individual components of that painting and add them all together to understand the meaning of the whole. Patients with lesions in the association areas may comprehend individual aspects of a painting but not the entire scene.

Luria's view that the cortical association regions have synthetic functions agrees with psychological theories that the intellectual functions mediated by these regions are constructional in nature and depend on advanced information processing capacity (Case, 1985; Piaget, 1954). Theoretical considerations suggest that the expanded size of the cortical association areas plays a fundamental role in the increased information processing capacities which render complex mental constructions possible (Gibson, 1988; 1990; in press).

Modern theories indicate that neural function occurs by means of parallel distributive processing mechanisms. This means at least three things. 1. Action, perception and cognition all reflect functioning neural networks as opposed to individual neurons functioning in isolation (McClelland and Rumelhart, 1986; Rumelhart and McClelland, 1986). 2. Neuronal percepts, actions and concepts are all broken into small component parts distributed among a number of neuronal networks. 3. Neural circuits function in parallel (simultaneously) rather than in sequence in order to produce complex perceptual, motor and cognitive behaviors.

Several considerations suggest that brain size may lead to expansions in parallel processing abilities. For one, increased cortical size correlates with increased numbers of cortical processing units or "modules" with the potential for parallel function. Further, we now know that the neocortical association areas contain multiple copies of visual, somatosensory, tactile and motor processing units (Kaas, 1987). Thus, the larger the association regions, the greater the duplication and multiplication of sensory and motor units, and the greater the numbers of sensorimotor units potentially firing simultaneously.

This finding is of interest given certain implications of neuronal network models. According to a number of such models, a single visual (or other sensorimotor) network may perceive many different objects depending on the pattern of neural activation. Thus, a single network might perceive a rose, a horse, a chimpanzee or an onion depending on the pattern of neuronal firing. However, each such network could perceive only one object at a time. In other words, a given neuronal network might perceive a chimpanzee or an onion, but it could not perceive a chimpanzee and an onion simultaneously as two discrete objects. Thus, a single network could not conceptualize the chimpanzee as acting upon the onion (Hinton, *et al.*, 1986). A perception of that nature would require at least two visual processing units firing in parallel.

This theory has important implications for human intelligence. If a single network could not perceive two discrete objects simultaneously, it could not conceptualize two socially interacting chimpanzees or groups of discrete but interacting objects such as a hammerstone and a nut, a knife, fork and piece of meat or a series of several discrete words. Thus, even the simplest social, tool-using and linguistic tasks appear to require duplication and multiplication of sensory networks. Similar considerations apply to the motor system. The greater the number of processing units, the greater the number of motor acts potentially processed at once.

The implications of parallel processing theories apply to sequential actions as well as to simultaneous perceptions. For instance, studies indicate that skilled actions, such as typing, demand the simultaneous activation of several motor impulses in advance, rather than the serial activation of one motor impulse after the other is completed (Rumelhart and McClelland, 1986). Thus, the numbers of acts which can be joined together as a part of one goal-oriented tool-using or linguistic sequence also depend on the numbers of sensory and motor processing units possessed and capable of firing in parallel.

As the neocortical association areas provide multiple sensory and motor processing units, the size of the neocortical association areas would be expected to correlate with the numbers of possible simultaneous and sequential perceptions of objects, words and motor actions. Thus, despite tendencies by some investigators to belittle the potential significance of brain size, the increased size of the association regions has almost certainly played a major role in the evolution of increased human information processing capacity.

Although this chapter focuses on intellectual tasks mediated by association regions, the implications of parallel processing mechanisms apply to other systems as well. Thus, for instance, Broca's speech area also enlarged in human evolution. One would predict that such enlargement would lead to greater oral discriminatory capacity and greater ability to plan sequences of oral actions.

The developmental canalization of language and object manipulation skills

All human populations use tools and language. Similarly, all physically and mentally normal human children learn at least one language, and all learn to use simple tools. Congenitally deaf humans deprived of language training invent their own sign language and syntactic rules (Goldin-Meadow and Mylander, 1991). Despite, however, their obvious tool-using and linguistic potentials, only some populations of apes use tools. None use symbolic languages. Apes raised in human homes do not invent their own languages. These considerations suggest that in addition to differing from apes in overall information processing capacities, humans differ from apes in the degree of canalization of their existing intellectual capacity in linguistic and technologi-cal directions. Judging by developmental studies, whatever helps canalize human behavior in these directions occurs very early. Most human infants speak their first words by the end of the first year of life or the beginning of the second. During the second year of life, they also develop a symbolic capacity, begin to combine words into short sentences and phrases and begin to use simple hammer and rake-like tools. From this perspective, it is interesting to compare the behavior of infant apes and humans. I was privileged to observe, for a period of several years, three chimpanzees who were being reared in a human home. The two males were eight and ten months of age respectively at the time my observations began; the female was four days old. The owners made every effort to raise these chimpanzees like humans – to play with them, vocalize in their presence, dress them, toilet train them, train them to use spoons, and provide them with human toys. The chimpanzees also frequently played with human children. I visited the home once a week, observed the animals' spontaneous behaviors and encouraged them to play with toys.

Despite these efforts, the chimpanzees' behavior differed radically from that described for human infants. For instance, from early infancy, humans engage in communicative behaviors which prepare them for social "conversation" and for instrumental use of language. Thus, beginning at six weeks of age, human infants spontaneously smile at other human faces. From an early age, they will also coo, wait for a maternal response and then vocalize again. From approximately six months of age, they will babble, both alone and in vocal exchanges with the mother.

The chimpanzees showed no such behaviors. Although the four-day-old chimpanzee would smile, if tickled under the chin and "fussed over", neither she nor any of the other chimpanzees ever developed a true social smile in response to a human face. None of the chimpanzees babbled, and none of them engaged in vocal turn-taking behaviors with me or other human care-givers. Nor have babbling and social smiling been reported for other apes, including home-reared apes.

The apes also differed from humans in their manipulative behavior. By four to eight months of age, human infants begin to engage in repetitive object–object manipulative behaviors which serve as precursors to both the causal use of objects as tools or constructional materials and the logical use of objects as steps to acquiring mathematical causality. Some of these repetitive actions fall in the realm of what Piaget calls secondary circular reactions or the repetition of actions in order to recreate interesting "spectacles" (Piaget, 1952). Thus, infants repeatedly shake rattles and kick mobiles. These secondary circular reactions apparently reflect and encourage the infant's developing understanding of physical causality.

By the second year of life, the infant's secondary circular reactions have matured into tertiary circular reactions or object–object and object–force–field manipulations which serve as "experiments in order to see" the effects of physical causality – such as speed, force, and height (Piaget, 1952). Tertiary circular reactions lay the cognitive foundations for tool use. During the second year of life, infants also manipulate spatial relationships between objects. Thus, towards the end of the second year, they will stack blocks or cups, place rings on sticks and solve simple puzzles.

Other early repetitive actions upon objects reflect emerging logico-mathematical understanding and classification skills. Thus, by six months of age, human infants repeat actions in a seriated fashion such as banging once, pausing, banging twice, pausing and then banging three times. By six months of age, they will also spontaneously position similar objects in small groups of two or three (Langer, 1980). During the second year, infants create larger groupings of objects and may create two or more groupings simultaneously, as for instance, in putting all of the blue objects in one group and all of the red objects in a second group. In addition, they may create groupings of different sizes and do so in serial order (Langer, 1986).

The chimpanzees behaved quite differently. Like human infants, they were very interested in objects. They rarely, however, exhibited secondary or tertiary circular reactions. In addition, they were mostly interested in single objects rather than in object–object relationships. I never saw them spontaneously place objects in classificatory groups. Even up to three years of age they had little interest in stacking blocks, banging objects together or engaging in other object–object relationships, such as placing rings on sticks or solving simple puzzles. The only object–object relationship task which seemed to intrigue them was hammering objects into pegboards and, then, not until the fourth year of life and only if I first placed the object in the proper opening.

While three chimpanzees are too few animals from which to generalize about the behavior of all apes, their behavior is not atypical of behaviors reported for other apes. Thus, other investigators have reported that young chimpanzees, gorillas and orangutans exhibit few, if any, secondary and tertiary circular reactions and show little interest in stacking blocks and other play with object–object relationships. Apes do develop interest in some of these activities eventually, but mainly after four to five years of age (Miles, personal communication; Russon, 1990; Vauclair and Bard, 1983).

Thus, although apes possess tool-using and symbolic capacities, the spontaneous behaviors of infant apes and humans differ dramatically in manipulative and communicative domains. These differences in infantile behaviors which help to canalize human cognitive development may be as important a component of the human cognitive niche as increases in adult information processing capacity.

Unfortunately, we lack information on the neurological basis of canalization behaviors. Some, such as increased babbling and increased manipulation of objects with respect to other objects, possibly reflect the enlargement of Broca's area and of the association areas, respectively. Other very early developing behaviors, such as the social smile and vocal turn-taking behaviors, may reflect structural changes or enlargements in early maturing subcortical areas. This question is open for further research.

Implications for the interpretation of the fossil record

These considerations have several important implications for our interpretations of the fossil record.

1. Increased size of the brain or of its component parts would have resulted in increased information processing capacity in those behavioral domains mediated by the enlarged regions. Just as brain size increased slowly, those aspects of language, tool use and social intelligence which depend on increased information processing capacity would also have evolved slowly.

2. As tool use is one aspect of complex social and technological networks in humans, it is misleading to judge the intelligence of fossil hominids by the form of their tools alone. Rather, the intellectual level of our ancestors is more likely

to have been reflected by the ways tools were used and by the complexity of the social-technological networks of which they were a part.

3.   As tools, language and social behavior are interdependent entities in hominids, no early hominid is likely to have reached modern levels of intelligence in any one of these domains without having reached it in all of them.

4.   As modern tool use, language and social intelligence depend not only upon overall adult information processing capacity, but also on the developmental canalization of tool use and language in human infants, linguistic and technological complexity depended on changes in infantile as well as in adult behavior.

5.   As linguistic and technological complexity depend on quantitative parameters of information processing capacity, it should be possible to determine logical sequences of cultural evolution based on the degree of information processing capacity necessary to support given sociotechnological systems.

An examination of the fossil record indicates that australopithecines who lived about 2 to 5 million years ago had slightly larger brains in comparison to body size than do the apes (Tobias, 1971). By 1.8 million years ago, the earliest members of the genus *Homo* appeared. In contrast to an average ape brain size of less than 400 cc., early *Homo* had a brain size in excess of 600 cc. In addition, early *Homo* exhibited expansions of Broca's speech area and of the parietal association areas. This suggests increased vocal capacities and increased information processing capacities. From 2 million years ago, brain size gradually increased until it reached (or exceeded) its modern levels in Neanderthals about 100000 years ago. A further change in brain form occurred, however, with the advent of modern *Homo sapiens sapiens* about 35000 years ago. At that time, the frontal and parietal lobes became more rounded and the occipital lobes less protuberant (Kochetkova, 1978).

Although no tools have been found with the earliest australopithecines, it is premature to conclude on that basis that they did not use tools or that their increased brain size brought no increase in intelligence or in tool-using capacity. Brains are metabolically very expensive organs. It is unlikely that brain size would have expanded without some functional reason. The ecology of these early hominids was quite different from that of the apes. A modern human gatherer with little more than a digging stick, a hammerstone and an ostrich egg-shell container would leave few or no remains that would be identified as tools and would seem no more intelligent than an ape, if intelligence were to be judged on the form of the tools alone. It is the complex sequence of preparing the digging sticks, gathering the food, cutting the food with tools, cooking it and dividing it equitably that renders the human gatherer's activities more intelligent and more demanding of information processing capacities than ape tool use. Thus, we should be asking what the minimal brain size increases in Australopithecines might have meant in terms of increased information processing capacity and, hence, in changes in the social and technological networks.

By the time of early *Homo*, hominids were producing and using sharp-edged stone tools. Since these tools look very similar to stone flakes left by chimpanzee nut-crackers, some conclude that they required no more intelligence than that possessed by chimpanzees (McGrew, 1987; Wynn and McGrew, 1989) and that early hominid brain advancements were not associated with increased tool making capacities (Wynn, 1988). Ongoing attempts to teach the bonobo, Kanzi, to make and use stone tools suggest, however, that apes may have difficulty producing acutely angled flakes. In addition, Oldowan tools were made by habitually right-handed individuals and chimpanzees lack this handedness (Toth and Schick, this volume). These factors suggest that early hominid activity may have demanded slightly more information processing capacity than that exhibited by apes.

Above and beyond this, however, intelligent tool use cannot be deciphered from the form of the tools alone. We have clear ecological evidence for major changes in life-style in early *Homo*. We know that early hominids used stone tools to cut soft tissues, probably tendons, from bone (Shipman, 1986). We also know that early hominids had regular stone caches where they presumably kept tools and reused them (Potts, 1984). These considerations point to differences between early *Homo* and the apes in methods of exploiting the environment and in social organization. The early hominid tool-using adaptations may well have required increased socialization of tool use and increased vocal or gestural communication capacities. Again, we should be asking in what way these hominids were making use of their increased information processing capacities and vocal skills.

Finally, it has been argued on the basis of tool form that *Homo erectus* possessed modern levels of intelligence (Wynn, 1979). As *Homo erectus* did not possess modern cranial capacities or modern brain form, it is highly unlikely that he possessed modern information processing capacities. Certainly the material remains of fossil *Homo sapiens* show signs of greater information processing capacity than those of *Homo erectus*. Thus, it is not until the appearance of essentially modern humans that we find evidence of tools constructed from several components, of the use of materials such as bone which demand complex processing methods, of the group exploitation of seasonally migratory prey, of art and of divisions of technical labor. While some of these behavioral advances may reflect advances in cultural, rather than in individual, information processing capacities, others such as long-term planning and flexibility of behavior are exactly what one would expect if frontal lobe function actually improved (Gibson, 1985).

Since no living animal possesses the technological adaptations and neurological organization of any of the fossil hominids which served as transition forms from ape-like ancestors to modern humans, it is not possible to model early hominid behaviors on any living animal (Tooby and DeVore, 1987). Rather, we must develop conceptual models of behavior based on our understandings of animal and human ecology, cognition and brain structure.

The information processing model presented here suggests that human developmental data can provide major clues to the evolution of cognition and social structure because information processing capacity increases in both development and in evolution. In particular, developmental data can provide clues to the complexity of technology, tool use, social behavior and mathematical thought possible with levels of information processing capacity intermediate between ape and human. No other source of data seems capable of providing such clues.

Parker and I have suggested, on the basis of Piagetian analyses, that early hominids could well have had a linguistic, tool-using and food-sharing system not unlike that of the modern mother–infant unit (Parker and Gibson, 1979; Gibson, 1983, 1988). Tool-using mothers, for example, may have extracted and processed foods using tools and then shared the food with offspring incapable of tool use themselves. Such food sharing may have selected for communication capacities similar to those of children just learning to talk – i.e., the capacity to request food or to note food or other significant objects in the immediate environment. Such an adaptation would have required only minimal advances in communicatory skills and social intelligence, and, hence, only minimal information processing capacity beyond that possessed by modern apes. Thus, it could have appeared very early in human evolution.

As information processing capacities increased, not only technological, but also social and linguistic capacities, would also have improved. On the grounds of Piagetian descriptions of language and play in children of four to five years of age, I previously suggested that a second step in the evolution of tool use and language could have involved on-the-spot cooperative efforts of small groups of hominids, as, for instance, in cooperative shelter building or cooperative divisions of labor in scavenging or hunting foods (Gibson, 1988). This would have demanded greater social and tool-using capacity than that present in modern apes, but much less than that possessed by modern human adults.

Only when information processing capacities reached levels equivalent to those of modern adolescents and adults would complex hunting and gathering strategies involving divisions of labor and the linguistic communication of

complex factual knowledge of ecology, animal behavior, and climatic fluctuations have become possible (Gibson, 1988). Thus, popular models suggesting
that early hominids mimicked either modern hunters or modern gatherers are
most unlikely to be correct.

In sum, information processing analyses suggest that the earliest hominids
may well have diverged from apes in human directions of increased tool-using,
linguistic and social capacity, but that language and social behavior continued
to increase in complexity until the human brain reached its modern size and
form. In addition, any change in human behavior which facilitated information
processing endeavors or accelerated their maturation would also have
increased technological, linguistic and social complexity. Such changes might
have included greater developmental canalization, improved phonemic production and perception, music, myth or other vocal methods of record keeping
and art, writing, and other methods of written record keeping.

References

Boesch, C. & Boesch, H. (1989). Hunting behavior of wild chimpanzees in the Tai
National Park. *American Journal of Physical Anthropology*, 78, 547–73.
Brown, R. (1980). The first sentences of child and chimpanzee. In *Speaking of Apes*,
ed. T. Sebeok & J. Umiker-Sebeok, pp. 85–101. New York: Plenum Press.
Byrne, R. & Whiten A. (1988). *Machiavellian Intelligence*. Oxford: Clarendon Press.
Case, R. (1985). *Intellectual Development: Birth to Adulthood*. New York: Academic
Press.
Dennett, D. (1988). The intentional stance in theory and practice. In *Machiavellian
Intelligence*, ed. R. Byrne & A. Whiten, pp. 180–202. Oxford: Clarendon Press.
Fortes, M. (1983). *Rules and the Emergence of Society*. Royal Anthropological
Institute Occasional Paper No. 39, 1983.
Gardner, R. A., Gardner, B. T. & Van Cantfort, T. E. (1989). *Teaching Sign
Language to Chimpanzees*. Albany: State University of New York Press.
Geschwind, N. (1965). Disconnection syndromes in animals and man. *Brain*, 88,
237–94.
Gibson, K. R. (1983). Comparative neurobehavioral ontogeny: the constructionist
perspective in the evolution of language, object manipulation and the brain. In
Glossogenetics. The Origin and Evolution of Language, ed. E. DeGrolier, pp. 41–
66. New York & Paris: Harwood Academic Publishers.
Gibson, K. R. (1985). Has the evolution of intelligence stagnated since Neanderthal
Man? In *Evolution and Development*, ed. G. Butterworth, J. Rutkouska & M.
Scaife, pp. 102–114. Brighton: Harvester Press.
Gibson, K. R. (1988). Brain size and the evolution of language. In *The Genesis of
Language; a Different Judgement of Evidence,* ed. M. Landsberg, pp. 149–172.
Berlin: Mouton de Gruyter.
Gibson, K. R. (1990). New perspectives on instincts and intelligence: Brain size and
the emergence of hierarchical mental constructional skills. In *"Language" and
Intelligence in Monkeys and Apes: Comparative Developmental Perspectives*, ed.

S. T. Parker & K. R. Gibson, pp. 97–128. New York: Cambridge University Press.

Gibson, K. R. (in press). The ontogeny and evolution of the brain, cognition and language. In *Handbook of Symbolic Intelligence*, ed. A. J. Lock & C. R. Peters. Oxford: Oxford University Press.

Goldin-Meadow, S. & Mylander, C. (1991). Levels of structure in a communication system developed without a language model. In *Brain Maturation and Cognitive Development: Comparative and Cross-Cultural Perspectives*, ed. K. R. Gibson & A. C. Petersen, pp. 315–44. New York: Aldine de Gruyter.

Goodall, J. (1986). *The Chimpanzees of Gombe*. Cambridge, MA: Harvard University Press.

Greenfield, P. M. (1991). Language, tools, and the brain: The ontogeny and phylogeny of hierarchically organized sequential behavior. *Behavioral and Brain Sciences*, 14, 531–95.

Greenfield, P. M. and Savage-Rumbaugh, E. S. (1990). Grammatical combination in *Pan paniscus*: Processes of learning and invention in the evolution and development of language. In *"Language" and Intelligence in Monkeys and Apes: Comparative Developmental Perspectives*, ed. S. T. Parker & K. R. Gibson, pp. 540–78. New York: Cambridge University Press.

Hinton, G., McClelland, J. L. & Rumelhart, D. E. (1986). Distributed representations. In *Parallel Distributed Processing, Vol. 1. Foundations*, ed. D. Rumelhart & J. McClellan, pp. 77–109. Cambridge, MA: The MIT Press.

Ingold, T. (1988). Notes on the foraging mode of production. In *Hunters and Gatherers I: History, Evolution and Social Change*, ed. T. Ingold, D. Riches, & J. Woodburn, pp. 260–85. Oxford: Berg.

Ingold, T. (in press). Social relations, human ecology and the evolution of culture: an exploration of concepts and definitions. In *Handbook of Symbolic Intelligence*, ed. A. Lock and C. Peters. Oxford: Oxford University Press.

Inhelder, B. & Piaget, J. (1958). *The Growth of Logical Thinking from Childhood to Adolescence*. New York: Harper and Row.

Jerison, H. (1973). *Evolution of the Brain and Intelligence*. New York: Academic Press.

Kaas, J. H. (1987). The organization of the neocortex in mammals: Implications for theories of brain function. *Annual Review of Psychology*, 38, 129–51.

Karmiloff-Smith, A. (1979). *A Functional Approach to Child Language*. Cambridge: Cambridge University Press.

Kochetkova, V. (1978). *Paleoneurology*. New York: John Wiley & Sons.

Kohlberg, L. (1984). *The Moral Development of the Child. Vol II. The Psychology of Moral Development*. San Francisco: Harper & Row.

Langer, J. (1980). *The Origins of Logic: Six to Twelve Months*. New York: Academic Press.

Langer, J. (1986). *The Origins of Logic: One to Two Years*. New York: Academic Press.

Luria, A. (1966). *Higher Cortical Functions in Man*. New York: Basic Books.

McClelland, J. & Rumelhart, D. (1986). *Parallel Distributive Processing, Vol. 2: Psychological and Biological Models*. Cambridge, MA: The MIT Press.

McGrew, W. (1987). Discussion presented at Fourth International Conference in Ecological Psychology, Trieste, Italy.

Miles, H. L. (1983). Two-way communication with apes and the evolution of language. In *Glossogenetics: The Origin and Evolution of Language*, ed. E. DeGrolier, pp. 201–10. New York & Paris: Harwood Academic Publishers.

Milner, B. (1967). Brain mechanisms suggested by studies of the temporal lobes. In *Brain Mechanisms Underlying Speech and Language*, ed. F. L. Darley, pp. 122–45. New York: Grune and Stratton.

Parker, S. T. (1985). A social-technological model for the evolution of language. *Current Anthropology*, 26, 617–39.

Parker, S. T. & Gibson, K. R. (1977). Object manipulation, tool use and sensorimotor intelligence as feeding adaptations in Cebus monkeys and great apes. *Journal of Human Evolution*, 6, 623–41.

Parker, S. T. & Gibson, K. R. (1979). A model of the evolution of language and intelligence in early hominids. *The Behavioral and Brain Sciences*, 2, 367–407.

Passingham, R. E. (1975). Changes in the size and organization of the brain in man and his ancestors. *Brain, Behavior and Evolution*, 11, 73–90.

Piaget, J. (1952). *The Origins of Intelligence in Children*. New York: W. W. Norton.

Piaget, J. (1954). *The Construction of Reality in the Child*. New York: Basic Books.

Piaget, J. (1955). *The Language and Thought of the Child*. London: Routledge & Kegan Paul.

Potts, R. (1984). Home bases and early hominids. *American Scientist*, 72, 338–47.

Premack, D. (1988). "Does the chimpanzee have a theory of mind?" revisited. In *Machiavellian Intelligence*, ed. R. Byrne & A. Whiten, pp. 160–79. Oxford: Clarendon, Press.

Reynolds, P. C. (1983). Ape constructional ability and the origin of linguistic structure. In *Glossogenetics: The Origin and Evolution of Language*, ed. E. DeGrolier, pp. 185–200. New York & Paris: Harwood Academic Publishers.

Rumelhart, D. & McClelland, J. (1986). *Parallel Distributed Processing, Vol. 1: Foundations*. Cambridge, MA: The MIT Press.

Russon, A. (1990). The development of peer social interaction in infant chimpanzees: Comparative social, Piagetian and brain perspectives. In *"Language" and Intelligence in Monkeys and Apes: Comparative Developmental Perspectives*, ed. S. T. Parker & K. R. Gibson, pp. 379–419. New York: Cambridge University Press.

Shipman, P. (1986). Scavenging or hunting in early hominids: theoretical framework and tests. *American Anthropologist*, 88, 27–43.

Teleki, G. (1973). *The Predatory Behavior of Wild Chimpanzees*, New Jersey: Bucknell University Press.

Tobias, P. V. (1971). *The Brain in Hominid Evolution*. New York: Columbia University Press.

Tooby, J. & DeVore, I. (1987). The reconstruction of hominid behavioral evolution through strategic modeling. In *The Evolution of Human Behavior: Primate Models*, ed. W. Kinsey, pp. 183–238. New York: State University of New York Press.

Vauclair, J. & Bard, K. (1983). Development of manipulations with objects in ape and human infants. *Journal of Human Evolution*, 12, 631–45.

Whiten, A. & Byrne, R. (1988). The manipulation of attention in primate tactical deception. In *Machiavellian Intelligence*, ed. R. Byrne & A. Whiten, pp. 211–23. Oxford: Clarendon Press.

Wynn, T. (1979). The intelligence of later Acheulean hominids. *Man*, 14, 379–91.

Wynn, T. (1988). Tools and the evolution of human intelligence. In *Machiavellian Intelligence*, ed. R. Byrne & A. Whiten, pp. 271–84. Oxford: Clarendon Press.

Wynn, T. & McGrew, W. (1989). An ape's eye view of the Oldowan. *Man*, 24.

Part IV

Perspectives on development

Part IV Introduction

Beyond neoteny and recapitulation: New approaches to the evolution of cognitive development

KATHLEEN R. GIBSON

Developmental frameworks were integral to nineteenth century evolutionary thought. In particular, Haeckel's biogenetic law, "ontogeny recaptitulates phylogeny", influenced generations of scholars. With the discovery of numerous exceptions, however, the biogenetic law fell into disrepute, and most evolutionary scholars eschewed developmental studies. As a result, the modern synthetic theory of evolution placed minimal emphasis on developmental mechanisms (Løvtrup, 1981). This extreme case of "throwing the baby out with the bath water" contributed to the poverty of theories of mental evolution, because all complex adult human behaviors reflect long developmental trajectories, and considerable learning.

Recognizing this lacuna, primatologists in the 1970s began to place primate cognitive and neural development within evolutionary frameworks (Chevalier-Skolnikoff, 1971, 1974, 1977; Fishbein, 1976; Gibson, 1970, 1977; Parker, 1973, 1977; Parker & Gibson, 1977), and three works appeared suggesting that the ontogeny of language recapitulates its phylogeny (Bates, *et al.*, 1979; Lamendella, 1976; Parker & Gibson, 1979). Nearly simultaneously, Stephen Gould, a paleontologist, published his now classic work, *Ontogeny and Phylogeny* (1977). Gould's work, in particular, set the framework which guides discussions of ontogeny to the present day. It was Gould's singular contribution to note that changes in developmental rates, heterochrony, have potentially far reaching effects on adult morphology and behavior. Gould also resurrected the concept of recapitulation, not as a law, but as one of several possible heterochronic outcomes. Finally, he popularized the view that adult human morphology and behavior reflect neoteny, a particular form of heterochrony, characterized by slow developmental rates and a permanent juvenilization of form.

Since the publication of *Ontogeny and Phylogeny*, Gould's views of neoteny and recapitulation have engendered considerable debate, especially among those with major interests in cognitive evolution (for reviews, see Gibson, 1991, in press). These debates have set the stage for serious academic investigation of developmental events and for a renewed appreciation of the centrality of developmental approaches for evolutionary theory. As a consequence, ontogenetic perspectives have become the rule, rather than the exception, among serious scholars of cognitive and linguistic evolution (Borchert & Zihlman, 1990; Butterworth, *et al.* 1985; Gibson, 1983, 1988, 1990, 1991, in press; Gibson & Petersen, 1991; Greenfield, 1991; Greenfield and Savage-Rumbaugh, 1990; Lock, 1983; Miles, 1983; Parker, 1984, 1985, 1987; Parker and Gibson, 1990, Ragir, 1986).

In accord with this modern tradition, developmental frameworks figured prominently in the conference discussions which have given rise to this volume, and in at least six of the conference papers (in this volume by Gibson, Goldin-Meadow, Langer, Lock, and Parker & Milbrath and also Greenfield, 1991). This section contains three of these contributions, representing a range of application of modern behaviorally-oriented developmental frameworks.

Langer's comparisons of early human infantile development with that of capuchin monkeys and macaques reveals species-specific differences in sequences of developmental events. His explicitly constructionist perspective indicates that cognitive skills, such as understandings of basic physical causality, simple logic, and classification, are essential prerequisites for the development of language and other representational systems. In turn, representation and symbolism are hierarchical mental constructions which integrate skills derived from earlier infantile cognitive accomplishments.

As noted by Langer, capuchin and macaque infants engage in causal object manipulations similar to those of human infants, although the monkeys do so at ages that are relatively later with respect to their growth cycles, and they incorporate fewer objects and fewer sets of objects into their manipulative schemes. The monkeys, however, do not engage in classificatory and logical object manipulations until they are nearly grown. Thus whereas human infants, at critical points in the growth cycle, engage simultaneously in causal, logical and classificatory behaviors, monkeys do not. Given their large brain sizes and advanced cognitive skills, humans can integrate each of the simultaneously occurring behaviors into representational systems. Even if monkeys had the requisite neural capacity (which they probably do not, see Gibson, this volume), they would lack representational systems because of these developmental disparities.

Langer's explication of heterochronic developmental events (which follows parallel frameworks by Antinucci, 1990 and Vauclair and Bard, 1983) has critical implications for both neoteny and recapitulation theories of human cognitive evolution. According to neoteny theory, human intelligence results from delayed development and a consequent permanent immaturity of the brain. Langer's work and that of other scholars, however, indicates that human development is actually accelerated in critical cognitive dimensions in comparison to that of apes and monkeys (Antinucci, 1990; Gibson, 1991; Vauclair & Bard, 1983). At the same time, this work suggests that earlier views of cognitive evolution, as primarily a matter of terminal addition (Parker & Gibson, 1979), were also overly simplistic. Rather, the development of new behaviors at early stages in development helps channel later developing cognitive events (see also Gibson, 1990, this volume).

Lock's summary of human developmental data indicates that, among small children just beginning to speak, gesture and vocalization are intertwined, and that many early communications are conducted in both modalities. In addition, as children develop concepts of objects and of causality, they incorporate these concepts into their vocabularies. Lock concludes, however, that despite the modern temporal association between the development of object concepts and their linguistic counterparts, tool-use and causal object manipulations by fossil hominids do not imply the incorporation of concepts of causality into early hominid communication systems. Indeed, we already know this from comparative data. Many animals possess these concepts, but have neither gestural nor vocal languages. Thus, Lock's data suggest that object concepts and understanding of causality are necessary, but not sufficient, prerequisites for language as we know it today.

Finally, Lock wisely cautions against drawing major conclusions about phylogeny from the ontogeny of a single species. Gould may have propelled recapitulation theories to new respectability, but he did not resurrect the biogenetic law. Recapitulation may, as Gould suggests, be a common evolutionary event, but its occurrence can only be inferred from comparative data, not assumed from the outset. Also, as Lock notes, and as Parker and Milbrath implicitly accept, human language and cognition are reinvented by human children under the guidance of adults. Early hominids would have lacked such guidance. Consequently, even hominids with all of the requisite cognitive skills may not have possessed fully-developed language.

In contrast to Lock, Parker and Milbrath present a rather strongly recapitulationist argument for the evolution of planning. Their views, however, are not based on presumptions of recapitulation, but on decades of comparative work by Parker, Gibson, and Chevalier-Skolnikoff, indicating that apes and monkeys reach levels of Piagetian intelligence and mental constructional capacities equivalent to those of human infants and children. The planning capacities that Parker and Milbrath discuss reflect these mental constructional skills. In addition, they argue cogently that if certain behaviors are prerequisite to the development of language or other cognitive skills in human children, they must also have been prerequisite to that development in our hominid ancestors.

Parker and Milbrath's paper explicitly links modern human language, technology, social behavior, and cognition through the medium of social planning. They indicate that the emergence of language and of advanced planning capacities required a cognitively-based integration of many separate, precursor behaviors. Prior to the emergence of this integrative complex, planning would have been rudimentary in nature. Once it was established, the stage was set for the emergence of elaborated communicative and social systems, such as those characteristic of the Upper Paleolithic.

Taken as a group the chapters in this section show how, by focusing on specific behaviors, we can move beyond assumptions of neoteny and recapitulation to make more direct comparisons of actual developmental schedules (see also Gibson & Petersen, 1991; Parker & Gibson, 1990). Such data can also help us determine which behaviors serve as essential precursors to language and other cognitive skills.

References

Antinucci, F. (1990). The comparative study of cognitive ontogeny in four primate species. In *"Language" and Intelligence in Monkeys and Apes: Comparative Developmental Perspectives*, ed. S. T. Parker & K. R. Gibson, pp. 157–71. Cambridge: Cambridge University Press.

Bates, E., Benigni, L., Bretherton, I., Camaioni, L., & Volterra, V. (1979). *The Emergence of Symbols: Cognition and Communication in Infancy*. New York: Academic Press.

Borchert, C. M. & Zihlman, A. L. (1990). The ontogeny and phylogeny of symbolizing. In *The Life of Symbols*, ed. M. LeC. Foster & L. J. Botscharow, pp. 16–44. Boulder: Westview Press.

Butterworth, G., Rutkowska, J., & Scaife, M. (1985). *Evolution and Developmental Psychology*. Brighton: Harvester Press.

Chevalier-Skolnikoff, S. (1971). The ontogeny of communication in *Macaca speciosa*. Ph.D. thesis, University of California at Berkeley.

Chevalier-Skolnikoff, S. (1974). *The Ontogeny of Communication in Stumptailed Macaques (*Macaca Arctoides*). Contributions to Primatology.*, vol. 2. Basel: S. Karger.

Chevalier-Skolnikoff, S. (1977). A Piagetian model for describing and comparing socialization in monkey, ape, and human infants. In *Primate Bio-Social Development*, ed. S. Chevalier-Skolnikoff & F. E. Poirer, pp. 159–87. New York: Garland Publishing.

Fishbein, H. D. (1976). *Evolution, Development and Children's Learning*. Pacific Palisades, CA: Goodyear.

Gibson, K. R. (1970). Sequence of myelinization in the brain of *Macaca mulatta*. Ph.D. thesis, University of California at Berkeley.

Gibson, K. R. (1977). Brain structure and intelligence in macaques and human infants from a Piagetian perspective. In *Primate Bio-Social Development*, ed. S. Chevalier-Skolnikoff & F. E. Poirer, pp. 113–57. New York: Garland Publishing.

Gibson, K. R. (1983). Comparative neurobehavioral ontogeny: The constructionist perspective in the evolution of language, object manipulation and the brain. In *Glossogenetics: The Origin and Evolution of Language*, ed. E. de Grolier, pp. 52–82. New York & Paris: Harwood Academic Publishers.

Gibson, K. R. (1988). Brain size and the evolution of language. In *The Genesis of Language: A Different Judgement of Evidence*, ed. M. Landsberg, pp. 149–72. Berlin: Mouton de Gruyter.

Gibson, K. R. (1990). New perspectives on instincts and intelligence: Brain size and the emergence of hierarchical mental constructional skills. In *"Language" and Intelligence in Monkeys and Apes: Comparative Developmental Perspectives*, ed. S. T. Parker & K. R. Gibson, pp. 97–128. Cambridge: Cambridge University Press.

Gibson, K. R. (1991). Myelination and behavioral development: A comparative perspective on questions of neoteny, altriciality and intelligence. In *Brain Maturation and Cognitive Development: Comparative and Cross-Cultural Perspectives*, ed. K. R. Gibson & A. C. Petersen, pp. 29–63. Hawthorne, NY: Aldine de Gruyter.

Gibson, K. R. (in press). The ontogeny and evolution of the brain, cognition and language. In *Handbook of Symbolic Intelligence*, ed. A. J. Lock & C. R. Peters. Oxford: Oxford University Press.

Gibson, K. R. & Petersen, A. C. (1991). *Brain Maturation and Cognitive Development: Comparative and Cross-Cultural Perspectives*. Hawthorne, NY: Aldine de Gruyter.

Gould, S. J. (1977). *Ontogeny and Phylogeny*. Cambridge, MA: Harvard University Press.

Greenfield, P. M. (1991). Language, tools and the brain: the ontogeny and phylogeny of hierarchically organized sequential behavior. *Behavioral and Brain Sciences*, 14, 531–95.

Greenfield, P. M. & Savage-Rumbaugh, E. S. (1990). Grammatical combination in *Pan paniscus*: processes of learning and invention in the evolution and development of language. In *"Language" and Intelligence in Monkeys and Apes: Comparative Developmental Perspectives*, ed. S. T. Parker & K. R. Gibson, pp. 540–78. Cambridge: Cambridge University Press.

Lamendella, J. T. (1976). Relations between the ontogeny and phylogeny of language: A neorecapitulationist view. In *Origins and Evolution of Language and Speech*, ed. S. R. Harnard, H. D. Steklis, & J. B. Lancaster, pp. 396–412. New York: New York Academy of Sciences.

Lock, A. (1983). "Recapitulation" in the ontogeny and phylogeny of language. In *Glossogenetics: The Origin and Evolution of Language*, ed. E. de Grolier, pp. 255–74. New York & Paris: Harwood Academic Publishers.

Løvtrup, S. (1981). Introduction to Evolutionary Epigenetics. In *Evolution Today: Proceedings of the Second International Congress of Systematic and Evolutionary Biology*, ed. G. Scudder G. E. & J. L. Reveal, pp. 139–44. Pittsburgh: Hunt Institute for Botanical Documentation, Carnegie-Mellon University.

Miles, H. L. (1983). Two way communication with apes and the origin of language. In *Glossogenetics: The Origin and Evolution of Language*, ed. E. de Grolier, pp. 201–10. New York & Paris: Harwood Academic Publishers.

Parker, S. T. (1973). Piaget's sensorimotor series in an infant macaque: The organization of nonstereotyped behavior. Ph.D. thesis, University of California at Berkeley.

Parker, S. T. (1977). Piaget's sensorimotor period series in an infant macaque: a model for comparing unstereotyped behavior and intelligence in human and nonhuman primates. In *Primate Bio-Social Development*, ed. S. Chevalier-Skolnikoff & F. E. Poirer, pp. 43–112. New York: Garland Publishing.

Parker, S. T. (1984). Playing for keeps. In *Play in Animals and Humans*, ed. P. K. Smith, pp. 271–93. New York: Blackwell.

Parker, S. T. (1985). Higher intelligence as an adaptation for social and technological strategies in early *Homo sapiens*. In *Evolution and Developmental Psychology*, ed. G. Butterworth, J. Rutkowska, & M. Scaife, pp. 83–100. Brighton: Harvester Press.

Parker, S. T. (1987). A sexual selection model for human evolution. *Human Evolution*, 2, 235–53.

Parker, S. T. & Gibson, K. R. (1977). Object manipulation, tool use and sensorimotor intelligence as feeding adaptations in cebus monkeys and great apes. *Journal of Human Evolution*, 6, 623–41.

Parker, S. T. & Gibson, K. R. (1979). A model of the evolution of language and intelligence in early hominids. *Behavioral and Brain Sciences*, 2, 367–407.

Parker, S. T. & Gibson, K. R. (1990). *"Language" and Intelligence in Monkeys and Apes: Comparative Developmental Perspectives*. Cambridge, England: Cambridge University Press.

Ragir, S. (1986). Retarded development: The evolutionary mechanism underlying the emergence of the human capacity for language. *The Journal of Mind and Behavior*, 6, 451–68.

Vauclair, J. & Bard, K. A. (1983). Development of manipulations with objects in ape and human infants. *Journal of Human Evolution*, 12, 631–45.

12

Human language development and object manipulation: Their relation in ontogeny and its possible relevance for phylogenetic questions

ANDREW LOCK

Communicative development

The developmental elaboration of human communication is outlined here as being a continuous process. This continuity should not be taken as implying one underlying set of processes or mechanisms. Nor should the use of the notion of "stages" be misunderstood. There is, for example, a period in which children predominantly use single words. It is *convenient* to refer to this as "the one-word stage". It is not, however, a stage with a definable beginning or end, but merges with other productions before and after it; it is not an homogeneous stage, for there are changes going on within it; and it is probably not founded on a single underlying set of "mechanisms". It is a stage for expositional convenience only, as are all the stages given here.

Stage 1: Precommunication (0–9 months)

The precommunication stage is that period in which human infants convey information to others by default. In the terminology of Vygotskian psychology, their abilities are intermental: that is, constituted in the social relationships between them and their cultural milieu. Infants in this period give no evidence of intending to communicate, but only communicate by virtue of the fact of their being within a socio-cultural human group, in which it is impossible *not* to communicate. This is "communication-in-the-eye-of-the-beholder" only. Thus an infant might yell out of frustration, yet be treated as if he or she had communicated "Help me".

The importance of this stage has been argued (e.g. Lock, 1980) to lie in inducting the infant into a set of meaningful social episodes with adults that provide a structure within which the infant can discover the meanings its activities are being accorded. Inchoate motives are structured with respect to particular practices, so that they can be pursued in a goal-oriented manner. Thus "discomfort", for example, is structured into particular "discomfort-removing" strategies, yielding a culturally-patterned set of motivations or goals for action.

These developments may be described as providing the child with the resources to uncover and control the implications of its "being-as-a-child". Thus early discomfort could be said to have a value for the infant ⟨whatever this bodily state is, I do not want it⟩. As the child is relieved of discomforts, it is given information that allows it to uncover and structure the implication of this, ⟨I want something else⟩, as ⟨I want that⟩. The dependent nature of human "being-as-a-child" is such that ⟨I want that⟩ has the further implication ⟨you do something to get me that⟩. But, in this first stage, infants only accomplish a structuring of their motives, and do not come to control this additional implication – thus their actions remain pre-communicative.

Stage 2: Pre-symbolic intentional communication (9–15 months)

Around nine months of age, infants begin to give evidence of controlling the implication in their actions of ⟨you do something to get me that⟩: that is, they *address* messages to other actors (see Figure 12.1). They begin to use a number of communicative gestures, these often being actions that are "lifted" from their original direct manipulation of the world (see Table 12.1). In addition to this direct origin, it is likely that some of an infant's repertoire is developed by imitation of parental models. There are marked individual differences in the adoption of these gestures, both in relation to their order and frequency of appearance, and in whether and when they are accompanied by vocalizations (e.g. Zinober & Martlew, 1986). The ontogenetic origins of pointing are less clear than for the other gestural categories (Lock *et al.*, 1990), but it appears to be a gesture with a universally similar form.

In addition to these physical gestures, the infant develops vocal counterparts to them (see examples given by Dore, 1986). Various categorization systems have been proposed for these vocalizations (see Nelson & Lucariello, 1986: 61–3, for a brief review). Common to these systems are the views that these vocalizations are not truly symbolic, are very tied to specific contexts, and are not phonetically structured. The complementarity between these vocalizations and the gestures that are used alongside them is apparent in the empirical findings of Bates *et al.* (1979), who conclude these systems are equivalent:

> At this point in development, the only difference between the two domains is in the modality of expression. . . . we see no evidence to suggest that a 13-month-

Fig. 12.1. Pre-symbolic intentional communication in the gestural mode by a 12-month-old female infant. Failing to reach a desired object (1), she vocally attracts her mother's attention (2) and establishes eye contact (3) at which point her mother says "Yes?", whereupon the infant directs her mother's attention (4) by pointing. That this pointing is functioning as a demand in this situation may be inferred by the mother from the quality of the attention–getting cry at (2).

old is in anyway biased toward the development of vocal language as opposed to gestural language (ibid: 177).

Grieve & Hoogenraad (1979) have characterized these early forms as a means of sharing *experiences* rather than *meanings:* that is, they are not fully referential. This shift to the referential, symbolic domain is accomplished in the next stage.

This present stage is also open to the sort of implicational analysis introduced earlier. It would seem that what is occurring is not a shift *beyond* the level of implication the child has already come to control, but an elaboration *at* the level already attained. That is, motives and situations are being more clearly discriminated by the infant, and this is reflected in the specificity of the means used to communicate them. A general ability to convey, say, ⟨you do something to get me that⟩ is being differentiated into ⟨you pick me up⟩, or ⟨you give me X⟩, where X can be specified by a specific gesture, vocalization or by pointing, while still having linked with it the entire message: that is, communicative "items" have holistic meanings at this point.

Table 12.1. *A possible classificatory scheme for pre-verbal communicative gestures*

	Function	Origin	Variation	Vocalization
Expressive (e.g. arm-waving, smiling, clapping, banging feet)	To convey emotional states, ±	Stylisation of early rhythmic movements	Depend on infant's 'temperament'?	Accompanies from 10 months
Instrumental (e.g. pick-up, handshake/nod, open palm)	To control the behaviour of another person	From the act of doing the thing itself	Cultural variation in form, perhaps, but universal?	Accompanies from 10 months – convey same meaning
Enactive (e.g. symbolic play – pantomime)	To represent actions and attributes of people and objects	From the technical act itself	Variable at level of dyad	13–21 months
Diectic (e.g. pointing, reaching)	To isolate an object from its general context	?	Probably universal	14–24 months, possibly at 9 months?

Source: Based on data from Zinober and Martlew, 1986.

Stage 3: Symbolic referential communication (15–24 months)

1. Reference and predication

The transition to symbolic, referential communication is poorly understood. It involves the establishment of vocalizations as *names*. Vocalizations become less tied to contexts, and, apparently, more to objects.[1] A "naming explosion" has often been reported, and is taken as evidence that a child has gained the insight into the general principle that things have names. There occurs, then, not just an association of a sound or gesture with an object, but a generalizable "understanding" of the relation between sounds and objects, a relation that holds over differing objects and contexts.

A number of situations have been described that doubtless feed into the establishment of names as "names". Many of these will be specific to particular cultures; none of them can be posited as necessary pre-requisites in their own right. The common factor amongst them is that they all establish ritual formats which serve to increase the saliency of the small number of elements varying across them. In "naming games" the format is generally for the adult to highlight an object, and the child to provide a response: different objects require different responses, but within ritualized formats. Objects similar in adult eyes can be substituted for each other while requiring the same response, so that sounds are not linked to, say, just one picture of a dog, but many different forms of dog; and can be substituted in other game formats, a straw, say, being used to function as a comb. Gestures or vocalizations used as demands also support aspects of referential symbolization, in that what is being demanded can be temporally displaced from the present.

Productive vocabulary grows slowly during the early part of this stage. Fifty words is often taken as the point at which the "naming explosion" may be expected (Nelson, 1973; Kagan, 1981), meaning that the infant can have a fairly large repertoire of functional vocalizations prior to the establishment of a "linguistic" referential system. Linguists such as Bloom (1973) and Dore (1986) credit the first signs of linguistic organization as late in the one-word period, when the child is able to produce words that relate syntagmatically to another element in the context; that is, in using a word communicatively the child "presupposes shared knowledge of an aspect of context (which could have been made explicit), but ... focuses on another" (Dore, op.cit: 36).

There are marked individual differences in the rate of vocabulary development. Nice (1926/27) tabulated 47 published vocabularies of 24-month-old infants, the average being 328 words, with a range from five to 1212 words. In a more recent study of 30 children Bates, Bretherton & Snyder (1988) report a mean vocabulary at 20 months of 142 words, with a range from one to 404 words. There is little evidence of words being phonologically organized along the lines of the adult system during this period. Rather, they are organized holistically in contrast to each other and not yet analysed into their component phonemes (Ferguson & Farwell, 1975).

Along with the attainment of reference at this age come some profound changes in what the child apparently refers to. Nelson & Lucariello (1986: 80) summarize these:

> during the first half of the second year the event representation remains unanalyzed in terms of specific concepts of objects, actions and actors. It is only during the second half of the second year, in general, that discrete concepts are differentiated from the whole.

Thus, once children have abstracted the general principle of word meaning from the various contexts in which it is embeddedly presented to them, they re-embed words into diverse contexts that allow them to abstractively create new meanings for them, and progress to predication.

In terms of the implicational framework already introduced, early words are being proposed as having holistic meanings, such that when a child says "down", for example, as it puts an object down, it means all of ⟨I am putting that object down⟩, and is not coding just one item of this meaning. The separate components coded in the adult language need to be constructed: actions, in English, need to be constructed as separate from agents and patients: where "kick" might originally mean ⟨I am kicking the ball⟩, it has to come to just represent the act of kicking; similarly for "ball" uttered in the same circumstances. Once these differentiations are made, then the child is moving itself beyond the realm of reference to that of predication; not just identifying something as a ball, but identifying a ball and saying something about it (see Lock, 1980: 148–76, for a detailed account of these developmental changes).

The point to be drawn here is that this stage of language development is one in which some quite fundamental aspects of the language system are being constructed. To call it a "one-word stage", as is often done in the literature, would obscure this. The child enters this stage with a communicative repertoire that is slowly re-worked into a linguistic one based on reference and predication, and in this re-working some major conceptual achievements on which language can be further elaborated are accomplished. At the same time, however, it should be recognized that this symbolic referential and predicative system, while pre-grammatical, is effective in specifying a range of intentions and meanings. The infant is able to convey implied meanings of a far greater range and power than it is able to render or control explicitly.

2. First combinations

There are two further reasons why, for present purposes, the notion of a "one-word stage" is inappropriate. First, many early "words" are in fact produced in combination with gestures: for example, naming is often accompanied by pointing; and requests are often produced with their gestural forms – "up" with outstretched arms, for example. Zinober & Martlew (1986) note that, for what they term "instrumental gestures" at least (see Table 12.1), there is a developmental progression in the form of word–gesture combination. Prior to 21 months their subjects used gestures in a supportive role to words, in that word and gesture appeared to have the same meaning. From 21 months, however, they were used in complementary roles. For example, a child says "Book" while placing mother's hand on a pile of books:

> the utterance "book" gives the name of the object being negotiated. The gesture indicates who shall perform the action (the mother) and what action is required (picking up the book). Together [word] and gesture communicate "you pick the book up, mummy" (ibid: 200).

Gestures thus continue to accompany words across the predication hurdle, and, indeed, Michael & Willis (1968) have catalogued 12 gestures that commonly signal messages such as "go away" and "come here" among 4–7 year olds[2].

Second, some word combinations do occur quite early in many children's productions. On the one hand there appear "rote" productions. These are not in any sense based on the application of a rule system, but are more "formula-like", and do not serve as productive templates for new combinations. On the other hand, as the conceptual distinctions noted above come to be made, they allow the construction of utterances in which these distinctions are represented: "Daddy pipe" (possessor), "more book" (recurrence), "all gone stick" (non-existence) appear to be pre-propositional combinations, as opposed to later appearing post-propositional ones such as "Daddy sit" (agent), "baby cry" (experiencer) (examples from Leonard, 1976). Thus, before some uses of words emerge as single-word utterances, other uses are being combined together, obscuring further the separation of a one- from a two-word stage.

None of the above combinational categories appear to be based on any knowledge of syntax, and in modern jargon, if there is a grammar module in the human linguistic system, and the data on adult speakers indicates there may well be, there is no evidence that it plays a role in establishing early "grammatical" abilities (see Bates *et al.*, 1988). It seems more likely that these early combinations are ordered on the basis of perceptual, semantic and pragmatic considerations (see Lempert, 1984, for a concise review), but exactly what is going on at this time remains unclear.

Stage 4: Grammar (2–5 years)

In the previous stage, children immersed in different language communities proceed in very similar ways (see Gentner, 1982, for a review). From the age of two years, however, language specific factors play a more dominant role. While grammatical development is incomplete at the age of five years, the bulk of the parent language's grammatical constructions emerge between these ages, starting with inflectional morphemes marking past tense or plural, and going through toward the full range of syntactic operations, such as (depending on target language), relativization, coordination, pronominalization and ellipsis. While learning is not error free, first language learners make relatively few errors in constructing complex grammars on the basis of a limited range of data. The errors that do occur have provided a rich source of data.

The patterning of errors suggests to some theorists (e.g., Davis, 1987; Macken, 1987; Roeper & Williams, 1987) that two learning processes are operating, one linked to morphological learning, and the other to the acquisition of major syntactic structures and principles. The first is a probabilistic type of learning, whereby the child masters the irregularities of the morphological system (e.g., a general rule for the formation of passives in English is to add *-ed* to the verb, but this does not hold for some verbs such as *break* or *sit*: acquisition of these is related to their frequency of occurrence in the target language); the second is a form whereby quite fundamental principles of grammatical organization are acquired on the basis of quite minimal experience.

It is this second form of "learning" that is at the root of claims regarding the unique (and innate) status of human linguistic abilities. While Chomsky's name is most often linked with this argument, much of its power comes from Gold's demonstration (1967) that

> only those languages which constitute finite sets of sentences are learnable without the use of both grammatical and ungrammatical examples (appropriately labelled) in the learners' input data (Johnson, Davis & Macken, in press).

Natural human languages consist of theoretically infinite sets of sentences. In addition, there is no evidence that children receive ungrammatical exemplars to guide them in learning which are the correct syntactic forms in their target language. These two "facts" lead "to the curious conclusion that, under the observed conditions of human language acquisition, human languages are unlearnable" (Johnson *et al.*, ibid).

Four contenders offering a way out of this paradox have been proposed in the literature. The "motherese hypothesis" argues that the data are presented to the language learner in an ordered fashion, from least to most complex. There is some evidence that this occurs, but not that it is a sufficient source on its own. Second, the child can be hypothesized as a "calculating machine", working out the probability of occurrence of a given exemplar, and inferring the ungrammaticality of a missing exemplar by its absence. The two-stage models referred to above effectively incorporate this hypothesis to account for some of the morphological learning that occurs, but, again, this hypothesis does not appear to be sufficient for explaining the major facts of syntactic acquisition.

The third hypothesis is to credit the child learner with some innately given knowledge of language structure which constrains the process of learning. The forms hypothesized for this knowledge have come in various guises. Presently, the favoured version is known as the "principles-and-parameters" theory (e.g. Hyams, 1986; Roeper & Williams, 1987). The child's task, in this view, is not so much to learn a language, but to use its "input data" to choose between which of an already specified set of possible languages it is dealing with. The child is conceived as having access to a set of options that a language can fit, and on the basis of very minimal data chooses between a set of pregiven rules as the principles by which the particular language it is hearing is structured. Present evidence of "real life" language learning by the child in this syntactic phase is not incompatible with this hypothesis, but whether it is compatible with reality is another matter (see below). This form of theorizing is generally embedded in what is termed the "modularity thesis" (e.g. Fodor, 1983). In this, human mental operations are regarded as compartmentalized, with sets of them being independent of others, and dedicated to specific tasks. There are thus claimed to be a set of operations available solely for language, that have their own separate history. If this is the case, then, as noted above, there is no evidence that this module is in operation prior to this stage, raising the issue of the continuity of the underlying process of development.

The fourth hypothesis is partly motivated by this question of continuity. It proposes that language is constructed in response to the child's communicative "needs", and tends to argue not for modularity, but for the role of common cognitive principles as operative in the task of constructing the child's abilities across domains. This hypothesis has a number of intuitive attractions. Perhaps because of the role Piagetian theory has played in the formative education of most contemporary developmental psychologists, we have a predilection for constructive accounts of development (of one sort or another), and tend only reluctantly to fall back on innatist explanations as a last resort, hoping that with new data we can get rid of them. The evidence supporting constructive approaches to language development is quite good, until one gets to the issue of grammar. There are not, at present, convincing developmental accounts of the construction of grammar out of earlier elaborated systems, and there thus does appear, as of present scholarship, to be a discontinuity at this point in the ontogeny of language. Reviewing these last two hypotheses in an evolutionary context, Johnson *et al.* (op. cit.) conclude that the third:

> provides an elegant solution to the learnability paradox, accounts for the rapidity, robustness, and comparatively errorless nature of language acqui-sition, and – supplemented with an appropriate theory of the learner – accords well with the empirical data from first language acquisition. However, ... the autonomous approach leads to rather implausible conclusions about human linguistic evolution, because it is a discontinuity view of both language development and evolution. ... But the autonomous linguist's view is that we know more about grammatical competence than about prehistory or evolu-tionary theory and, thus, an implausible evolutionary scenario is preferable to an implausible theory about language development.

This issue will be returned to in the conclusion.

Stage 5: Post-syntactic development (5 years onward)

The learning of a native language is by no means complete with the acquisition of the major syntactic structures. The main developments still to occur are, crudely, the matching of the child's grasp of syntax with the child's grasp of the real world. Thus, for example, children have difficulty in understanding passive sentences, tending to pay more attention to the ordering of the words in the sentence than their apparent grammatical relations: thus "The dog was chased by the cat" is likely to be misinterpreted.

The development of tool-use

There is a surprising dearth of recent studies on the development of human tool-use, as opposed to understanding the properties of objects and the space they occupy. Detailed normative data are presented by Gesell (e.g., 1954), and these form the basis of many developmental assessment tests. Two points that may be drawn from this data concern physical maturation and the role of imitation in development.

First, both the components of object manipulation and their co-ordination into functional activities only begin to achieve adult status around five to six years of age (see, for example, illustrative material presented by Gesell, op.cit.). The activities of grasping and releasing objects, and the embedding of these into larger activities such as placing objects on top of each other (to build a tower of blocks, say) or to throw an object, have a maturational component. In addition there is a subtler issue of behavioural use: crudely, the difference between what, in a testing situation, an adult is able to demonstrate a child can do, and what, in real life, a child does do. The behavioural sophistication of neonates, for example, has only been recently uncovered. But such demonstrations rely on sophisticated adult techniques. It seems likely that these early capacities have previously been overlooked because the infant was not provided with situations in which such capacities had any use. In addition to having a maturational component, then, behavioural skills also have a motivational component to their developmental elaboration. These points have an impact on how ontogenetic information may be extrapolated to the phylogenetic arena (drawn out below).

Second, Gesell's data (op. cit.) demonstrate a lag in constructional activities between the child's abilities to imitate the construction of adult models and the ability to copy the finished product. A child can reproduce an adult construction from imitating the adult's actions of constructing before being able to reconstruct a completed model (and reproduction is even more difficult if the model is not visually present at the time of the child's building). This introduces another complication into the reading of the cognitive capacities of the child from the evidence of its productions: an object that is, for example, symmetrical in three dimensions, implying a capacity for image manipulation, can be reconstructed without the use of such an ability via the imitative reproduction of a string of actions[3].

Despite being thin on the ground, however, there are three strands of contemporary investigation of relevance here: the use of objects in symbolic play; the relation of cognitive abilities used and formed in relation to actions on the world of objects and those that might be involved in the elaboration of language abilities; and the relation of constructive play patterns to the (syntactic) patterns of emerging language.

Symbolic play with objects

Objects can be used in conventional activities in ways that are not conventionally appropriate for their use; a hair dryer as a gun, a spoon as a comb, a saucepan as a mirror, for example. Most of the naturalistic studies report such play as emerging around two years of age (e.g. Inhelder *et al.*, 1972; Nicolich, 1977), later than other forms of pretend play, such as pretending to be asleep[4]. This activity is taken to imply that the child has developed representational abilities, and is supportive of the Piagetian claim that the capacity for symbolism emerges across a number of domains simultaneously, rather than being specific to language. Specific parallels between object use and emerging language have been observed in the early stages of development. Bates *et al.* (1979) propose developmental stages for object play and language that are strikingly similar. It is difficult to see how this area of research could be directly applied phylogenetically, but it raises the more specific issue of whether there is, in development, some common underlying ability that feeds into both object strategies and language emergence.

The relation between cognition and language development

Investigations of the relation between cognition and language development represent, by-and-large, refinements of Piaget's original claim that symbolism arises out of the attainment of representational intelligence at the end of the sensorimotor period (stage 6), and, as such, is relatively modality-free, with language being just one of the media in which it is exhibited. This very *general* claim appears to hold up: objects are used symbolically in much the same way as early vocalizations (e.g. Musatti & Mayer, 1987). By contrast, Piaget's more *specific* claim that achieving stage 6 levels of intelligence is a pre-requisite for the emergence of language has gained little support. What has emerged, however, is a more detailed picture of the inter-relations between specific abilities in the two domains at this period.

Gopnik & Meltzoff (1986) offer what they term a "specificity hypothesis". This argues that, rather than stage 6 intelligence as a whole being a precursor to the emergence of language, there will be very specific links between cognition and language. Thus they predicted and confirmed a close coupling between: the developmental mastery of object permanence and the acquisition of disappearance words; and between insightful tool use to reach a goal and the acquisition of success/failure words. At the same time, no cross-coupling between abilities in the two domains was detected. A second study (1987) obtained a similar relation between the ability to categorize objects (that is, to spontaneously sort a presented pile into similar groupings) and the "naming explosion" referred to above.

Bates and her co-workers (1979, 1988) have put forward a view of the relations between cognition and early language from data gathered from three cohorts of children, two in America and one in Italy, between the ages of nine and 28 months. Both studies employ a correlational approach, abstracting out from a plethora of measures what goes with what in development at any particular point in time, and what measures at one time predict the appearance of abilities at a later time. The original study covered the period from nine to 13 months, and established three packages of developing skills that they argue to be pre-requisites for the emergence of language symbols. The second study covered the range 13 to 28 months in an attempt to follow the contribution of these packages in the move from linguistic symboling to linguistic patterning.

The three skills isolated, and claimed to be pre-requisites for symbolic development, in the sense that each was both independent of the others and required to pass some (unspecified) threshold of elaboration before providing the foundation for linguistic symbols to be constructed, were: conventional gesturing that implicates objects beyond the infant's body (as opposed to ritualized "showing off"); means–end analysis (as indexed by tool-use); and imitation (both of actions *per se* and actions upon objects). Means–end analysis and imitation abilities are indexed through the child's performance on standard developmental tests, and do not yield as precisely specifiable a relationship with language as do the measures used by Gopnik & Meltzoff (op. cit.), but, nonetheless, they provide evidence for what Bates *et al.* (op. cit.: 128 ff) term a "local homology" or "skill-specific" model of development:

> a package of related structures, capacities that are implicated in the development of linguistic and nonlinguistic symbols. There is also supporting evidence from abnormal language development, and from comparisons across species, suggesting that the same capacities that are present when language emerges are absent when language fails to emerge (ibid: 316).

In addition, there are indications in their data that the relationships that hold do so only at certain "sensitive" periods. One way of interpreting this is that, for example:

> tool use at 9 months involves a capacity that feeds into the 13-month discovery that things have names. Beyond that point, however, "more" tool use capacity may be irrelevant to "more" naming (Bates *et al.*, 1979: 367–8).

This interpretation leads to their presenting a "threshold model", applicable to both the phylogenetic and ontogenetic emergence of symbols:

> once certain critical threshold levels were reached in each of these three domains [communicative intent, tool use, and imitation], it was possible for the same three capacities to join in the service of a new function, the symbolic capacity (ibid).

In the more recent study (1988), data were not available regarding the early establishment of communicative intent, but the remaining underlying skills of analysis and rote reproduction were found to continue to play a constructive role in language development, with the addition of an unspecified language "comprehension" factor. Again, these skills were found to make varying contributions at different points in development. In addition, no evidence emerged for a split between lexical and grammatical development, meaning that the same processes operating in the elaboration of vocabulary are also operative in the establishment of early grammar (thus, if there is any specifically linguistic knowledge brought to bear in language development – as implied by the autonomous model – it only plays a role after this point).

The emergence of combinatorial skills

Combinatorial abilities appear and develop between around 12 to 30 months of age, and do so in most areas of the child's life: language, from one, to two, to multiword utterances; motor imitation; pretence play; peer social interaction; and problem solving. A number of studies have documented parallel developments during the second year of life in numerous of these domains: between language development, combinatorial abilities and decentration in symbolic play; problem-solving tasks such as obeying directions, building towers, and replacing puzzle pieces; self-awareness, problem-solving and representational skills; and peer social interaction, imitation, and language (see Brownell, 1986, 1988, for reviews of the literature). It is thus possible that these convergent developments derive from some common source[5]: perhaps the maturation of a general combinatorial ability; something related to the possession of symbolic representation; the attainment of predication; or to changes in the memory and information processing capacities of infants.

As with the very specific relations between cognitive abilities and language use found by Gopnik & Meltzoff (op. cit), there are indications of some very local relationships among these various parallel developments. For example, Brownell (1988) presents intercorrelations of measures of imitation, language and peer interaction overtures for children aged 20–27 months. While there is a common pattern of relationships between the various imitation and language measures, only manipulative imitation shows any relation to peer overtures. Crudely, if initiating interaction with a peer is thought of as an attempt to manipulate their actions, similar in nature to manipulating a non-social object, then this is another relation falling under the "specificity" or "local homology" hypotheses[6].

With respect to structural (grammatical) properties of language, Goodnow & Levine (1973) pointed out that the act of copying a design could be described in terms of grammatical rules, suggesting direct parallels between the organization of language and action. However, a large number of activities are so describable[7], and it may just be that complex activities in general tend to be hierarchically organized, and thus that there is only an analogical relationship between their structure and that of language[8]. In general, though, the problems involved in establishing relationships between ontological domains pale beside those inherent in establishing them between ontogeny and phylogeny.

Relating ontogeny and phylogeny

The "recapitulation saga" indicates that there can be no simple reading from ontogeny to phylogeny. Recent formulations suggest that a mosaic of heterochronic changes may be postulated as underlying many evolutionary changes, with selective pressures operating at numerous points in ontogeny (cf. Gould, 1977). This view proposes a dissociation between ontogenetic rates of growth, development and maturation of organic systems. This dissociation may have recapitulatory, progenic, neotenic or hypermorphotic results. Rather than forms and structures evolving by the direct action of selective forces upon them, selection can operate on the ontogenetic processes that give rise to those forms and structures. Neoteny is claimed as the major process responsible for human evolution.

However, while such an approach rehabilitates comparisons between ontogeny and phylogeny in the morphological field, it is not directly applicable to the cultural elaboration of human action and "knowledge". Neoteny clearly plays little role in this sphere (cf. Gibson, in press), for human knowledge does not remain infantile during ontogeny. Variations on the themes of recapitulation and common constraint have been advanced as explanations for the observation, originally made by Pascal but present in Piaget's views, that the elaboration of human knowledge appears as if it were conducted by an individual mind learning continuously throughout history.

Explanations along these lines have, to date, been bedevilled by the lack of a clear formulation of the relation between language and thought/knowledge. At its worst, this lack of clarity leads to the racist excesses of Victorian evolutionary sociology – the "primitive mind of the poor illiterate savage" syndrome. Yet modern scholars can still, often implicitly, fall into the trap set by their accepting, on the one hand, that structurally all extant human languages are equally complex ("evolved"), and, on the other, that the practices of different extant cultures apparently instantiate different levels of knowledge ("unequally evolved"). Hypothetico-deductive reasoning, Piaget's final stage, does not appear to be universally attained: how is equality in one sphere to be equated with inequality in another, without invoking the racist bogey, or claiming that some adult cultures are founded on the intellectual achievements of Western children?

The point being made here is that, often unreflectively, issues of culture, biology and child development are conflated in Western cultural rhetoric, because some fundamental distinctions are not made. If ontogenetic information is to inform an understanding of the evolution of human abilities, it *must* be recognized, as a starting point, that three different spheres of change are operating simultaneously in human evolution, and that feedback loops hold between them. Thus:

1.   The skills that make language possible have to be assembled to allow a language system to be formulated;
2.   A language system is unlikely to be initially formulated fully-fledged, but will be elaborated over time within cultural practices. This elaboration may involve the recruitment of additional skills over and above those required for its original impetus;
3.   The cultural practices that sustain language will themselves be changed by the possibilities opened by the possession of language, feeding back to the way in which the language system is further elaborated, and thus, in turn, feeding back to the skills required for the acquisition of what might be far removed from the originally created system. Additionally, reversing this perspective, the skills that provide for the earlier establishment of language will themselves act as constraints on whether and how the task of elaboration is carried out, as will the cultural practices that language both supports and is embedded in at any particular time.

Ontogeny itself has also to be brought into this melange. The prolonged nature of human infancy makes it a period that will selectively accrue adaptations to itself, with the result that the ontogenetic process will not map directly to the phylogenetic one. For example, where in a recapitulationist climate the phenomenon of neonatal reflexive walking might be explained as a vestige of an adaptation for locomotion in a former environment (e.g. Jersild, 1955), it may well be an adaptation to the particular conditions of ontogeny in which it appears: Prechtl (1984), for example, considers the reflex as one that allows the neonate to turn in the womb, and Ianniruberto (1985) has suggested it plays a role in engaging the neonatal head in the cervix just prior to birth. Such adaptations as may accrue in ontogeny necessarily complicate any direct reading from one domain (development) to another (evolution).

Again, changes in the relative timing of developmental events can have consequences for other, apparently unconnected events. Vauclair (1982) notes that the timing of locomotion in ape and human infants leads to the creation of markedly different modes of object exploration. Apes use all four limbs to locomote towards an object and then initially prehend it orally via outstretched lips, whereas immobile human infants need to interpose their hands as instruments with which to get objects to their mouths. This hand–mouth coordination is maintained once the human infant becomes mobile, rather than being superseded by the ape pattern, resulting in the adoption of quite different object exploration strategies in the two species at later stages in development and consequently contributing to markedly different adult outcomes in object skills.

The force of these points is that there can be no direct reading of human development for its evolutionary information. The temporal correlation of events in ontogeny will directly yield little about their possible temporal correlation in phylogeny. That, for example, we find relations between specific object oriented activities and language in early development (Meltzoff, 1988) is not, of itself, a ground for inferring that these skills evolved together: the processes that underlie the two activities may have separate origins and histories, so that phylogenetic evidence for the presence of one skill gives no grounds for assuming the presence of the other. For this, additional arguments need to be marshalled. Again, the possibility that language development is a discontinuous process does not imply that this is the case for its evolution: infants may now be adapted to acquire complex linguistic systems as a result of the previous elaboration of complex linguistic systems to constitute their developmental environment, their phylogenetic elaboration occurring by different means to those now used in ontogeny. In addition, infants develop within the arena of many cultural support systems, and once these are withdrawn, development is hindered[9]. The possession of the full wherewithal for language is no guarantee of the production of language: ontogeny occurs by guided reinvention, a different situation to that of original invention. Finally, much of the phylogenetic record of human behaviour was produced by the actions of adults, and there is little reason to assume that the products of prehistoric adult activities will be directly comparable to the ontogenetic products of modern infancy and childhood. In sum, with respect to language, it must be remembered that the ontogenetic problem – e.g., how children come to be able to acquire syntactic language – is not the phylogenetic problem – e.g., how language came to have a syntactic structure for children to acquire.

If, then, there can be no simple reading from development to evolution, what progress can be made in using knowledge of development in pursuit of an evolutionary scenario? This same question can be posed with respect to the use of comparative and cross-cultural psychological data. The following points provide a minimum framework within which to proceed:

1. There are a suite of separable/dissociable skills contributing to the cognitive abilities underlying human symbolic abilities.
2. These underlying skills can be evolved at independent rates.
3. The evolution of these underlying skills is partly independent of the elaboration of the systems they support.
4. The elaboration and cultural conservation of the systems these underlying skills make possible provides: 1. an environmental substrate that scaffolds and

thus enhances the use that can be made of those same underlying skills; and 2. an environment in which the skills that have provided for their elaboration can be further acted on by processes of selection.

5.   The rate and direction of these systems' elaborations are largely prompted by broad ecological factors. These act to provide the conditions that allow the properties of the systems themselves to be flushed out in a transcendental manner to "bootstrap" new "objects" for both old and newly required underlying skills to be elaborated and to operate on.

Ontogenetic information can play a role in filling out some of this framework, but only to the extent of hypothesizing possible scenarios, and thus posing questions on which different areas of study could bring evidence to bear.

Notes

1 Yet in becoming context-free, they also, in a sense, become object-free, in that the phenomenon of word extension occurs: a vocalization that was previously used in only one context with respect to one object can be used in other contexts, and with other, "similar" objects.

2 Pettito (1987) argues that prelinguistic gestures are radically discontinuous with language, on the grounds that deictic prelinguistic pointing gestures do not facilitate the acquisition of signed linguistic deixis in deaf children (performed by pointing at another to signal second person pronoun). It is not clear why locative deixis should be claimed to facilitate person deixis, and her general claim of discontinuity may be a result of comparing an early emerging gestural form with what would be viewed in this present treatment as a much later emerging symbolic form. However, her point does raise the general issue that the inter-relation of gesture and early words portrayed here may reflect a temporal co-existence of two very differently based systems, and that continuity cannot be assumed. The view presented here is more a continuity-with-restructuring one, but the jury must really be regarded as still out on this issue.

3 Note also that Gesell's data, in common with more recent investigations, concern *additive* constructions through the arrangement of pre-existing component parts, whereas, until the Upper Palaeolithic, human tools were made by subtractive techniques (cf. Wynn, this volume).

4 Bates *et al.* (1979) report the behaviour at 13 months, the discrepancy probably due to differences in the criteria used in their study as compared to others (McCune-Nicolich, 1981).

5 Note, however, that the crucial empirical experiments have not been conducted: that is, if these parallel developments are dependent on some common underlying structure, then stimulation in one area should lead to developments in other areas as a result of induced changes in the underlying structure.

6 A similarity that may fuel evolutionary speculations about the relations of language, tool-use and social activity in the present "Machiavellian" climate (see, for example, Byrne & Whiten, 1988).

7 For example, Cleveland & Snowdon (1982) argue that the sequencing of calls in cotton-top tamarins requires a simple phrase grammar to account for it. Peters (1972) has written a grammar for rhesus monkey copulation.

8 See also Gibson (this volume), and Greenfield (e.g., 1978, 1991).

9 The most extreme example of this actually comes from the field of comparative psychology – the ape language programs. In these, the power of the human cultural support systems is amply demonstrated, enabling animals that possess quite unadapted repertoires to use their cognitive abilities to attain levels of symbolic

functioning far beyond those that simian cognition can elaborate via simian support systems (see, for example, Savage-Rumbaugh and Rumbaugh, this volume).

References

Bates, E., Benigni, L., Bretherton, I., Camaioni, L. & Volterra, V. (1979) *The Emergence of Symbols: Cognition and Communication in Infancy.* New York: Academic Press.

Bates, E., Bretherton, I. & Snyder, L. (1988) *From First Words to Grammar: Individual Differences and Dissociable Mechanisms.* New York: Cambridge University Press.

Bloom, L. (1973) *One Word at a Time: The Use of Single Words before Syntax.* The Hague: Mouton.

Brownell, C.A. (1986). Convergent developments: cognitive-developmental correlates of growth in infant/toddler peer skills. *Child Development*, 57, 275–86.

Brownell, C.A. (1988). Combinatorial skills: converging developments over the second year. *Child Development* 59, 675–85.

Byrne, R. W. & Whiten, A. (eds.) (1988). *Machiavellian Intelligence: Social Expertise and the Evolution of Intellect in Monkeys, Apes and Humans.* Oxford: Oxford University Press.

Cleveland, J. & Snowdon, C.T. (1982). The complex vocal repertoire of the adult cotton-top tamarin (*Saguinus oedipus oedipus*). *Zeitschrift fur Tierpsychologie*, 58, 231–70.

Davis, H. (1987). The acquisition of the English auxiliary system and its relation to linguistic theory. Unpublished Ph.D. thesis, University of British Columbia.

Dore, J. (1986). Holophrases revisited: their "logical" development from dialog. In *Children's Single-Word Speech*, ed. M. Barrett, pp. 23–58. Chichester: Wiley.

Ferguson, C. & Farwell, C. (1975). Words and sounds in early language. *Language*, 31, 419–39.

Fodor, J.A. (1983). *The Modularity of Mind.* Cambridge, MA: M.I.T. Press.

Gentner, D. (1982). Why nouns are learned before words: Linguistic relativity versus natural parsing. In *Language Development. Vol. 2: Language, Thought and Culture*, ed. S. Kuczaj II, pp. 301–34. Hillsdale, NJ: Lawrence Erlbaum Associates.

Gesell, A. (ed.) (1954). *The First Five Years of Life: A Guide to the Study of the Pre-School Child.* London: Methuen.

Gibson, K. R. (1990). New perspectives on instincts and intelligence: Brain size and the emergence of hierarchical mental constructional skills. In *"Language" and Intelligence in Monkeys and Apes: Comparative Developmental Perspectives*, eds. S. T. Parker & K. R. Gibson, pp. 97–128. Cambridge: Cambridge University Press.

Gibson, K. R. (in press). The ontogeny and evolution of the brain, cognition and language. In *A Handbook of Human Symbolic Evolution*, ed. A. J. Lock & C. R. Peters. Oxford: Oxford University Press.

Gold, E.M. (1967). Language identification in the limit. *Information and Control*, 10, 447–74.

Goodnow, J. & Levine, R. (1973). "The grammar of action": sequence and syntax in children's copying. *Cognitive Psychology*, 4, 82–98.

Gopnik, A. & Meltzoff, A. N. (1986). Relations between semantic and cognitive development in the one-word stage: the specificity hypothesis. *Child Development*, 57, 1040–53.

Gopnik, A. & Meltzoff, A.N. (1987). The development of categorization in the second year and its relation to other cognitive and linguistic developments. *Child Development*, 58, 1523–31.

Gould, S. (1977). *Ontogeny and Phylogeny*. Cambridge, MA: Harvard University Press.

Greenfield, P.M. (1978). Structural parallels between language and action in development. In *Action, Gesture and Symbol: The Emergence of Language*, ed. A. J. Lock, pp. 415–45. London: Academic Press.

Greenfield, P. M. (1991). Language, tools, and brain: The ontogeny and phylogeny of hierarchically organized sequential behavior. *Behavioral and Brain Sciences*, 14, 531–95.

Grieve, R. & Hoogenraad, R. (1979). First words. In *Language Acquisition*, ed. P. Fletcher & M. Garman, pp. 94–104. Cambridge: Cambridge University Press.

Hyams, N. (1986). *The Acquisition of Parameterized Grammars*. Dordrecht: Reidel.

Ianniruberto, A. (1985). Prenatal onset of motor patterns. Paper presented to Conference on Motor Skill Acquisition in Children, Nato Advanced Study Institute, Maastricht, Netherlands.

Inhelder, B., Lezine, I., Sinclair-de-Zwart, H. & Stambak, M. (1972). Les debuts de la fonction symbolique. *Archives de Psychologie*, 41, 187–243.

Jersild, A. (1955). *Child Psychology*. 4th. edn. London: Staples Press.

Johnson, C., Davis, H. & Macken, M. (in press) Symbols and structures in language acquisition. In *A Handbook of Human Symbolic Evolution*, ed. A. J. Lock & C. R. Peters. Oxford: Oxford University Press.

Kagan, J. (1981). *The Second Year of Life: The Emergence of Self-Awareness*. Cambridge, MA: Harvard University Press.

Lempert, H. (1984). Topic as starting point for syntax. *Monogr. Soc. Res. Child Dev.* Serial No. 208, Vol. 49.

Leonard, L. B. (1976). *Meaning in Child Language*. New York: Grune and Stratton.

Lock, A. J. (1980). *The Guided Reinvention of Language*. London: Academic Press.

Lock, A. J., Young, A. Y., Service, V. & Chandler, P. (1990). Some observations on the origins of the pointing gesture. In *From Gesture to Language in Hearing and Deaf Children*, ed. V. Volterra & C. J. Erting, pp. 42–55. Berlin: Springer-Verlag.

Macken, M. (1987). Representation, rules and overgeneralization in phonology. In *Mechanisms of Language Development*, ed. B. MacWhinney, pp. 367–98. Hillsdale, NJ: Lawrence Erlbaum Associates.

McCune-Nicholich, L. (1981). Toward symbolic functioning: structure of early pretend games and potential parallels with language. *Child Development*, 52, 785–97.

Meltzoff, A. N. (1988). Imitation, objects, tools, and the rudiments of language in humans. *Human Evolution*, 3, 45–64.

Michael, G. & Willis, F. (1968). Development of gestures as a function of social class, education and sex. *Psychological Record*, 18, 515–19.

Musatti, T. & Mayer, S. (1987). Object substitution: its nature and function in early pretend play. *Human Development* 30, 225–35.

Nelson, K. (1973). Structure and strategy in learning to talk. *Monogr. Soc. Res. Child Dev.* Serial No. 143, Vol. 38.

Nelson, K. & Lucariello, J. (1986). The development of meaning in first words. In *Children's Single-Word Speech*, ed. M. Barrett, pp. 59–86. Chichester: Wiley.

Nice, M.M. (1926/27). On the size of vocabularies. *American Journal of Speech*, 2, 1–7.

Nicolich, L. (1977). Beyond sensori-motor intelligence: assessment of symbolic maturity through analysis of pretend play. *Merril-Palmer Quarterly*, 23, 89–99.

Peters, C.R. (1972). Evolution of the capacity for language: a new start on an old problem. *Man n.s.*, 7, 33–49.

Pettito, L. A. (1987). "Language" in the prelinguistic child. In *The Development of Language and Language Researchers: Essays in Honor of Roger Brown*, ed. F. S. Kessel, pp. 187–221. Hillsdale, NJ: Lawrence Erlbaum Associates.

Prechtl, H.F.R. (1984). Continuity and change in early neural development. In *Continuity of Neural Function from Prenatal to Postnatal Life*, ed. H.F.R. Prechtl, pp. 1–15. Oxford: Blackwell.

Roeper, T. & Williams, E. (1987). *Parameters and Linguistic Theory*. Dordrecht: Reidel.

Vauclair, J. (1982). Sensorimotor intelligence in human and non-human primates. *Journal of Human Evolution*, 11, 257–64.

Zinober, B. & Martlew, M. (1986). The development of communicative gestures. In *Children's Single-Word Speech*, ed. M. Barrett, pp. 183–215. Chichester: Wiley.

13

Comparative cognitive development

JONAS LANGER

Common general purpose structures underlie logical cognition (e.g., classifying, ordering, and quantifying), physical cognition (e.g., the causality involved in tool construction and use), and language (see Langer, 1980, 1986, for details of this theoretical proposal). A general purpose structure that is central to both cognition and language is combinativity. Combinativity includes composing, decomposing, and recomposing operations. These operations construct fundamental elements, such as sets and series, without which cognition and language are not possible. To illustrate the generality of these combinativity structures, consider an aspect of composing. At least two objects must be composed with each other if: (a) they are to be classified as identical or different; and (b) a tool is to be used as a causal instrument to an end. So, too, at least two symbols must be composed with each other if they are to form a minimal grammatical expression.

Basic combinativity structures are found in human infants (Langer, 1980, 1986; see also replications in Peru with Aymara and Quechua Indian infants by Jacobsen, 1984, and autistic children by Slotnick, 1984). They are also found in the two nonhuman primate species that have been studied so far on this question, *Cebus apella* and *Macaca fasciculeris*, by the Antinucci (1989) group. As an illustration consider composing two or more objects into sets by bringing them into contact or close proximity with each other. While their growth curves diverge, all three primate species are quite productive in their rate of composing. At the youngest ages at which they have been tested, the mean rate of composing objects into sets is 4.90 per minute by human infants (age six months), 2.16 by cebus (age 16 months), and 2.29 by macaques (age 22 months). With increasing age the rate of composing increases a bit (during the first year) and then remains basically stable (during the second year) in human infancy; increases sharply in cebus; and does not change in macaques. Thus,

300

composing is part of the early as well as later behavioral repertoire of all three primate species.

Given that general combinativity structures are common to all three species, it is not surprising that similar elementary cognitions develop in all three as well. These include: (a) logical cognitions, such as composing objects into single classes only to produce qualitative identity (e.g., placing circles with circles and not squares); (b) arithmetic cognitions, such as exchanging objects within single sets of objects to produce quantitative equivalences within single sets only (e.g., placing a ring behind a spoon, taking the ring away, and substituting a doll in its place); (c) physical cognitions, such as causal dependency within single sets only (e.g., using one object as a tool to repeatedly push another object such that its displacements are a direct function of the pushing); and (d) other basic physical cognitions, particularly of space and objects, such as object permanence.

Cebus and macaques, however, do not progress to the level of cognitive development already attained by human infants during their second year (Langer, 1989, 1990). This fundamental comparative developmental difference is the result, in part, of divergence in the cross-species construction of the *elements* of their cognition. The elements of organisms' cognition constrain the level of their intellectual operations. Two features are crucial: the constancy and the power of the elements. Systematic logical cognition (such as a number system) and physical cognition (such as a system of causal experimenting) are not possible without constant given elements. Moreover, the more powerful the elements the more they permit organisms to construct new and progressively more advanced logical and physical knowledge.

As already noted, set construction (e.g., composing objects together) is a major source of cognitive elements for primates. Important indices of the constancy and power of the sets constructed are their size (i.e., the number of objects composed) and the temporal relations between the sets (i.e., isolated, consecutive, or overlapping). With increasing age, infants increasingly compose two or more sets of three or more objects in temporal relation to each other (e.g., two contemporaneous sets of objects in one-to-one spatial and numerical correspondence). In comparison, cebus and macaques remain limited to composing single sets of no more than three objects that are temporally isolated from each other.

Thus, the cognitive development of cebus and macaques is constrained by the limited elements they can construct. Even relatively mature cebus and macaques are limited to constructing single sets of no more than three objects, while human infants already begin to exceed these limits by constructing multiple related sets of increasingly numerous objects. The comparative consequence is that cebus and macaques are locked into developing no more than relatively simple cognitions, while progressive possibilities open up for children to map new and more advanced cognitions.

Recursive cognitive development in humans but not in cebus and in macaques

A second fundamental basis for cross-species comparison is the developing organization of cognition both within and between domains of knowledge. Human cognitive development is recursive. It already begins to be recursive during infancy (see Langer, 1986, 1988, in press, for more detailed expositions). Recursive cognitive development is marked by multistructural, multilevel, and multilinear change in the organization of knowledge.

Multistructural change includes progressive differentiation of the domains of knowledge (e.g., logical, numerical, causal and spatial). Multistructural change also includes progressive integration of cognitive domains. To illustrate, human infants begin to develop a logic of causal experimenting during their second year, thus integrating the physical and logical domains of cognition (Langer, 1985, 1986).

Recursive cognitive development begins to be multilevel towards the end of human infants' first year (Langer, 1980). Multilevel cognition is manifest when infants' initial elementary (first-order) cognitions begin to co-exist simultaneously with the onset of more advanced (second-order) cognitions. First-order cognitions are operations on single cognitive elements such as single sets or functions. Second-order cognitions are operations which hierarchically integrate two first-order cognitive elements such as two sets or two functions. This scheme is extended to third, fourth, etc., orders or levels of cognitive development. Note that because first-order cognitive forms become elements of second-order forms, first-order cognitions are prerequisite to the development of second-order cognitions, second-order cognitions are prerequisite to the development of third-order cognitions, and so on.

Multilinear change is defined as divergent development of both domains and levels of cognitions. With development, some cognitive structures become central; that is, they become regulatory and controlling. With development there is also progressive hierarchic integration of simpler within more advanced levels of cognition. In other words, development is not simply a pattern of progressive differentiation of domains and the linear proliferation of progressively higher-order levels of cognition. Instead, development comprises domain differentiation and specialization, centralization in which some structures serve regulatory and controlling functions, and hierarchic integration of lower-order cognitions by higher-order cognitions.

As indicated in the introduction, the initial contents of cognitive development are minimal elements, such as single sets of objects. The initial forms of cognitive development are simple structures, such as minimal classificatory and causal constructions (deployed on the contents or elements of cognition). Thus, the initial relation between the forms and contents of cognitive development is elementary: minimal cognitive constructions are mapped onto minimal elements of cognition.

A keystone of recursive development is progressive change in the relation between the forms and contents of cognition. This opens up the possibility of transforming forms (structures) into contents (elements) of cognition. Thus, the initial simple linear cognitions (e.g., minimal classifying) become potential elements of more advanced hierarchic cognitions (e.g., comprehensive taxonomizing). In passing I might note that recursive cognitive development is a precondition for the formation of all reflective cognition, including abstract reflection (Piaget, 1977) and metacognition (e.g., Astington, Harris, & Olson, 1989).

The elements of cognitive development are limited to contents such as single sets of very few objects in cebus and macaques. As already indicated we have found that older infants exceed this limit by generating multiple related sets of increasingly numerous objects. In addition, and perhaps more importantly, in humans the elements of cognitive development are not limited to contents such as actual sets of objects. By late infancy, the elements begin to be expanded to include forms of cognition (e.g., classifications, correspondences, and exchanges) as well as objects, sets, series, etc. Towards the end of their second year infants begin to map their cognitive constructions onto each other. For example, as described in greater detail in the next section, some infants compose two sets of objects in spatial and numerical one-to-one correspondence. Then they exchange equal numbers of objects between the two sets such that they preserve the spatial and numerical correspondence between the two sets. These infants map exchanges onto their correspondence mappings. Thereby they produce equivalence upon equivalence relations.

Thus, in their second year human infants begin to map their cognitions onto each other. By this recursive procedure, they generate more advanced (representational) cognitions where the elements of their cognitive mappings are as much other cognitive mappings as actual things. This is our empirical evidence for identifying the cognition of young human infants, cebus and macaques as first-order linear cognitions; while identifying the cognition that begins to develop in human infants during their second year, but not in cebus and macaques, as second-order hierarchical cognitions.

The idea is that recursive mapping is necessary in order to develop representational, reflective, and hierarchic cognition. Lacking this recursive procedure, cebus and macaques never develop these more advanced forms of cognition. Beginning to construct this procedure, infants originate these more advanced forms of cognition during their second year. As indicated later in the section on cognition and language, this has important ramifications for the level of symbolization or language that can be acquired.

Parallel cognitive development in humans but not in cebus and in macaques

Two kinds of cognition (i.e., special purpose structures) are mapped onto elements by both human infants and other primates: logical and physical cognitions. Following Piaget's (1972) usage, I have labelled logical mappings as operations (actually, proto-operations in human infancy and in cebus and macaques). Operations have properties of necessary and reversible part–whole relations (or progressively acquire these properties in the course of development). Also following Piaget's (Piaget, Grize, Szeminska, & Vinh Bang, 1977) usage I have labelled physical mappings as functions (actually, protofunctions in human infancy and in cebus and macaques). Functions construct means–ends dependency relations that are contingent (e.g., achieving effects or goals is contingently dependent upon causes or tools).

At the outset, logical and physical cognition share common initial elements (e.g., single sets of objects). With development, they continue, in part, to share some common elements (e.g., multiple sets of objects). More importantly, however, their elements progressively diverge. As already noted, recursive cognitive development implies that the elements of more advanced operations (e.g., second-order operations) also include simple logical operations (e.g., first-order operations). Likewise, the elements of more advanced functions (e.g., second-order functions) also include simple physical functions (e.g., first-order functions). Hence, the elements of advanced operations include elementary operations while the elements of advanced functions include elementary functions. The progressive divergence in the elements of operations and functions reinforce the differentiated structural developmental trajectories of logical and physical cognition.

Nevertheless, operations and functions develop along parallel trajectories in human infancy. Both first-order operations and functions develop during the first year of infancy. Both second-order operations and functions develop during the second year of infancy. A brief illustrative excursion into some of our findings may be helpful.

The development of *operations* during infancy is extensive. As early as age six months infants consistently compose objects into two-object sets (e.g., by placing a ring behind but not touching a spoon) and rarely into three-object sets. They also consistently recompose these sets into derivative sets (e.g., by displacing the spoon so it touches the ring). Some six-month-olds already begin to recompose their two-object sets by first-order exchange operations of replacing, substituting, and commuting. All these exchange operations construct quantitative equivalence within single sets.

Consider the development of substituting. Simple first-order substituting is limited to single, nonreversible two-object sets by one-third of infants at age six months (e.g., by taking away the ring and substituting a finger-doll in its place behind the spoon). By age 12 months all infants produce quantitative equivalence in two-object sets by reversible substituting (e.g., by going on to take away the finger-doll and put back the ring in its place behind the spoon). One-half of these infants also already substitute objects in three-object sets they have constructed. During infants' second year, first-order substituting that produces quantitative equivalence is expanded to single sets comprising larger numbers of objects (e.g., by age 18 months one-half of infants substitute in four-object sets).

At the same time, during their second year infants begin to generate second-order substituting. Thus, by age 24 months all infants compose two sets of objects in one-to-one correspondence such that the two sets are quantitatively equivalent (e.g., two stacks of three rings). Some infants then substitute equivalent numbers of objects between the two sets (e.g., exchange the two topmost rings in the two stacks for each other). These infants produce equivalences upon equivalences by mapping substituting upon correspondence.

Parallel development is found in the *functions* generated in human infancy. Consider the development of tool construction and use. As early as age six months infants consistently generate simple first-order causal functions or dependencies. On the one hand, infants construct and replicate effects that are direct functions of causes while observing the effects. For instance, one object, such as a block, is used as a tool with which to push another dependent object several times in succession while looking at the dependent object moving. On the other hand, infants anticipate and predict causal effects. For instance, they accurately use one object as a tool to block, stop, and trap another object that is moving in front of them (see von Hofstein, 1983, for similar findings on infants' catching behavior).

With increasing age, infants progressively construct dependent effects that are direct functions of independent causes. On the one hand, infants construct direct ordered functions. For instance, objects are used as tools to push other objects harder and harder while watching what happens. On the other hand, infants construct direct categorical functions. For instance, after discovering that cylinders roll away easily and blocks do not, infants only push the cylinders in front of them and not the blocks. We have formalized these means–ends relations as one-way ratio-like functions such as "Moving further is a function of pushing harder" and "Rolling objects is a function of pushing cylinders" (see Langer, 1986, pages 370–375).

At the same time, during their second year infants begin to generate second-order causal functions. They are more indirect analogical or proportion-like functions. To illustrate, older infants, like younger ones, use one object as a tool with which to push a second dependent object. But beginning at age 18 months, when the effect is that the dependent object rolls away, infants may also transform the tool into a means with which to block the dependent object from rolling further. Correlatively, infants thereby transform the end or goal from rolling to stopping. When the dependent object stops rolling infants transform the same tool back into a means with which to make the dependent object roll away again. And so on. Thus, older infants begin to covary causes with effects to form second-order proportion-like dependencies, such as "Moving is a function of pushing as stopping is a function of blocking."

Logical and physical cognition, we have seen, are independent special purpose cognitive structures. Nevertheless, they develop in parallel in human infancy from first- to second-order operations and functions. Independent but parallel development permits operations and functions to interact and influence each others' progress. Logical cognition progressively introduces elements of necessity and certainty into physical cognition; while physical cognition progressively introduces elements of contingency and uncertainty into logical cognition (Langer, 1985, 1986).

Most significant for our present concerns is the "operationalization" of causality. This includes the origins of a logic of experimenting by applying ordering and classifying operations to causal functions. This is prerequisite to the formation of even minimal tool systems (i.e., classes and orders of tools such as constructing sets of sticks graded by length for retrieving objects at different distances). As already noted, we have traced the origins of operationalizing causality to infants' first-order ratio-like constructions (e.g., "Moving further is a function of pushing harder"). According to this hypothesis, these are the initial constructive roots of hypotheticodeductive causality that begins to develop in adolescence (Inhelder & Piaget, 1958).

From the start of human ontogeny, then, logical and physical cognition comprise contemporaneous special purpose structures that become progressively interdependent while not losing their unique properties. Synchronic development facilitates information flow between different cognitive structures. Mutual and reciprocal influence between logical and physical cognition adds power to each and, perhaps most importantly, lends impetus to progressive transformation in each.

In comparison, the cognitive ontogeny of cebus and macaques is doubly limited. First, it is limited to the development of first-order constructions. Second, it is limited to asynchronic development. This restricts information flow between different cognitive structures. We have already spelled out the general features of the first limitation. We will now consider the general features of the second constraint.

Central physical cognitions (such as object permanence and causal relations) develop before central logical cognitions (such as classifying and substituting). The development of these physical functions is well underway or completed by the time of the developmental onset of logical operations. To illustrate, cebus complete their development of object permanence (up to stage 5) during their first year (Natale, 1989) and only begin to develop logical cognition during their second year (Spinozzi & Natale, 1989; Poti & Antinucci, 1989).

The development of causal cognition also antedates logical cognition. Simple first-order functions (such as using a support as a tool to obtain a goal object) develop by age nine months in cebus and 15 months in macaques (Spinozzi & Poti, 1989). More advanced first-order functions (such as using a stick as an instrument to get a goal object) develop by 18 to 20 months in cebus; and may never develop in macaques (Natale, 1989). Thus, simple first-order causality is well developed by macaques or completely developed by cebus by the time of onset of their logical cognition. Advanced first-order causality is well developed by cebus or nonexistent in macaques by the time of onset of their logical cognition.

In good part, then, physical and logical cognition develop consecutively in cebus and macaque ontogeny. Asynchronic cognitive developments are segregated from each other in time. Since logical and physical cognition are out of developmental phase, no direct interaction or ready information flow is possible between them.

Logical cognition begins to develop after causal cognition in cebus and macaque ontogeny. The development of their causal cognition is well underway by the age at which they begin to develop logical cognition. In comparison, logical cognition originates at the same age as causal cognition in human ontogeny; and they progress in tandem. Neither begins or ends before the other during childhood and adolescence. Hence, both are open to each other's influence and to similar environmental influences in human but not in cebus and macaque ontogeny.

The causal cognitions developed by cebus and macaques, including their tool construction and use, are not subject to the progressive influence induced by information flow from (a) contemporaneously developing logical cognitions and (b) environmental influences mediated or interpreted by logical cognitions. We have already seen in the previous section that their causal cognition is not recursive and therefore not reflective. Altogether, then, this is consistent with Visalberghi's report (this volume) that even when cebus use tools successfully, they do not show any understanding of the causal process by which they achieve success. Their tool making and use is not informed by their logical cognition, which only begins to develop after their causal cognitive development is finished or almost finished. This means that their tool construction and use necessarily remains rudimentary. In particular, forming tool systems is highly unlikely if not impossible. In comparison, progressive and increasing possibilities open up for new and more powerful tool construction and use in human ontogeny (e.g., Piaget, 1987).

Cognition and language

We start from two premises: cognition generates knowledge; symbols express meaning. Cognition, I suggested in the previous section, includes operations and functions that generate two kinds of knowledge, logical and physical knowledge. Symbolic processes, on this view, complement cognitive processes. Symbolic processes express or represent meaning based upon the knowledge generated by cognitive processes. The symbolic media used to express meaning range from gestural and iconic to linguistic and mathematical notation.

It might also be worth briefly reminding ourselves of some prime differences between symbols and tools. Tool construction and use is a physical causal phenomenon. As such, tools are always material means towards material ends. They are relatively permanent instruments that, in themselves, are not symbolic. Symbols, on the other hand, are expressive or notational phenomena. Their primary functions are to represent and communicate. When used instrumentally, symbols are used rhetorically, that is, as social means towards social ends. So you can make a request with symbols, not tools; and you can retrieve an object with tools, not symbols. Symbols are relatively transient events. They are not physical causes even when used rhetorically (i.e., as social causes).

So far as I can tell, only humans have linked together tool systems with symbol systems to form hybrids such as computers. We alone have constructed systems of symbolic instruments. Note that symbolic tools are essentially symbolic means to symbolic ends. Their instrumental functions are informational, not physical. You cannot use a computer to retrieve an object. Rather, a computer is useful to inform, instruct, etc., another (organism or robot) how to retrieve an object. A computer requires some form of physical instrumental intermediary that it instructs or informs in order physically to retrieve an object.

What then might be the relations between cognition and language in evolution and development? On the present view, cognition provides axiomatic properties necessary for any grammatical symbolic system, including language and mathematics. But symbolic systems also have special purpose properties not found within cognition *per se*. For example, semantic rules of selection and representation are autonomous and vary from one symbolic medium to another, such as language and mathematical notation (see Langer, 1986, chapter 19 for a fuller discussion). Language and mathematical notation are powerful heuristic media that multiply new phenomena (i.e., possibilities, considerations, problems, contradictions, gaps, etc.) upon which cognition may operate.

Determining the relations between cognition and language is a central problem for all major theories of cognitive development (Piaget, 1951; Vygotsky, 1962; Werner & Kaplan, 1963). Unlike previous theories, however, our proposals are not based upon ontogenetic data that confound cognitive with linguistic data. They are based upon data on the development of operations and functions that are independent of the data on the development of symbol formation, such as pretend routines and verbal utterances.

These data sets led us to conclude that cognitive development is equal to or greater than linguistic development during human infancy. This proposition takes into account our data on cognitive development (Langer, 1980, 1986) and the data on symbolic and linguistic development generated in our studies (Langer, 1980, 1982, 1983, 1986) and that reported in the literature (e.g., Bloom, 1970; Bowerman, 1978; Braine, 1963; Brown, 1973; and Maratsos, 1983). It is, of course, impossible to compare quantitatively cognitive with symbolic development since there is no common developmental metric that can measure both. Nevertheless, the data are rich enough to extract a set of qualitative generalizations:

1.   First-order cognition is well-developed during the second half of infants' first year when their symbolic behavior is extremely rudimentary and involves little more than signalling.

2.   Second-order cognition originates toward the end of infants' first year when their symbolic and linguistic productions begins to be substantial.

3.   Second-order cognition is well developed by the second half of infants' second year when their linguistic production is beginning to develop some power.

To the extent that cognition and language may inform each other's development during infancy, the predominant potential influence would therefore be from cognition to language. Since it leads ontogenetically, cognition can inform and constrain language. Since language lags behind ontogenetically during most of infancy, it is less possible for it to affect cognition. As language catches up with cognition by late infancy and early childhood, the influences between cognition and language may become more mutual.

Further, infants develop second-order cognition before they progress from rudimentary signalling to more advanced language marked by supple syntax and complex semantics. Second-order cognition is a necessary condition for infants to produce and comprehend arbitrary but conventional rules by which symbols stand for and communicate referents in grammatical forms. Second-order cognitions may well be axiomatic to grammatical formations in which linguistic elements are progressively combinable and interchangeable yet meaningful. For example, this is not possible without the second-order operation of substituting elements within and between two compositions (or sets) that, as we have seen, develops toward the end of infants' second year. The hypothesis is that second-order operations (of composing, decomposing, matching, commuting, substituting, etc.) provide the rewrite rules without which grammatical constructions are not possible.

We have also seen that cebus and macaques develop first-order but not second-order cognition. On the present view, they have therefore evolved the cognitive structures necessary for signalling. Cebus and macaques have not evolved the recursive hierarchical cognitive structures necessary for advanced grammatical language that human infants begin to develop in their second or third year (for related discussions see Bickerton, 1990 and Lieberman, 1991).

Signal systems are rudimentary symbolic systems. They are very poor systems for generating new phenomena for cognitive consideration. If, as seems to be the case, cebus and macaques are limited to signalling, then their symbolizing can only play a minor role in expanding their cognitive development. In comparison, when human infants begin to develop advanced language their symbolizing can play a progressive role in fostering their continuing cognitive development. With continuing symbolic development, most especially in mathematical expressivity in childhood and adolescence, new and ever more powerful possibilities are increasingly opened up for cognitive development. Obviously, for example, advanced tool construction and use (such as of cars and computers) is not possible without the prior development and application of both advanced logical cognition and mathematical symbolization.

Heterochrony

To account for the comparative divergences in primate cognitive and linguistic development reviewed in the previous sections we have been led back to the venerable evolutionary hypothesis of heterochrony (Langer, 1989; see also Gould, 1977, and McKinney & McNamara, 1991, on the concept of heterochrony). Cebus and macaque ontogeny progresses from unilinear growth of physical cognition to multilinear growth of physical and logical cognition. Human ontogeny is multilinear from the start. Hence, the age of onset for physical cognition is roughly the same in all three species; but the age of onset for logical cognition is accelerated in humans.

These data indicate phylogenetic displacement in the ontogenetic onset or timing of physical cognition relative to logical cognition in humans as compared to cebus and macaques. The realignment of the out-of-phase development of physical and logical cognitions in ancestral populations into contemporaneous developments in human infancy opens up the possibility of information flow from the start between these different cognitive structures. The result is progressive open-ended cognitive and linguistic development in humans but not in cebus and macaques. Data gathering on the cognitive development of chimpanzees (*Pan troglodytes*) by the Antinucci group will further test the generality of heterochrony as a mechanism of primate cognitive evolution. Preliminary analyses of the chimpanzee data support the heterochrony hypothesis (Langer, in press). Emphasizing the role of heterochrony in the evolution of progressively more advanced primate cognition does not, of course, mean excluding other factors such as brain size and organization considered in the chapters in Part III of this volume.

Conclusion

Data on the comparative cognitive and linguistic ontogeny of human (particularly infants and young children) and nonhuman primates (i.e., cebus and macaques) support five propositions:

1.   General purpose structures (e.g., composing, decomposing, recomposing, commuting, replacing, and substituting) are axiomatic to the development of both cognition and language.

2.   Special purpose structures (e.g., means–ends relations in the case of tool making and use) mark and differentiate cognitive and linguistic development.

3.   Logical (e.g., classifying) and physical (e.g., causality) cognition develop in parallel or synchrony in children but not in cebus and in macaques.

4.   The intellectual development of children is recursive and leads to their progressively representational cognition. Key features of recursive development (e.g., hierarchizing) are missing in cebus and in macaques. This precludes their developing advanced representation.

5.   Heterochrony accounts in part for the differential evolution of intelligence by human and nonhuman primates.

References

Antinucci, F. (ed.) (1989). *Cognitive Structure and Development of Nonhuman Primates*. Hillsdale, NJ: Lawrence Erlbaum.

Astington, J.W., Harris, P.L., & Olson, D. (eds.) (1989). *Developing Theories of Mind*. New York: Cambridge Univeristy Press.

Bickerton, D. (1990). *Language and Species*. Chicago: University of Chicago Press.

Bloom, L. (1970). *Language Development: Form and Function in Emerging Grammars*. Cambridge, MA: MIT Press.

Bowerman, M. (1978). Structural relationships in children's utterances: Syntactic or semantic? In *Readings in Language Development*, ed. L. Bloom, pp. 217–30. New York: Wiley.

Braine, M.D.S. (1963). The ontogeny of English phrase structure: The first phrase. *Language*, 39, 1–13.

Brown, R. A. (1973). *A First Language: The Early Stages*. Cambridge: Harvard University Press.

Gould, S. J. (1977). *Ontogeny and Phylogeny*. Cambridge, MA: Harvard University Press.

Inhelder, B. & Piaget, J. (1958). *The Growth of Logical Thinking from Childhood to Adolescence*. New York: Basic Books.

Jacobsen, T. A. (1984). *The Construction and Regulation of Early Structures of Logic. A Cross-Cultural Study of Infant Cognitive Development*. Unpublished doctoral dissertation. University of California at Berkeley.

Langer, J. (1980). *The Origins of Logic: Six to Twelve Months*. New York: Academic Press.

Langer, J. (1982). From prerepresentational to representational cognition. In *Action and Thought*, ed. G. Forman, pp. 37–63. New York: Academic Press.

Langer, J. (1983). Concept and symbol formation by infants. In *Toward a Holistic Developmental Psychology*, ed. S. Wapner & B. Kaplan, pp. 221–34. Hillsdale, NJ: Lawrence Erlbaum.

Langer, J. (1985). Necessity and possibility during infancy. *Archives de Psychologie*, 53, 61–75.

Langer, J. (1986). *The Origins of Logic: One to Two Years*. New York: Academic Press.

Langer, J. (1988). A note on the comparative psychology of mental development. In *Ontogeny, Phylogeny, and Historical Development*, ed. S. Strauss, pp. 68–85. Norwood, NJ: Ablex.

Langer, J. (1989). Comparison with the human child. In *Cognitive Structure and Development of Nonhuman Primates*, ed. F. Antinucci, pp. 229–42. Hillsdale, NJ: Lawrence Erlbaum.

Langer, J. (1990). Early cognitive development: Basic functions. In *Developmental Psychology: Cognitive, Perceptuo-Motor, and Neuropsychological Perspectives*, ed. C. A. Hauert, pp. 19–42. Amsterdam: North Holland.

Langer, J. (in press). From acting to understanding: The comparative development of meaning. In *The Nature and Ontogeny of Meaning*, ed. W. F. Overton & D. S. Palermo. Norwood, NJ: Lawrence Erlbaum.

Lieberman, P. (1991). *Uniquely Human*. Cambridge: Harvard University Press.

Maratsos, M. (1983). Some current issues in the study of the acquisition of grammar. In *Handbook of Child Psychology*, Vol. III, ed. P. H. Mussen, pp. 707–86. New York: Wiley.

McKinney, M. L. & McNamara, J. K. (1991). *Heterochrony: The Evolution of Ontogeny*. New York: Plenum.

Natale, F. (1989). Stage 5 object–concept. In *Cognitive Structure and Development in Nonhuman Primates*, ed. F. Antinucci, pp. 89–96. Hillsdale, NJ: Lawrence Erlbaum.

Piaget, J. (1951). *Play, Dreams and Imitation in Childhood*. New York: Norton.

Piaget, J. (1972). *Essai de Logique Operatoire*. Paris: Dunod.

Piaget, J. (1977). *Reserches sur L'Abstraction Reflechisante: Vol.1. L'Abstraction des Relation Logicoarithmetiques*. Paris: PUF.

Piaget, J. (1987). *Possibility and Necessity. Vol. 1. The Role of Possibility in Cognitive Development*. Minneapolis: University of Minnesota Press.

Piaget, J., Grize, J. B., Szeminska, A., & Vinh Bang (1977). *Epistemology and Psychology of Functions*. Dordrecht: Reidel.

Poti, P. & Antinucci, F. (1989). Logical operations. In *Cognitive Structure and Development of Nonhuman Primates*, ed. F. Antinucci, pp. 189–228. Hillsdale, NJ: Lawrence Erlbaum.

Slotnick, C. (1984). The organization and regulation of block constructions: A comparison of autistic and normal children's cognitive development. Unpublished doctoral dissertation, University of California at Berkeley.

Spinozzi, G. & Natale, F. (1989). Classification. In *Cognitive Structure and Development of Nonhuman Primates*, ed. F. Antinucci, pp. 163–88. Hillsdale, NJ: Lawrence Erlbaum.

Spinozzi, G. & Poti, P. (1989). Causality I: The support problem. In *Cognitive Structure and Development of Nonhuman Primates*, ed. F. Antinucci, pp. 113–20. Hillsdale, NJ: Lawrence Erlbaum.

von Hofstein, C. (1983). Catching skills in infants. *Journal of Experimental Psychology: Human Perception and Performance*, 9(1), 75–85.

Vygotsky, L. S. (1962). *Thought and Language*. New York: Wiley and MIT.

Werner, H. & Kaplan, B. (1963). *Symbol Formation*. New York: Wiley.

14

Higher intelligence, propositional language, and culture as adaptations for planning

SUE T. PARKER AND CONSTANCE MILBRATH

Introduction

In this chapter we argue that capacities for higher intelligence, language and technology co-evolved in homininae in part as adaptations for planning. Specifically, we argue that the evolution of this complex culminated in *Homo sapiens* in higher intelligence, propositional language, and culture.

We propose to use this planning model in the context of a larger approach to reconstructing the evolution of planning which involves the following steps: 1. using logically constrained sequences of development in various domains in human infants and children as guidelines for understanding the sequence of evolution of these abilities; 2. placing human cognitive, linguistic, and technological abilities in a comparative context relative to those of other primate species with the help of these and other frameworks; 3. inferring the most likely abilities of the common ancestor of apes and humans from these comparative data; 4. inferring the most likely time of appearance of new abilities in various human ancestors from archeological data, using Goodenough's concepts of culture and language; 5. deducing the adaptive significance of these new abilities on the basis of analogy with living humans.

Planning theory as an integrative framework

What is needed to address these issues is a model which shows how human intelligence, technology, and language work together as an integrated functional whole, and how that whole relates to culture. We look to planning theory for this purpose because it provides an integrative framework that captures and reflects socially and ecologically relevant functions of mentality. Planning theory offers the virtue of temporal and spatial comprehensiveness, since it is applicable to the entire range of human cultural behavior irrespective of time and place. It is this aspect which renders it capable of explaining technological, motivational, intellectual, linguistic and social aspects of human behavior in an integrated fashion.[1]

To address the evolution of intelligence, technology, and planning through comparative studies a model is required which also reveals the continuities between the earliest reflexive actions of human infants and the highly elaborated declarative plans of adults, and by implication, the continuities between the simplest goal directed activities of nonhuman primates and the most complex plans of adult humans. DiLisi's developmental planning model offers this heuristic advantage.

Planning theory is a branch of cognitive science which concerns the nature of plans, and planning processes. Contrasting approaches to planning theory can be traced back to two seminal books, one by Miller, Galanter, & Pribram, *Plans and The Structure of Behavior* (1960), the other by Newell & Simon, *Human Problem Solving* (1972). Miller, Galanter, & Pribram define plans as hierarchical processes that control the order and sequence of operations to be performed in solving a problem. According to their problem-solving approach, strategies and heuristics for problem-solving form the nucleus of planning behavior. This approach emphasizes continual feedback and evaluation of sub-goal attainment as the planning sequence advances. In contrast, Newell & Simon separate planning heuristics from the planning process. They focus on the constraints the process of plan formulation imposes on planning procedures. Plan formulation involves a formal analysis of the planning domain, and in the most successful plans, formulation is entirely anticipatory to action.

The Miller, Galanter, & Pribram approach equates planning skills with problem-solving strategies, algorithms or plans-in-action (Greeno, 1978; Hayes-Roth & Hayes-Roth, 1979; Kreitler & Kreitler, 1987), while the Newell & Simons approach emphasizes the representational and anticipatory nature of plans (Pea, 1982; DiLisi, 1987; Kagan, 1984). In some cases this latter approach focuses on the meta-cognitive aspects of planning, knowing "that" (or why) or "when" a plan needs to be formulated (Brown, 1978; Brown & DeLoache 1978; Flavell, 1985).

The two approaches described above could equally well be distinguished by the degree of emphasis they place on what Anderson (1980) has called procedural as against declarative knowledge: In the case of planning based on procedural knowledge, planning is embedded in the operations employed in problem-solving. "Knowing how" to make a tool is an example of procedural knowledge which also constitutes procedural planning. While preceded by intention, procedural plans are represented primarily in action schemes, which, if interrupted, can be modified on the spot. In contrast, planning based on declarative knowledge involves symbolic representation which allows anticipation of consequences, and modification of planning sequences before plan execution. Declarative planning is more flexible in that modifications can be made by substituting new parts for old parts and/or interchanging sequences in anticipation of differing results. Declarative plans can also be consciously described while procedural plans (until they are embedded in and accessed through declarative plans) are largely unconscious and uncommunicable. Both procedural and declarative knowledge are integral to plan execution in adults: Declarative knowledge allows the planner to recognize the need for a plan, and to articulate or communicate a plan, while procedural knowledge helps the planner execute a specific plan more efficiently and effectively.

The development of planning abilities

It would be useful for the purposes of this model to know the stages of development of problem-solving and planning abilities and their relationship to intellectual and linguistic development. Various developmental approaches have emerged in recent years: Case (1985) and Siegler (1986), for example, follow the development of problem-solving by reformulating Piaget's stages in information processing terms. Richard DiLisi (1987), in contrast, traces a developmental sequence which begins in procedural modes and evolves into increasingly abstract declarative modes. It is this sequence which we use as a basis for our model.

DiLisi argues that during development, the planning process becomes increasingly distant from plan execution, requiring more protracted periods of plan formulation. At the same time planning becomes increasingly flexible until it is regulated entirely in anticipation. DiLisi distinguishes the following four sequentially developing types of planning.

Type 1 planning, planning in action, is characterized by context-bound functional procedures that are immediately executed. It includes no representational component beyond awareness of the goal. Some of these early procedural plans are reflexive while others are skilled motor sequences which have been automaticized. They are organized around such specifically defined goals as sucking or grasping. As sensorimotor schemes develop, behavior becomes increasingly voluntary and differentiated, and practical problem-solving strategies begin to appear.

Type 2 planning, planning of action, is characterized by mental anticipation of behavioral sequences as well as goals, and by representation of and communication of plans. Plan formulation is now differentiated from plan execution, but it is still temporally contiguous with execution. Because young children are incapable of deliberate, conscious and self-reflective planning, regulation of type 2 planning depends on external feedback whereby plans are locally guided by specific contexts and goals, as for example, when a child adjusts the position and movements of his hand on the hammer and his hand on the nail in response to the dynamics of the interaction between the hammer and the nail as he tries to drive a nail into wood.

Type 3 planning, planning as a strategic representation, is more highly differentiated because planners at this stage are aware of the need for a plan, the process of plan formulation, and plan execution. They are capable of mentally representing multiple courses of action, anticipating the consequences of undertaking each possible plan, and evaluating and revising (regulating) a plan prior to executing it. Thus they are capable of entertaining multiple goals. A child at this stage might, for example, consider several alternative ways of going to a friend's house, evaluating the probabilities of encountering a bully along each potential route, while simultaneously considering the desirability of passing by the ice cream store as against the comic book store.

Type 4 planning, planning as an end in itself, is characterized by such complete differentiation of the planning process from the execution context that plans can be constructed as ends in themselves rather than as means to a goal. At this stage problems and contexts can be completely hypothetical, and plans purely representational. Plans become altogether declarative. At this level, for example, the adolescent or adult might plan an entire architectural design for an imaginary village on another planet. At this stage planning becomes pure invention.

DiLisi's type 1 planning, planning in action, includes poorly differentiated schemes characteristic of Piaget's first three stages of the sensorimotor period, as well as schemes characteristic of stage 4 that are differentiated from their goals. During stage 4 of the sensorimotor period, at about eight months of age, infants begin to represent goals independently of, and prior to, their actions. Practical anticipation and selection of procedures, therefore, are possible only after stage 4 (Piaget, 1952).[2]

DiLisi's type 2 planning, planning of action, which includes not only procedural plans but the beginning of declarative plans (insofar as plans can now be communicated), develops in conjunction with preoperational abilities (Inhelder & Piaget, 1964). During the symbolic subperiod of preoperational development, from about two to five years of age, children begin to speak in sentences, to draw pictures, and to play make-believe games. Symbolic mental representation allows children to anticipate and regulate their actions prior to their execution. This opens the way for re-working a plan on subsequent trials, that is, for regulation through subsequent feedback.

DiLisi's type 3 planning, planning as strategic representation, develops in conjunction with concrete operational abilities (Inhelder & Piaget, 1964). During the concrete operations period, from about six to twelve, children develop the capacity for mental reversibility, which allows them to mentally anticipate the consequences of a number of actions. These and other abilities allow them to understand hierarchical classification systems, number systems, projective spatial concepts, and games with rules (Piaget, 1966). Increasingly effective and complex planning is a natural outgrowth of the child's emerging abilities to conceptualize, to evaluate, and to regulate applied procedures, i.e., to mentally test, revise, and re-test plans before applying them.

During the early concrete operations period from six or seven years of age, children become capable of empirical abstraction, i.e., trial-and-error coordination of their actions (Piaget, 1974). Knowledge derived from empirical abstraction allows them to generalize from experience and to anticipate future possibilities. Re-working and modifying plans continues to be done primarily through feedback forms of regulation. We characterize planning at this level as type 3A planning.

During the late concrete operations period, from nine or ten years of age, children become increasingly capable of reflexive abstraction (Piaget, 1974), i.e., inferential coordination of procedures on concrete objects. At this stage children's understanding of causal relationships begins to supplant trial-and-error as a more powerful regulator of behavior. Regulation of actions becomes proficient when the child's understanding of necessary relationships ("necessity") imposes constraints on actions (Piaget, 1985: 1987). Children know "how," "why" (or "that") and "when" to do things, and can represent this knowledge propositionally in sequences that constitute complex plans. Modification of plans can now precede execution entirely, i.e., through feedforward regulation. Planning is now fully declarative (as well as procedural). We characterize planning at this level as type 3B planning.

DiLisi's type 4 planning, when planning becomes an end in itself, and plans can be solely declarative and generated for their own sake with respect to hypothetical unrealizable goals, develops in conjunction with formal operations (Inhelder & Piaget, 1958). At the stage of formal operations from 11 or 12 years of age, children become capable of systematically testing the causal role of variables in an event, and of systematically using units of measurement. At this age they can understand the concept of universal rules of a game. They also become capable of meta-cognitive reflected abstraction, i.e., of reflecting on their own cognitive functioning. At this stage reflexive coordinations themselves become objects of thought.

While we recognize that defining planning as a continuum from the simplest forms of organized behavior to the most complex could dilute the meaning of the concept, we take the view that the heuristic advantages of defining planning in this manner outweigh the disadvantages. Although we follow DiLisi's formulation here, we do not insist that the earliest stages of planning be called true planning; they could just as well be called proto-planning. We do maintain, however, that procedural plans, i.e., plans situated in actions (see Suchman, 1987), provide a key transition in evolution because, once transformed through a signalling system or through symbolic means, they gain power by virtue of being communicable and increasingly flexible.

It should also be noted that lower stages of planning continue to operate in certain spheres and contexts throughout the life span, thus, lower stages coexist with higher stages of planning in older children. The current stage represents, then, the highest level at which the child operates, but does not imply the absence of planning characteristic of lower stages.

Linguistic event representation: A model for the transition from procedural to declarative planning

The evolutionary relationship between language and planning can be approached through a discussion of the relationship of language development, intellectual development, and event representation. Event representation includes "... representations of objects, persons, and person roles, and sequences of actions appropriate to a specific scene. In other words, it includes specific social and cultural components essential for carrying through a particular activity. If children understand to any degree the ongoing activities of their world ... they must have formed some sort of representations of these activities" (Nelson, 1983; p.135).

According to Nelson's model, children's linguistic representations take the form of "scripts" for familiar recurrent activities. "Scripts" have the following characteristics (Nelson & Gruendel, 1986): 1. social contexts; 2. sequential organization; 3. temporal, spatial and causal structures which dictate which elements must be represented; 4. goal-directed organization; 5. decomposability into smaller units of activity; and 6. generalized content.

Accounts of children's first utterances indicate that a word, e.g., "bath", denotes the entire sequence associated with bathing: going to the bathroom, taking off clothes, etc. (Nelson, 1983). A generalized account of the event is symbolically represented in the word. Similarly, early sentences often take the form of "scripts" recounting in sequential order such routine events as going to bed; grammatically, these accounts typically involve the "timeless present" verb form and the general "we" or "you" subject form, e.g., "... then you put on your jammies, and you get in bed, and you say 'night-night.'" Such "scripts" are highly generalized and consistent from child to child, and show little developmental change (Nelson & Gruendel, 1986). In contrast, episodic accounts of specific memories show significant developmental change through childhood. Nelson suggests that the early linguistic "scripts" reflect the organization of some underlying event representation crucial to subsequent language and conceptual development (Nelson, 1983; Nelson & Gruendel, 1986).

Young children also represent events through such nonlinguistic symbolic means as pretend play (Bretherton, 1984) and drawings (Piaget, 1962). These other forms of event representation may also play important roles in the development of declarative planning, e.g., as precursors to representations of spatial relations.

It is significant in this context that these early appearing, developmentally stable, general event representations symbolically encode routine procedures, or to phrase it differently, seem to be global representations of "how to" do something. In other words, the child's first utterances codify simple routine procedures.

Before new procedures can be codified, various parts of the generalized event representation must be conceptualized independently, i.e., they must be differentiated, disembedded, and decontextualized. Nelson (1983) proposes two processes of decontextualization: 1. the child uses one part of the account (e.g., a particular food item) in a number of different "scripts," (e.g., shopping, eating, preparing) and 2. the child uses several accounts as frames in which different parts or items can be inserted (e.g., several food items are inserted in the eating script). As the parts of the event become decontextualized they are abstracted from the whole as conceptually independent units. Abstraction opens the possiblity of recombining parts into new procedures, and of modifying old procedures by inserting new parts. Procedural accounts therefore become increasingly enriched.

Language is a vehicle for intentional expression characteristic of the transition from procedural to declarative planning because it helps the child to "fix and isolate her goals", and hence to sustain longer sequences of action and to engage in joint activities (Pea, Mawby & MacKain, 1982; p.2). We suggest that the transition from plans as procedures to declarative plans develops with linguistic event representation from the codification of a procedure, to the decontextualization of its parts, resulting in increased potential for plan modification and regulation.

Relationships among language, intelligence, technology, planning and culture

Language is a multi-level symbolic system comprised of three interlocking subsystems: the sound system, the meaning system, and the grammatical system, all of which operate within pragmatic social and physical environments (Parker, 1985a): "A language or dialect, then is composed of a number of systems of varying degrees of autonomy, these systems being articulated in a particular way. Changes in any one of these systems or change in the manner of their articulation will result in a somewhat different language" (Goodenough, 1981; pp. 19–20). Propositional language is the language involving correct use and understanding of such logical connectives as "because", and is therefore necessary to describing causality and logic.

Language is critical to declarative planning: because 1. language is a medium for declarative planning; 2. language classifies and names objects and routines involved in their use, and encodes and communicates the group's store of knowledge concerning technology and the organization of tasks; 3. it is a means for self-regulation during planning (Vygotsky, 1978), including self-regulation through coordination of procedures; 4. language is a means for communicating plans to others, regulating and coordinating their behaviors as potential agents or objects of plans; 5. it involves other people as co-planners or commentators who provide social feedback on planning (Goodnow, 1987).

As with language and declarative planning, the relationship between language and intelligence is a close one, which has been briefly addressed above. This topic has been widely explored by psycholinguists (e.g., Bates, 1979; Slobin, 1973) as well as developmental psychologists (e.g., Vygotsky, 1978; Piaget, 1962). It is also hotly debated by Piagetians and Chomskians but such debate is beyond the scope of this chapter.

The relationships between technology and language, and between technology and intelligence, have received less attention, and remain more controversial. A few investigators have related stone tool production techniques to Piagetian spatial concepts (Parker & Gibson, 1979; Wynn, 1979, 1989), and others have related Upper Paleolithic notation systems to temporal concepts (Marshack, 1972; Conkey, 1980). Some archeologists believe that the finished artifact reveals little or nothing about the processes of manufacture (see Wynn, 1985; Davidson and Noble, this volume). Other archeologists have interpreted data on nonverbal transmission of stone tool production techniques as evidence that there is no necessary relationship between language and technology. Using a developmental approach to this issue, Greenfield (1991) has related hierarchical object manipulation strategies to syllable manipulation strategies in human infants.

We suggest that necessary relationships among language, technology and symbolic intelligence are implicit in the planning and event representation models: Goal directed actions by agents using tools (instruments), and the transformed recipients of their actions (patients), are encoded in the grammar of event representation. This formulation implies an ontogenetically early relationship among these three elements beginning in the symbolic subperiod of preoperations. Insofar as these are logically necessary relations, we can infer that they also occurred phylogenetically (Parker & Gibson, 1979; Lock, 1983; Parker, 1985a,b; Gibson, 1986).

When we speak of language and planning, we can hardly avoid discussing their relationship to culture. Of the various definitions offered by cultural anthropologists (e.g., Kroeber & Kluckhohn, 1952), we prefer Goodenough's cognitive definition of culture as ". . . standards for deciding what is, standards for deciding what can be, standards for deciding how one feels about it, standards for deciding what to do about it, and standards for deciding how to go about it" (Goodenough, 1963, pp. 258–9; cited in Goodenough, 1981). Whereas some anthropologists include artifacts in the definition of culture, Goodenough explicitly distinguishes culture from artifacts, referring to the material manifestations of culture as "*cultural artifacts*" (Goodenough, 1981; p. 50). We contend that the classification of artifacts or institutions in or out of the definition of "culture" is unimportant so long as we recognize that these products of culture constitute significant environments to which individuals adapt.

Goodenough enumerates the contents of culture as follows: "forms, propositions, beliefs, values, rules and public values, recipes, routines and customs, systems of customs, and meaning and function" (Goodenough, 1981; p.63). All these contents play important roles in declarative planning. Rules **are** plans for directing and limiting the scope of possible behaviors; recipes and routines are the means or procedures for achieving goals; and beliefs and customs and values set the goals that are to be achieved. According to Goodenough, language is culture, and makes culture possible by encoding its contents: "A language, then provides a set of forms that is a code for other cultural forms" (Goodenough, 1981; pp. 65).

As should be clear from Goodenough's definitions, language and culture are closely related in that language is both a form of culture and a medium of culture. A language community generates, transforms, selects, stores, and transmits plans and procedures. In other words, a language community functions like a computer bulletin board interactively expanding and updating the available database. Language thereby removes planning from the pre-human domain of individual actions which are coordinated with or modeled after those of other animals through social facilitation and/or local and stimulus enhancement (Galef, 1988), and into the cultural domain.

Finally, we argue that culture in Goodenough's terms depends upon the presence of at least some cohorts of individuals who display the capacity for propositional language, abstract reasoning, and type 3B and 4 hypothetical-deductive planning. Culture implies symbolic inter-coordination of anticipated actions among participants, with subsequent modification and elaboration in anticipation of execution. Conversely, the existence of hypothetical-deductive planning necessarily implies culture.

Two ends of the comparative scale

Having postulated a human planning complex involving declarative planning, propositional language, operational intelligence, and culture, and having identified the stages of development of planning abilities in human children, we are prepared to survey planning abilities from a comparative perspective. Data on the planning abilities of our closest living relatives provide a necessary basis for reconstructing the evolution of the planning complex.

It seems clear that whereas virtually all animals act in a purposeful manner, most animals have at their disposal a rather limited set of planning rules and procedures, which at best allow them to plan a few steps in advance in response to a rather limited set of problems. Mammals in general have access to a larger repertoire of actions to use in service of their goals, and higher primates in particular have quite an extensive repertoire of manipulatory schemes that they can use in service of both subsistence and reproductive goals. Though the degree to which their strategies involve planning is more difficult to assess, type 1 context-bound planning-in-action seems to be common in monkeys. Evidence of higher level planning seems to be found primarily among the great apes.

Most observers of chimpanzees agree, for example, that these apes engage in some planful behaviors. Kohler's description of insightful problem solving with spatially dislocated tools (Kohler, 1917/27), for example, suggests type 2 incipient representation of means, and possibly type 3A mental anticipation of actions. Since Kohler's time various investigators have published compelling evidence that chimpanzees engage in planning: e.g., Savage-Rumbaugh, Rumbaugh, & Boysen's (1978) experimental demonstration that chimpanzees can communicate requests for tools appropriate to a problem-solving task; Boesch & Boesch's (1984) data on the collection and transport of stones to distant anvil sites preparatory to cracking open nuts, and their description of maternal teaching of tool use (Boesch, this volume); DeWaal's (1983) description of the chimpanzee Luit's political strategy of consistently punishing associations between his competitor and females. Even chimpanzees, however, are limited in their ability to expand, update, and communicate their database of procedures. Despite the existence of regional differences in tool use, chimpanzees do not qualify as cultural animals (Tomasello, 1990). Indeed, as we elaborate below, only *Homo sapiens* is a fully cultural animal in Goode-nough's sense.

Although little research has been devoted to this topic, most anthropologists would agree that modern humans in all societies, however primitive their subsistence patterns, display type 4 planning and hence culture. If we extend our temporal range through history and into prehistory, we find that all Upper Paleolithic "cultures" yield evidence suggestive of these capacities.

Given the enormous gap between modern humans and great apes in intellectual, linguistic, and hence planning abilities, students of the evolution of human cognitive abilities naturally look to paleontological and archeological data for clues to the capacities of our various human ancestors.[3]

Paleontological and archeological clues to planning abilities

It is generally agreed that the transition from Middle to Upper Paleolithic tool cultures marks a major shift in hominine adaptation. After this transition the archeological remains of our ancestors meet criteria for cultural animals implicit in Goodenough's definition of culture: imposed form, standardization, regionalization, and rapid historic change in tools (e.g., Braidwood, 1975; Stringer, 1989). In Europe for example, the Upper Paleolithic is characterized by the invention of the bow and arrow, a great emphasis on blades, and the development of bone, ivory, and antler implements apparently associated with greater dependence on reindeer (Gamble, 1986). Significantly, these features are associated with cave art and statuary, notation systems (Marshack, 1972), and seasonal aggregations for the harvesting of animal resources (Conkey, 1980). This period was followed by geographic expansion into Australia and Siberia (Whallon, 1989).

Regionalization and seasonal aggregation in turn seem to imply increased population densities, formalized reciprocal relations among neighboring groups, i.e., alliance networks, and a socially organized division of labor (as opposed to separation of labor). All of these features in turn imply systematic information exchange among local groups, and hence, fully-fledged language, planning and prediction (Whallon, 1989).

Unfortunately, the temporal framework for this transition is unclear. Whereas the transition to the Upper Paleolithic in Europe occurs at about 35000 years ago, some African archeologists see signs of this transition in Africa beginning as early as 80000 years ago (Mellars & Stringer, 1989). Likewise, the nature and timing of the morphological transition from archaic to anatomically modern *Homo sapiens* is disputed. Whereas the transition in Europe is dated at 35000 years ago, paleontological evidence from Africa suggest the appearance of modern looking *Homo sapiens* dating from around

150000 years ago or earlier. This interpretation, which suggests an African origin for modern humans, has been given greater credence in light of racial cladograms based on mitochondrial deoxyribonucleic acid (mDNA), which also suggest the origin of modern races from a single African population 200000 years ago (Stoneking & Cann, 1989; Rouhani, 1989).

Both the African fossil and the mDNA data seem to support the single origin as opposed to the multiple origin model for the evolution of anatomically modern *Homo sapiens*. Recent evidence of long term concurrent habitation of modern *Homo sapiens* and Neanderthals in Israel also supports this interpretation (Stringer, 1989; 1990).

Significantly, anatomical differences in the chin and base of the skull between these two groups seem consistent with the hypothesis that while anatomically modern *Homo sapiens* possessed the articulatory apparatus for modern language (Lieberman, 1984), Neanderthals were less able to articulate speech sounds, though recent evidence from finds of the hyoid bone fails to support this interpretation (Arensburg *et al.*, 1989). Endocasts of the two groups suggest that Neanderthals fail to show the brain morphology characteristic of anatomically modern *Homo sapiens*. Specifically, they may be more primitive in their frontal and parietal lobes, which are precisely the regions implicated in declarative planning (Gibson, 1985; 1988; contra Holloway, 1985). Recent reinterpretations of archeological data on Neanderthals moreover argue that these creatures did not engage in rituals or bury their dead (Gargett, 1989).

If these archeological and paleontological interpretations are correct, they would support the notion that full declarative type 3B and/or 4 planning first appeared in anatomically modern *Homo sapiens* (whether at 80000 years or at 35000 years ago). This seems to imply that these creatures were capable of late concrete or formal operational reasoning.

If we accept this reconstruction of *Homo sapiens*, and the reconstruction of the earliest homininae (putatively *Australopithecus afarensis*) as bipedal small-brained creatures who were not yet manufacturing worked stone tools (Johanson & Edey, 1981), we have the two ends of the continuum of hominine cognitive evolution, with a chimpanzee-like ancestor at one end, and anatomically modern humans at the other end. The australopithecines, including *A. afarensis* and their descendants, *A. robustus* and *A. boisei* may have, like living great apes, been capable of at least type 2 and possibly type 3A planning. See Figure 14.1 for a family tree of the two lineages of homininae reflected in the two genera, *Australopithecus* and *Homo*.

This reconstruction sets some parameters, suggesting that the capacity for type 3A planning consolidated and ramified in human ancestors between *Australopithecus afarensis* and *Homo sapiens*.

It gives us no clues, however, concerning the rates of evolution of cognitive

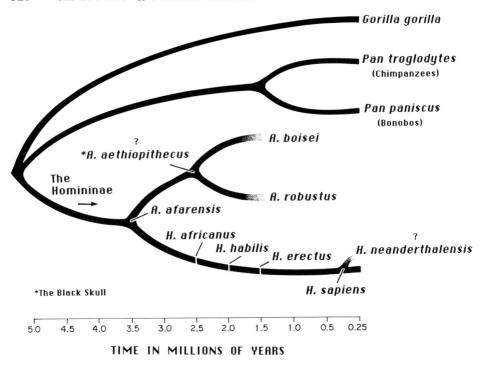

Fig. 14.1. A possible hominine phylogeny.

abilities within and among domains from early preoperations through concrete operations. Developmental logic suggests that each series or domain may have evolved in its own developmental sequence; developmental evidence tells us that the sequences develop concurrently in modern human infants and children. These approaches give us no definite clue to the rates of evolution in different domains prior to the emergence of mature language and culture and of modern humans. For this we must rely upon reconstructions from archeological data, however problematic their interpretation may be.

Archeological evidence suggests that larger brained *Homo habilis* (circa 2.5 million years ago) were the earliest tool manufacturers. The picture is complicated by new evidence of primitive features in the forelimbs of *Homo habilis* and the newly raised possiblity that more than one species of *Homo* lived concurrently (Simons, 1989). The long persistence of static, regionally invariant tool (Oldowan) traditions associated with these creatures, who had brains of approximately 700 cc, suggest that they had neither propositional language (as opposed to simpler forms of language) nor culture in Goodenough's sense, and were probably restricted to type 2 or 3A planning.

Likewise, archeological evidence associated with *Homo erectus* (circa 1.6 to 0.3 million years ago), the putative descendent of *Homo habilis*, and putative ascendant of *Homo sapiens*, suggests long persistence of their more advanced Acheulean or handax tool culture, and hence the absence of fully-fledged language and culture. On the other hand, even the simple biface tools they manufactured and the simple shelters they apparently constructed imply the capacity for late preoperational or early concrete operational topological and euclidean spatial concepts (Parker & Gibson, 1979; Wynn, 1979; 1989), and at least type 2 planning and possibly type 3A planning involving mental anticipation, and some form of symbolic communication of plans temporally contiguous with their execution. Likewise, the gathering of materials and shaping them to any kind of standard suggests the capacity for detailed imitation and fairly complex procedural planning.

This brief overview suggests that formal operational reasoning (Gibson, 1985; 1988; Parker, 1985b), full declarative planning, and the associated capacity for culture probably emerged during the latest stage in human evolution, i.e., in the transition from anatomically primitive to anatomically modern *Homo sapiens* sometime between 200 000 and 35 000 years ago. This is not to say, however, that more primitive forms of reasoning and planning ceased to exist; rather, more primitive forms must have co-existed with the new abilities, as they do in humans today.

The adaptive significance of declarative planning

This reconstruction of the emergence of declarative planning alongside the emergence of anatomically modern *Homo sapiens* raises the question of what new kinds of selection pressures occurring during this period could have shaped the transition. Reconstruction of Upper Paleolithic selection pressures necessarily involves ethnographic analogies with living humans, however subject to error the analogic process might may be. Use of ecological and evolutionary models is one important means for strengthening analogic inference (Wylie, 1985). Also crucial to reasoned use of ethnographic analogies is a recognition of the diversity of Paleolithic habitats and subsistence strategies (Yellen & Harpending, 1972), and of the diversity of kinship and residence patterns among historically known populations of hunter–gatherers (e.g., Murdock, 1949; Fox, 1967; Service, 1962; Coon, 1977).

A central challenge in reconstructing Upper Paleolithic selection pressures is to identify new circumstances coinciding with the transition from archaic to modern *Homo sapiens* which would have favored individuals with advanced planning abilities. Climatological shifts attendant on glacial episodes may have been one such circumstance. Expansion into highly variable and seasonal habitats with unstable resources may have been an associated circumstance

(Whallon, 1989; Gamble, 1986). As groups moved into diverse habitats they must have depended on new technologies, and on new routines for employing these technologies. Likewise, they must have experimented with new social forms for exchanging locally scarce resources. In other words, the challenges of inhabiting such regions must have favored the invention of innovative physical and social solutions (Gamble, 1976).

On the technological side, inhabitants of regions with high seasonal variation would have benefited from planning/inventing new kinds of shelters, clothing, implements for seasonal hunting (e.g., bows and arrows), food harvesting, transport, preparation and storage techniques, devices for reckoning annual cycles in relation to fish, game, and other resource runs, devices for mapping terrain and resources, devices for navigating in long distance travel similar to those used by modern hunter–gatherers (Mason, 1895/1966; Coon, 1977; Parker, 1985b), as well as devices for communicating through items of material cultural (Gamble, 1976; Gibson, personal communication 1991).

On the social side, inhabitants of such regions would have benefited from planning/inventing social alliances to be contracted and maintained during the periods of seasonal aggregation in which larger populations would have congregated. Such aggregations are reconstructed by archeologists (Gamble, 1976; Conkey, 1980), and are common among historic hunter–gatherer populations in all areas of the world (Mauss, 1950/1979; Service, 1962; Coon, 1977; Lee, 1979). Large scale regional movements associated with such migratory herd animals as reindeer, for example, seem to have appeared in the Upper Paleolithic in Europe (Conkey, 1980).

Formation of alliances may have become a major adaptation of anatomically modern *Homo sapiens* as they moved into variable habitats (Whallon, 1989; Gamble, 1982): Such alliances between families have the singular advantage of evening out temporal and regional fluctuations in key resources, and of providing coalitions for resource harvest and defense (Wobst, 1976; Gamble, 1976). This alliance model is consistent with the way bands are formed among living hunter–gatherers, for whom the band has been defined as a "temporally unstable aggregation(s) of families with links of kinship or friendship functioning primarily to smooth the daily variations in exploitative success of its members by foodsharing" (Yellen & Harpending, 1972).

Among historic human societies such alliances generally occur in the context of marriage and resource exchanges based on kinship. Typically such alliances have been established and maintained during periods of seasonal aggregation when feasting, trading, rituals, fighting and sexual activities are at a high pitch (e.g., Mauss, 1950/1979; Coon, 1977; Lee, 1979). Among the !Kung San, for example, "Marriage brokering was another of the activities bringing together large groups of San. Because of the small size of local groups and the frequent disparities in sex ratios, parents had to look far afield in arranging a marriage for their son or daughter. ... The large winter camps gave families the opportunity of casting a wider net in seeking a spouse" (Lee, 1979; p. 366).

We suggest that the mental capacities associated with the cultural planning complex arose through both natural selection and social selection operating on Upper Paleolithic populations as they settled in unstable and variable habitats: Natural selection would have acted directly by favoring individuals with planning skills related to improved resource acquisition and survival, as for example in inventing technologies. Social selection, in contrast, would have acted indirectly by favoring those individuals whose planning skills helped them acquire more mates, and to influence the behavior of offspring, kindred, friends, and followers, as for example in inventing and using systems of kinship terms, and of marriage and resource exchange (Parker, 1984; 1985 a& b).[4] In other words, we suggest that in the Upper Paleolithic, both natural and social selection favored type 3B and 4 planning, thereby integrating linguistic, technological, and intellectual skills into a single cultural planning complex.

Acknowledgements

We thank the Drs. Kathleen Gibson and Tim Ingold, the Organizers of the Conference and the editors of this volume, for their comments and suggestions. We also thank the Wenner-Gren Foundation, Dr. Sydel Silverman, and Ms. Laurie Obbink and other staff members for sponsoring and hosting an intellectually exciting conference, and the participants of that conference for the ferment they engendered. We also thank the following individuals for helpful comments on the manuscript: Drs. Roy Pea, Margie Purser, Anne Russon, and Shirley Silver. We also thank Phyllis T. Spowart for drawing Figure 1.

Notes

1 Although Piagetian frames are valuable for identifying and comparing cognitive abilities in various domains (physical including causality and space; logical–mathematical including classification, number, etc.), these frames address these abilities structurally and without reference to their adaptive contexts. Considered individually, various of these cognitive structures suggest such specific adaptive functions as stone tool manufacture (Parker & Gibson, 1979; Wynn, 1979; 1989), aimed missile throwing, and allocation of resources (Parker & Gibson, 1979). Such frames, however, do not provide an integrated functional pattern, and therefore offer only modest guides for understanding the possible relationships among cognitive domains in our human ancestors.

2 We differ from DiLisi in distinguishing type 1A planning characteristic of sensorimotor series stages 1–4 from type 1B planning characteristic of sensorimotor

series stages 5 and 6. Type 1A planning constitutes procedural planning which unfolds in the actions themselves, while type 1B planning involves incipient representation of means. Piaget himself did not focus on planning in a social context, and so far few planning theorists have used Piagetian frames. Therefore the empirical evidence for these developmental changes is limited. Research by Kreitler & Kreitler (1987), however, is consistent with both DiLisi's developmental typology, and his Piagetian interpretations of this typology.

3 Given current estimates of the highest cognitive abilities of our closest relatives, chimpanzees and gorillas, in Piagetian terms (see Dore & Dumas, 1987; Parker, 1990 for reviews), we conclude that the post-sensorimotor period abilities characteristic of the late preoperations, concrete operational, and formal operational periods must have originated after humans and chimpanzees diverged from their common ancestor (Parker & Gibson, 1979). If we map the pattern of occurrence of various sensorimotor abilities in various primate species we see a mosaic pattern which suggests that independent selection pressures have operated in the evolution of various sensorimotor series.

4 Social selection is a general term embracing forms of selection which are mediated by conspecifics (e.g., sexual selection, kin selection, parental manipulation, and reciprocal altruism), as contrasted with natural selection which is mediated by other species and abiotic forces (Trivers, 1971; Alexander, 1974; West-Eberhard, 1983).

References

Alexander, R. (1974). The evolution of social behavior. *Annual Review of Ecology and Systematics*, 4, 325–85.

Anderson, J. R. (1980). *Cognitive Psychology and Its Implications*. San Francisco: Freeman.

Arensburg, B., Tillier, A. M., Vandermeersch, B., Schepartz, L. & Rak, Y. (1989). A Middle Palaeolithic human hyoid bone. *Nature*, 338, 758–60.

Bates, E. (1979). *The Emergence of Symbols*. New York: Academic Press.

Boesch, C. & Boesch, H. (1984). Mental maps in wild chimpanzees: An analysis of hammer transports for nut cracking. *Primates*, 25, 160–70.

Braidwood, R. J. (1975). *Prehistoric Men*. 8th edn. Glenville, IL: Scott, Foresman and Company.

Bretherton, I. (1984). Introduction: Piaget and event representation in symbolic play. In *Symbolic Play*, ed. I. Bretherton, pp. 3–41. New York: Academic Press.

Brown, A. (1978). Knowing when, where and how to remember: A problem of metacognition. In *Advances in Instructional Psychology*, vol. 1, ed. R. Glaser, pp. 76–166. Hillsdale, NJ: Erlbaum.

Brown, A. & DeLoache, J. S. (1978). Skills, plans and self-regulation. In *Children's Thinking: What Develops?*, ed. R. S. Siegler, pp. 73–96. Hillsdale, NJ: Erlbaum.

Case, R. (1985). *Intellectual Development: Birth to Adulthood*. New York: Academic Press.

Conkey, M. (1980). The identification of prehistoric hunter-gatherer aggreggation sites: The case of Altamira. *Current Anthropology*, 21, 609–29.

Coon, C. S. (1977). *The Hunting Peoples*. New York: Penguin.

DeWaal, F. (1983). *Chimpanzee Politics*. New York: Harper & Row.

DiLisi, R. (1987). A cognitive-developmental model of planning. In *Blueprints for Thinking*, ed. S. Friedman, E. Scholnick & R. Cocking, pp. 79–109. Cambridge: Cambridge University Press.

Dore, F. Y. & Dumas, C. (1987). Psychology of animal cognition: Piagetian studies.

Psychological Bulletin, 102(2), 219–33.

Flavell, J.H. (1985). *Cognitive Development*. Englewood Cliffs, NJ: Prentice-Hall.

Fox, R. (1967). *Kinship and Marriage*. Harmondsworth: Penguin.

Galef, B. G., Jr. (1988). Imitation in animals: History, definitions, and interpretation of data from the pyschological laboratory. In *Social Learning: Psychological and Biological Perspectives*, ed. T. Zentall & B. G. Galef, Jr., pp. 3–28. New York: Erlbaum.

Gamble, C. (1976). Interaction and alliance in Palaeolithic society. *Man*, 17, 92–107.

Gamble, C. (1986). *The Palaeolithic Settlement of Europe*. Cambridge: Cambridge University Press.

Gargett, R. (1989). Grave shortcomings: The evidence for Neanderthal burial. *Current Anthropology*, 30(2), 157–90.

Gibson, K. R. (1985). Has the evolution of intelligence stagnated since Neanderthal man (commentary on Parker) In *Evolution and Developmental Psychology*, ed. G. Butterworth, J. Rutkowska, & M. Scaife, pp. 102–14. Brighton: Harvester Press.

Gibson, K. R. (1986). Cognition, brain size and extraction of embedded foods. In *Primate Ontogeny and Social Behaviour*, ed. J. C. Else & P. C. Lee, pp. 93–105. New York: Cambridge University Press.

Gibson, K. R. (1988). Brain size and the evolution of language. In *The Genesis of Language*, ed. M. E. Landsberg, pp. 149–73. New York: Mouton deGruyter.

Goodenough, W. (1981). *Language, Culture and Society*. 2nd edn. Menlo Park: Benjamin Cummings.

Goodnow, J. (1987). Social aspects of planning. In *Blueprints for Thinking*, ed. S. L. Friedman, E. K. Scholnick & R. R. Cocking, pp. 179–202. Cambridge: Cambridge University Press.

Gould, R. (1980). *Living Archeology*. New York: Cambridge University Press

Greenfield, P. (1991). Language, tools, and brain: The ontogeny and phylogeny of hierarchically organized sequential behavior. *Behavioral and Brain Sciences*, 14, 531–95.

Greeno, J.G. (1978). A study of problem solving. In *Advances in Instructional Psychology*, vol. 1, ed. R. Glaser, pp. 13–75. Hillsdale, NJ: Erlbaum.

Hayes-Roth, B. & Hayes-Roth, R. (1979). A cognitive model of planning. *Cognitive Science*, 3, 275–310.

Holloway, R. (1985). The poor brain of "Homo sapiens Neanderthalensis": See what you please. In *Ancestors: The Hard Evidence*, ed. E. Delson, pp. 319–24. New York: Alan Liss.

Inhelder, B. & Piaget, J. (1958). *The Growth of Logical Thinking from Childhood to Adolescence*. New York: Basic Books.

Inhelder, B. & Piaget, J. (1964). *The Early Growth of Logic in the Child*. London: Routledge & Kegan Paul.

Johanson, D. & Edey, M. (1981). *Lucy*. New York: Werner Books.

Kagan, J. (1984). *The Nature of the Child*. New York: Basic Books.

Kohler, W. (1917/1927). *The Mentality of Apes*. New York: Vintage Press.

Kreitler, S. & Kreitler, H. (1987). Conceptions and processes of planning: the developmental perspective. In *Blueprints for Thinking*, ed. S. L. Friedman, E. K. Scholnick & R. R. Cocking, pp. 179–204. Cambridge: Cambridge University Press.

Kroeber, A. & Kluckhohn, C. (1952). *Culture: A Critical Review of Concepts and Definitions*. Papers of the Peabody Museum of American Archeology and Ethnology. Cambridge, MA: Harvard University.

Lee, R. B. (1979). *The !Kung San.* New York: Cambridge University Press.

Lieberman, P. (1984). *The Biology and Evolution of Language.* Cambridge, MA: Harvard University Press.

Lock, A. (1983). "Recapitulation" in the ontogeny and phylogeny of language. In *Glossogenetics: The Origin and Evolution of Language*, ed. E. deGrolier, pp. 255–73. Paris: Harwood Academic Press.

Marshack, A. (1972). *The Roots of Civilization.* New York: McGraw Hill.

Mason, O. T. (1895/1966). *The Origins of Inventions.* Boston: M.I.T. Press.

Mauss, M. (1950/1979). *Seasonal Variations of the Eskimo.* London: Routledge & Kegan Paul.

Mellars, P. & Stringer, C. (1989). Introduction. In *The Human Revolution*, ed. P. Mellars & C. Stringer, pp. 1–14. Princeton: Princeton University Press.

Miller, G., Galanter, R. & Pribram, K. (1960). *Plans and the Structure of Behavior.* New York: Holt, Rinehart & Winston.

Murdock, G. P. (1949). *Social Structure.* New York: Free Press.

Nelson, K. (1983). The derivation of concepts and categories from event representation. In *New Trends in Conceptual Representation: Challenges to Piagetian Theory*, ed. E. Scholnick, pp. 129–149. Hillsdale, NJ: Erlbaum.

Nelson, K. & Gruendel, J. (1986). Children's scripts. In *Event Knowledge*, ed. K. Nelson, pp. 21–45. Hillsdale, NJ: Erlbaum.

Newell, A. & Simon, H. (1972). *Human Problem Solving.* Englewood Cliffs, NJ: Prentice-Hall.

Parker, S. T. (1984). Playing for keeps. In *Play in Animals and Humans*, ed. P. K. Smith, pp. 271–93. Oxford: Blackwell.

Parker, S. T. (1985a). A social technological model for the evolution of language. *Current Anthropology*, 26(5), 617–39.

Parker, S. T. (1985b). Higher intelligence as an adaptation for social and technological strategies in early *Homo sapiens.* In *Evolution and Developmental Psychology*, ed. G. Butterworth, J. Rutkowska, & M. Scaife, pp. 83–100. Brighton: Harvester Press.

Parker, S. T. (1990). Origins of comparative developmental evolutionary studies of primate mental abilities. In *"Language" and Intelligence in Monkeys and Apes: Comparative Developmental Perspectives*, ed. S. T. Parker & K. R. Gibson, pp. 3–64. New York: Cambridge University Press.

Parker, S. T. & Gibson, K. R. (1979). A developmental model for the evolution of language and intelligence in early hominids. *Behavioral and Brain Sciences*, 2, 367–408.

Pea, R. D. (1982). What is planning development the develpment of? In *Children's Planning Strategies: New Directions for Child Development*, vol. 18, ed. D. L. Forbes, & M. T. Greenberg, pp. 5–27. San Francisco: Jossey-Bass.

Pea, R. D., Mawby, R. & MacKain, S. (1982). World-Making and World-Revealing: Semantics and Pragmatics of Modal Auxiliary Verbs During the Third Year of Life. Paper presented at the Seventh Annual Boston U. Conferences on Child Language Development.

Piaget, J. (1952). *The Origins of Intelligence in Children.* New York: W. W. Norton.

Piaget, J. (1962). *Play, Dreams and Imitation in Childhood.* New York: W. W. Norton.

Piaget, J. (1966). *The Moral Judgment of the Child.* New York: Free Press.

Piaget, J. (1970). *Structuralism.* New York: Harper.

Piaget, J. (1974). *Success and Understanding.* Cambridge, MA: Harvard University Press.

Piaget, J. (1985). *The Equilibration of Cognitive Structures: The Central Problem of Intellectual Development*. Chicago: University of Chicago Press.

Piaget, J. (1987). *Possibility and Necessity*, vols. 1 & 2. Minneapolis, MN: University of Minnesota Press.

Rouhani, S. (1989). Molecular genetics and the pattern of human evolution: plausible and implausible models. In *The Human Revolution*, ed. P. Mellars & C. Stringer, pp. 47–60. Princeton: Princeton University Press.

Savage-Rumbaugh, S., Rumbaugh, D. & Boysen, S. (1978). Symbolic communication between two chimpanzees (*Pan troglodytes*). *Science*, 201, 641–4.

Service, E. (1962). *Primitive Social Organization*. New York: Random House.

Siegler, R. (1986). *Children's Thinking*. New York: Prentice-Hall.

Simons, E. (1989). Human origins. *Science*, 245, 1343–50.

Slobin, D. (1973). Cognitive prerequisites of grammar. In *Studies of Child Language Development*, ed. D. Slobin & C. Ferguson, pp. 175–206. New York: Cambridge University Press.

Stoneking, M. & Cann, R. (1989). African origin of human mitochondrial DNA. In *The Human Revolution*, ed. P. Mellars & C. Stringer, pp. 17–29. Princeton: Princeton University Press.

Stringer, C. (1989). The evolution of modern humans: A comparison of the African and non-African evidence. In *The Human Revolution*, ed. P. Mellars & C. Stringer, pp. 123–53. Princeton: Princeton University Press.

Stringer, C. (1990). The emergence of modern humans. *Scientific American*, December, 98–104.

Suchman, L. A. (1987). *Plans and Situated Actions*. New York: Cambridge University Press.

Tomasello, M. (1990). Cultural transmission in the tool use and communicatory signaling of chimpanzees. In *"Language" and Intelligence in Monkeys and Apes: Comparative Developmental Perspectives*, ed. S. T. Parker & K. R. Gibson, pp. 274–310. New York: Cambridge University Press.

Trivers, R. (1971). The evolution of reciprocal altruism. *Quarterly Review of Biology*, 46, 35–57.

Vygotsky, L. S. (1978). *Mind in Society*. ed. M. Cole, V. John-Steiner, S. Scribner, & E. Souberman. Cambridge: Harvard University Press.

West-Eberhard, M. (1983). Sexual selection, social competition, and speciation. *Quarterly Review of Biology*, 58, 155–83.

Whallon, R. (1989). Elements of cultural change in the later Palaeolithic. In *The Human Revolution*, ed. P. Mellars & C. Stringer, pp. 433–53. Princeton: Princeton University Press.

Wobst, H. M. (1976). Locational relationships in Paleolithic society. *Journal of Human Evolution*, 5, 49–58.

Wylie, A. (1985). The reaction against analogy. *Advances in Archeological Method and Theory*, vol. 8, pp. 63–111. New York: Academic Press.

Wynn, T. (1979). Intelligence of later Achuelean hominids. *Man*, 14, 371–91.

Wynn, T. (1985). Piaget, stone tools and the evolution of human intelligence. *World Archeology*, 17, 32–43.

Wynn, T. (1989). *The Evolution of Spatial Competence*. Illinois Studies in Anthropology. No. 17. Chicago: University of Illinois Press.

Yellen, J. & Harpending, H. (1972). Hunter–gatherer populations and archeological inference. *World Archeology*, 4, 244–52.

Part V

Archaeological and anthropological perspectives

Part V Introduction

Tools, techniques and technology

TIM INGOLD

Archaeologists have established, with a fair degree of certainty, that around 1.5 million years ago ancestral populations of *Homo erectus* began making a kind of flaked stone implement known in the trade as the 'Acheulean hand axe'. Of a size that fits neatly into the palm of a modern human, these implements are immediately striking – by comparison with earlier stone tools – on account of their elegant bilateral symmetry and overall regularity of form. The designation 'hand axe' is unfortunate, however, as no-one really knows what they were used for: suggestions range from picks and cleavers to frisbee-like projectiles (see Calvin in this volume, Chapter 10). In formal rather than functional terms, they are more accurately called 'bifaces'. Their discovery in sites covering the three continents of Africa, Europe and Asia, over a period spanning more than a million years, poses one of the most profound enigmas known to archaeological science. Condensed in the 'Acheulean problem' are almost all of the issues tackled by contributions to this section, three of which – by Toth and Schick, Davidson and Noble, and Wynn – raise the problem explicitly.

The issues are these: How can we account for such a degree of formal standardization, over such vast expanses of space and time? Can this be taken as evidence for the operation of cultural norms? Did the persistence of the artefact form entail a process of cultural transmission, and did this process depend on verbal communication? Did the objects conform to a plan or representation in the mind of the tool-maker? Do they tell us anything about the nature and evolution of hominid sociality? Might they have had communicative or semiotic as well as technical functions?

To the first question, both Wynn and Toth and Schick, on the one hand, and Davidson and Noble on the other, present totally contradictory answers. There were, say Toth and Schick, clear cultural norms for making particular shapes of bifaces, a view echoed by Wynn in his comment that since there is no obvious way in which these shapes were constrained by function, they must have been dictated by what he calls 'community standards' integral to any kind of cultural tradition. Yet it has to be admitted that if this is so, the tradition involved was of a kind quite without parallel in the archaeological and ethnographic record of anatomically modern humans. For even the most conservative traditions known from this record, supported by a fully evolved language capacity and the whole panoply of institutions that this makes possible, possess nothing even approaching the stability of the Acheulean form. In the light of this problem, Davidson and Noble propose a radical alternative. They argue that there were no strictly cultural norms or standards until the advent of the Upper Palaeo-lithic, coinciding with the arrival of modern humans, and that the shapes of bifaces can be fully accounted for by the mechanical constraints of the manufacturing process. This process, moreover, may not even have been geared to the production of bifaces: it is possible – Davidson and Noble suggest – that bifaces are residual cores left after successive removals of flakes. The flakes, then, were meant to be used, not the cores.

Let me follow up these alternative answers, since they lead to further questions of critical significance. Suppose that the Acheulean form *does* attest to the operation of normative standards. How, then, were these standards maintained and transmitted? Given suitable raw materials in the local environ-ment, each generation could learn from the last through demonstration and imitation, a pedagogic process which – as Boesch shows elsewhere in this volume (Chapter 7) – occurs even among chimpanzees. Wynn argues that the skills of making and using tools are very generally acquired through apprenti-ceship, that is through practice on the job, in which each novice acquires the 'knack' for him- or herself through repetition and rote memorization (see also Lave 1990). This requires neither language nor any other advanced cognitive capability. Nevertheless, making a biface is by no means easy and requires a great deal of practice, as Toth and Schick have found in experimental attempts to replicate the manufacturing process. Whether or not the ease of acquisition of this skill would be significantly enhanced, had the novice the benefit of verbal instruction from a tutor, has still to be determined; it could be argued, however, that the Acheulean form owes its stability, at least during the earlier periods of its occurrence, to the *limitations* of *Homo erectus*'s cognitive and linguistic capabilities. In other words, these hominids had the ability to replicate an established design, but not to conceptualize novel forms in advance of their realization. As Bloch has recently argued (1991: 193), it is not necessary for the effective transmission of everyday practical knowledge that it be expressible in verbal form, but this *may* be a condition for creative innovation.

This very limitation, Toth and Schick suggest, may also account for the otherwise anomalous *absence* of Acheulean implements from the earliest sites in Europe and Asia. For to reach these continents from their original area of distribution in Africa, hominid populations had to traverse regions in which the kind of raw material needed for making bifaces was not available. There was consequently a hiatus in the inter-generational process of apprenticeship, for novices had no material on which to work. The fact that Acheulean forms appear in later periods throughout both Europe and Asia, in association with populations of archaic *Homo sapiens*, suggests to Toth and Schick that these hominids now possessed the ability to hold on to an image of cultural form, and to communicate this image across the generations by means of verbal instruction, *independently of its physical demonstration*. That is to say, they would have been capable of teaching in the narrow sense defined by Peter Wilson: 'an activity that rests on forms without reality beyond themselves and prepares the learner without requiring him to experience the object taught' (Wilson 1980: 145). This is undoubtedly a crucial capacity shared by all modern humans, and possibly by their archaic predecessors as well, but its invocation in the context of the 'Acheulean problem' is not without its difficulties. For one thing, it places extraordinary demands on memory and oral tradition; moreover there are plenty of examples in the ethnographic record to attest that entire cultural traditions may be lost *despite* the possibility of verbal transmission. And for another, the reappearance of the Acheulean in Europe and Asia need not have been due to its maintenance across an intervening bottleneck. Hominids with the capacity to maintain an oral tradition over many generations would have been equally capable of reinventing a tradition previously forgotten.

The answer proposed by Davidson and Noble to the Acheulean problem neatly disposes with the entire issue of cultural standards and their conservation. Bifaces arise simply from the fact that hominids were making (and breaking) flake tools. Other issues, however, figure centrally in their approach, of which I wish to focus on two in particular. First, what is the role of planning in tool manufacture? Secondly, what is the ontological status of the 'tool' or 'artefact'?

It has, of course, traditionally been part of the archaeologist's job to identify and classify the forms of prehistoric stone tools. But it is a mistake, Davidson and Noble argue, to suppose that these forms and the constituent operations in their manufacture, retrospectively reconstructed by the analytically minded archaeologist, were envisioned by the original toolmakers as mental plans or images, in advance of their realization in the material. And it is doubly erroneous to infer from this supposition that the toolmakers possessed linguistic abilities. The mistake arises from what Davidson and Noble call the 'finished artefact fallacy', the assumption that the objects which the archaeologist recovers represent final forms corresponding to the self-consciously articulated, prior intentions of their one-time makers. If one flake is removed at a time from a stone core, and if the object is to use the flake, then the 'depth' of planning need not extend beyond the production of each flake, and certainly does not extend to a presumed final form of the core. Indeed the 'end' may come, if at all, at the point when the attempt to remove yet another flake from a much used core causes the latter to fracture so as to be of no further use at all. Likewise a single flake may be repeatedly used and reduced, until it is eventually discarded. To call this discarded form a 'finished artefact' is rather like saying that the 'finished pencil' is one that – through repeated use and resharpening – has been reduced to a stub. The real breakthrough marking the advent of modern human cognitive and linguistic abilities is indicated, according to Davidson and Noble, by the appearance in the Upper Palaeolithic of a quite different set of toolmaking techniques, involving cutting and grinding rather than flaking, of material such as bone and ivory as well as stone. For grinding *is* a process whereby preconceived forms are imposed on raw material, forms which are constrained little if at all by the technique or by the properties of the material itself.

To question the idea that the form of a material object is the realization of a plan in the mind of a maker is to cast doubt on the very status of the object as an 'artefact'. Virtually by definition, an artefact is a thing that results from the shaping of naturally given raw material to a preconceived cultural standard (e.g. Goodenough 1981: 50, cf. Parker and Milbrath, this volume Chapter 14). The idea of completion is therefore inherent in the notion of the artefact: it is finished once the object has been brought into conformity with the standard. Whatever happens to it beyond that point belongs to the phase of 'using' rather than 'making' – indeed it is only by invoking a notion of finality that these phases can be marked off from one another at all. Yet any object may, in the course of its existence, undergo any number of alternating phases of manufacture and use. Thus the remains of old pieces of equipment may form the stock out of which new ones are built, fulfilling in them quite different functions, in successive operations of the kind that Lévi-Strauss (1966) classically called *bricolage*. How, then, do we distinguish between the use of an old object and the making of a new one? Am I writing with my pencil, or manufacturing a blunt stick for use as a probe by grinding down the point? Am I using a core to make flakes or making a core for use as a hand axe? The answer can only be given in relation to the project of an agent, depending on whether the modifications undergone by the object belong to its present use or are preparatory to an intended future use (in the latter case, of course, the archaeologist might expect to find evidence of wear resulting from that use). It follows that the quality of 'being an artefact' no more inheres in the object itself than does the quality of 'being a tool'. As regards the latter, Ingold shows that it entails the conjoining of the object to the technical skill of a user, whose presence is presupposed when we call the object a tool of a certain kind. Likewise, 'being an artefact' depends on its conjunction to an intended project. In short, both the instrumentality and the artefactuality of objects are conditional upon the situational contexts of their engagement in practical activity, and as objects endure over time, having histories of such engagement, so their status can change. Perhaps it makes no sense to ask whether or not the Acheulean biface was an artefact, or even a tool. It has, in the course of time, been many things to many people.

Davidson and Noble have drawn our attention to the 'finished artefact fallacy'; Reynolds, in his contribution to this section, introduces what he calls the 'great tool-use fallacy'. This lies in the assumption that the use of tools is individual rather than social in nature. The theme of the fundamentally social organization of technical activities is central for both Reynolds and Ingold, and there is much agreement between them. Ingold shows how, in Western thought (and in anthropological theory), practical, tool-using activity came to be linked to the individual side of a pervasive conceptual dichotomy between individual and society. Reynolds, for his part, shows that it is precisely the 'social structure of manual skill' that differentiates human tool-use from that of non-human primates, which remains essentially individualistic despite the highly social nature of these species (see also Gibson, this volume Chapter 11). Both insist that the domains of social and technical activity are not separate yet interdependent, but rather consubstantial – at least in the context of non-industrial societies. In these societies, technical relations *are* social.

Only when it comes to modern industrial societies do the two authors appear to diverge. Reynolds argues that the intimate, face-to-face 'task group' (see Helm 1968, for an earlier formulation of this concept), which organizes technical activities in small-scale societies, can be found operating in much the same way in industrial contexts as well. Ingold, on the other hand, claims that the transition to industrial capitalism and the rise of mechanized production have been accompanied by depersonalization and deskilling, and a consequent withdrawal of technical forces from the domain of social relations, leading to the modern institutionalized separation of technology and society. In fact, both Ingold and Reynolds may be right. Ingold is concerned, above all, with how modern ideas of technology and the machine have shaped the way practices – including those of co-operation – are *discursively represented*, and these representations may not necessarily conform to experience (though he implies that they often do). Reynolds is more concerned with the actual experience of social-technical co-operation in the workplace, yet significantly he chooses his examples from situations of product development rather than subsequent mass production.

Starting from the observation that human tool-use almost invariably involves 'heterotechnic' co-operation, in which several persons work together in different yet complementary tasks, Reynolds launches a highly original argument, under the title of the 'theory of complementation', which links this kind of co-operation with what he calls 'polylithic' constructional abilities in the field of toolmaking, and syntax in the field of language. A polylith is a construction of several components, linked by joins, that can be rotated along any axis without falling apart. 'Polypods', by contrast, are multi-component constructions that hold together by virtue of their arrangement in a gravitational field. The suggestion is that heterotechnic co-operation, polylithic construction and syntactical communication all rest on the same formal cognitive principles. If Reynolds is right, then ancestral hominids may have had a polypod technology far more advanced than anything known among contemporary non-human primates, whilst yet lacking even the simplest polyliths (and, presumably, the co-operative and linguistic abilities of modern humans). The Acheulean biface, one supposes, could have fitted in well with such a technology. Can we, then, infer that its makers lacked the kind of sociality manifested as heterotechnic co-operation?

For Ingold, a major concern is to distinguish between the concepts of technique and technology, the former signifying the embodied skills of human agents, the latter comprising the operational principles embodied in the external apparatus of production. He believes that by uncritically allowing the concept of technology, which is of modern Western provenance, to structure our thinking about human tool use in general, and even about the tool use of non-human animals, we may be led to assume that technique is a property of the tools themselves rather than a personal accomplishment of their users. Toth and Schick, too, warn that one cannot read off the level of skill in a population from the record of its tools, or judge artifice from artefacts, and the same point is apparent from McGrew's comparison (1987 and Chapter 6, this volume) of the tool-kits of chimpanzees and Tasmanian Aboriginal hunter–gatherers, where the superficial similarity hides a fundamental difference between chimpanzees and humans in the way the tools are incorporated, for the latter, into the subjective and intersubjective experience and know-how of their makers-cum-users. Ingold argues that an exclusive focus on the tools can lead us to view technical evolution as a process of complexification when what is involved, in reality, is an *objectification* of technique in the form of technology, much as the technique of speech is objectified in the form of systems of writing. Reynolds shares the view that it is the ideology of industrialism that constructs Aboriginal hunter-gatherers as people with 'simple' technology, and – with Ingold – he insists on distinguishing 'technology' from 'tool-use'. For Reynolds, however, technology refers not to an objectified system of mechanical principles but to the social organization for co-operation that intervenes between the wider structures of society and the tool-user in person.

The acquisition of technical skill is a key theme of Wynn's contribution to this section. His emphasis is on the mundane, everyday practices which we learn by apprenticeship and are liable to repeat, with more or less the same format and sequencing, on innumerable occasions throughout our lives. In a passing comment, however, Wynn puts his finger on a source of much misunderstanding in discussions of tool use. He notes that the relevance of 'object manipulation' to 'tool use' is by no means established. For readers who imagine these to be the same thing, this observation might seem surprising. Surely, using tools entails the manipulation of objects. But Wynn's concern is really to differentiate object-manipulation as a technique of *puzzle-solving* from what might be called technical *artistry* – the skill that inheres not in figuring out what to do, but in the doing itself. In the puzzle-solving approach to tool use, everything is seen to depend on the ability to devise solutions – in the form of plans for subsequent execution – which incorporate the manipulation of detached objects as a means to achieve the desired goal. For example, in Visalberghi's experiments with capuchin monkeys (this volume, Chapter 5), the solution lies in using a stick to extract a nut from a tube. Once the solution is arrived at, the execution is relatively straightforward – as unproblematic, say, as moving a chess piece from one square to another across the board. This puzzle-solving approach naturally lends itself to the kind of modelling of cognitive processes, exemplified in this volume by Parker and Milbrath (Chapter 14), according to which every episode of action can be resolved into a mental operation of plan-construction followed by a purely physical event of behavioural execution. It is doubtful, however, whether an approach of this kind can help us to understand how humans can successfully accomplish such tasks as throwing a lasso, riding a bicycle or playing a musical instrument. Here the problems are, in a sense, already solved, yet these tasks, as we all know, are 'easier said than done'. It will not do, as Bourdieu has remarked (1977: 96), to regard such practices, which are characteristic of everyday life, as 'no more than "executions" . . . or the implementation of plans'. There is more to human activities (including those of speaking) than the behavioural transcription of mental constructions. Moreover as both Reynolds and Ingold recognize, planning is itself an activity carried out by people in a social context; it is not the achievement of an isolated, self-contained intellect encased within the body of the organism, as cognitive science is inclined to suppose. The same, of course, is true of memory.

Finally, what do the contributions to this section tell us about the relations between tool use and language? On this point, Wynn's argument is essentially a negative one. He approaches the issue by thinking of both tool behaviour and speech as layered systems that can be analysed on a number of levels. Thus tool behaviour can be considered in terms of the biomechanical constraints imposed by the executive organs of the body (analogous to those imposed on speech by the structures of the vocal tract and ear), the ways in which elementary movements are linked into sequences (analogous to the syntactical organization of speech) and the organization of knowledge (analogous, perhaps, to the semantic organization of language). But with regard to the latter two levels, the comparison with language reveals fundamental differences. Wynn argues that action sequences in tool behaviour, insofar as they follow an established routine, are *not* generated by rules of syntax but are merely 'strings-of-beads' produced by chaining one action to the next. As to the organization of knowledge, he shows that while tool users do bring together ideas about form, function and material into larger conceptual 'constellations', these differ from the semantic paradigms of language firstly in that they are tied to particular contexts of use, and secondly in that they are idiosyncratic – specific to their individual constructors rather than common to a technical community.

For Wynn, then, tools and tool behaviour afford few if any grounds for inferences about language. For Davidson and Noble, tools from the Upper Palaeolithic onwards, which reveal planning and 'second order intentionality', provide positive indications of the emergence of language along with anatomically modern humans. Toth and Schick are inclined to credit the earlier, archaic *Homo sapiens* with language, but go further to suggest that the preferential right-handedness evident from the earliest (Oldowan) tools of *Homo habilis*, dating from almost two million years ago, may have represented a preadaptation for linguistic abilities. For Reynolds, the clearest evidence for language would appear in the form of polylithic constructions. From such a range of disparate views it is not possible to draw firm conclusions. Let me close this introduction, however, by returning to the Acheulean biface. For whatever it was used for, this use must have involved an element of skill, and skill, in turn, is an index of personal identity. Both Wynn and Ingold point out that for this reason, the very use of tools constitutes a statement about identity, and therefore has a communicative or semiotic as well as a purely technical aspect. In some cases, the semiotic aspect can become dominant, to the extent that the 'tool' itself becomes practically useless. For the contemporary experimental archaeologist-turned-flint-knapper, the ability to make a biface is certainly an important aspect of identity within the (scientific) community, even if he or she has no idea how to use it. Perhaps this was so for people in the past as well.

References

Bloch, M. (1991). Language, anthropology and cognitive science. *Man* (N.S.), 26, 183–98.

Bourdieu, P. (1977). *Outline of a Theory of Practice*. Cambridge: Cambridge University Press.

Goodenough, W. H. (1981). *Language, Culture and Society* (2nd edn.). Menlo Park: Benjamin Cummings.

Helm, J. (1968). The nature of Dogrib socioterritorial groups. In *Man the Hunter*, ed. R. B. Lee and I. DeVore, pp. 118–25. Chicago: Aldine.

Lave, J. (1990). The culture of acquisition and the practice of understanding. In *Cultural Psychology: Essays on Comparative Human Development*, ed. J. W. Stigler, R. A. Shweder and G. Herdt, pp. 309–27. Cambridge: Cambridge University Press.

Lévi-Strauss, C. (1966). *The Savage Mind*. London: Weidenfeld & Nicolson.

McGrew, W. C. (1987). Tools to get food: the subsistants of Tasmanian Aborigines and Tanzanian chimpanzees compared. *Journal of Anthropological Research*, 43, 247–58.

Wilson, P. J. (1980). *Man, the Promising Primate* (2nd edn.). New Haven, CT: Yale University Press.

15

Early stone industries and inferences regarding language and cognition

NICHOLAS TOTH AND KATHY SCHICK

Introduction

Attempting to infer levels of cognitive and communicative complexity in early hominids from analyzing the prehistoric technological record is akin to trying to reconstruct the social organization and sexual habits of early hominids by analyzing assemblages of fossil bones from palaeontological localities. Making larger-scale, inductive inferences regarding levels of hominid cognition (or social organization) is problematic since our only models for these are drawn from modern human and non-human species, which are themselves the products of thousands or millions of years of evolution along unique trajectories and of a myriad of selective processes. Thus the accuracy of such inferences from the evidence at hand is largely dependent upon one factor: how closely the extinct forms resemble modern analogs which can be carefully analyzed from a comparative perspective. For early hominids that lived in the late Pliocence and early Pleistocene, our modern analogs are not that good.

The challenge to palaeoarchaeology is to identify what patterns of material culture in the prehistoric record have implications for intelligence and language (or proto-language). While the evolution of brain functions proceeds along Darwinian lines, with changes in gene frequencies leading to changes in phenotypic traits that can be advantageous, deleterious, or neutral in given environments, the evolution of technology (and much of culture) proceeds along Lamarckian lines, according to which acquired traits are inherited through learning. This means that it is possible for technologies to change literally overnight if innovations that improve a prehistoric hominid's adaptive capacity are invented or received by diffusion from other areas. It is also possible that technology was one of the cultural catalysts which accelerated the pace of brain expansion and cognitive development (constituting a "bio-cultural feedback system", Washburn, 1960).

One should always assume that the level of technological sophistication can only reliably indicate a *minimum* level of intelligence and language that must have already been reached or passed; it is quite possible that, throughout human evolution, populations were capable of more sophisticated types of tools and material culture, had there been a need for them in their adaptive strategies or had accumulated cultural knowledge allowed for such innovations. *Homo sapiens sapiens* hunter–gatherers in the late Pleistocene and early Holocene exhibited an astonishing range of stone tool assemblages, some very simple and some extremely complex. Yet, so far as we know, all of these groups possessed the same basic level of brain function and linguistic complexity which characterizes our present species.

Modern human thought processes (including aesthetic expression and symbolic forms of communication) are unique in the animal world, and have probably been in existence only for the last 40 000 to 100 000 years. For early *Homo sapiens sapiens* it is difficult to judge levels of cognitive ability: being anatomically modern *skeletally* does not necessarily mean, especially in the time period prior to 40 000 years ago, that hominids were fully modern *behaviorally* or even *cognitively*. What patterns of cognition and communication would have been like in more remote periods of time for archaic *Homo sapiens*, *Homo erectus*, *Homo habilis*, or the australopithecines is largely gleaned imperfectly through examining remains of the prehistoric material culture and palaeoneurological evidence, and from attempts to reconstruct vocal tracts based upon the basicranial anatomy of fossil forms.

This chapter will concentrate on prehistoric material culture as a potential source of information about the cognitive and communicative levels of extinct hominid taxa. We will take the existing archaeological record and experimental archaeological approaches as our points of departure, and suggest future lines of research which might shed more light on these problems.

Synopsis of early technologies

The earliest reliably dated stone artifacts are approximately 2.4 million years old and come from East African Rift localities. The world's earliest stone technologies are normally placed under the rubric of "Oldowan" or "Mode I" industries, characterized by such forms as battered percussors or hammerstones, simple cores made on cobbles or blocks, casually retouched flakes, and unmodified flakes and fragments (debitage). These industries are characteristic of the time range 2.4 to 1.5 million years ago at such sites as the Omo, Ethiopia; Gona (Hadar), Ethiopia; Koobi Fora, Kenya; and Olduvai Gorge, Tanzania. Sites that may be of a similar age, though not radiometrically dated, include the earlier part of the Casablanca sequence in Morocco and the South African cave sites of Sterkfontein and Swartkrans (Harris, 1983; Isaac 1984; Toth and Schick, 1986).

The earliest tools of East Africa would be contemporaneous with robust australopithecines (*A. aethiopicus*, evolving into *A. boisei* in East Africa and perhaps into *A. robustus/crassidens* in South Africa), as well as with at least one more gracile fossil form: *Australopithecus africanus* in South Africa and fossils represented by isolated teeth in East Africa that may be transitional into *Homo habilis*. By approximately 2.0 million years ago there is evidence of brain expansion in forms that are assigned to *Homo habilis*, although some researchers believe that this taxon actually comprises at least two species, a larger-brained form of *Homo* (e.g. KNM ER 1470) and a smaller-brained (and perhaps smaller-bodied) form (e.g. KNM ER 1813, STS Member 5, OH 62).

Susman (1988) has emphasized the possibility that robust australopithecines as well as early *Homo* may have made stone tools for a variety of tasks; while there is no *a priori* reason why *Australopithecus robustus/boisei* should have been tool-makers, it is also clear that wherever stone tools are found (at localities where fossil bone is preserved), the genus *Homo* is normally present as well. It will probably be impossible to demonstrate which of the co-existing hominid forms was responsible for a given assemblage of stone tools without decades of more research and repeated associations of hominid fossils with palaeolithic archaeological sites.

About 1.5 million years ago, there was a technological shift in parts of Africa towards producing forms called "handaxes", "cleavers", or "picks" (generically called "bifaces"), often on large flakes struck from boulder cores. These so-called "Acheulean" or "Mode II" industries appeared in many parts of Africa, then in the Middle East, Europe, and parts of the Indian subcontinent during the next 1.2 million years. There is a general tendency for these forms to become more refined through time, although there are many exceptions. In addition, the simpler "Mode I" types of artifacts are usually present in varying frequencies. The emergence of these new Acheulean artifact forms roughly coincides with the emergence of a new species, *Homo erectus* (KNM WT 15000; KNM ER 3733, 3883), and with the subsequent extinction of the robust australopithecines by about one million years ago.

An assessment of Oldowan technology

During the past decade, a detailed experimental archaeological and functional program has been carried out in East Africa, concentrating on the early stone age artifacts from Koobi Fora, Kenya (Isaac and Harris, 1978). Much of this research focused upon the nature of the Oldowan assemblages found between 1.9 and 1.4 million years ago in this study area, as well as upon early Acheulean technology (Toth, 1985 a,b; Toth, 1987; Schick 1987; Toth and Woods, 1989). Methodologically, this approach aimed to replicate the types of raw materials, inferred techniques, and presumed methods (strategies) used by early hominids in producing their artifacts.

Some of the basic conclusions about Oldowan technologies that were drawn from this study were:

1. Oldowan tools are technologically quite simple, and many or most of the so-called "core-tools" (choppers, discoids, polyhedrons, heavy-duty scrapers, etc.) appear simply to be discarded cores manufactured in the process of flake production. These core forms are not necessarily tools, nor do they necessarily correspond to "mental templates" held by early tool-making hominids. The final morphology of these core forms may be determined largely by the size, shape, and raw material of the rock used.

2. Although such Oldowan artifactual forms are technologically simple, some important principles of flaking stone had been mastered by two million years ago. These include (a) the ability to recognize acute angles on cores to serve as striking platforms from which to detach flakes and fragments, and (b) good hand–eye coordination when flaking stone, including the dexterity to strike the core with a hammerstone with a sharp, glancing blow. It would appear that a strong power grip, as well as a strong precision grip, was characteristic of early hominid tool-making populations at this time.

3. The nature of the artifact assemblages at these early stone age sites suggests that hominids were transporting a great deal of stone artifactual material around the landscape, and sometimes preferentially discarding it in areas of high concentration. These are the types of prehistoric occurrences that archaeologists tend to excavate (Schick, 1987; Toth, 1987). At Oldowan sites at Koobi Fora the later stages of flaking of cobble cores tend to be found, rather than the initial stages of core reduction.

4. These early tool-making populations appear to have been preferentially right-handed. This inference is based upon analysis of cortical flakes which indicates a preferential, clockwise turning direction of cores consistent with observations of modern right-hand knappers (Toth, 1985a). The implications of this observation are not entirely clear. Although consistently asymmetrical use of hands, paws, or feet seems to be rare or nonexistent among non-human animals (Collins, 1963; Warren, 1980), recent research has strongly suggested that there is asymmetrical specialization of cerebral hemispheres in non-human primates as well as in other species. It is possible that in the evolution of early hominid tool-making populations there was a more profound lateralization in the brain that may have been a pre-adaptation for other functions, including language (Calvin, 1983; Falk, 1980).

5. Spheroids, which some have regarded as intentionally-shaped tools used

as "bola stones", missiles, club heads, etc., are often simply well-curated hammerstones that naturally assume these globular, battered shapes through hours of accumulated use (Toth, in prep.).

One interesting question to ask is: were these early tool-making hominids essentially bipedal, savannah-dwelling chimpanzees, with the same basic cognitive and communicative systems as are observed in extant members of the genus *Pan*, who have been seen to make and use tools in in a variety of different situations (Boesch and Boesch, 1981, 1984; Hannah and McGrew, 1987; Kortlandt, 1986; McGrew, 1974, 1977; Nishida, 1973; Nishida and Hiraiwa, 1982; Sugiyama and Koman, 1979)? We suggest that this is probably not the case, for the following reasons:

1. The distances over which early hominids transported raw materials (and probably foodstuffs as exhibited by animal bones of a range of species, some with cut-marks and probable hammerstone fracture patterns) were greater than those normally covered by chimpanzees; these hominids repeatedly carried rocks from their geological sources up to several (sometimes more than ten) kilometers to a presumed activity area (Leakey, 1971; Hay, 1976). Chimpanzees, on the other hand, rarely if ever transport tools or foodstuffs over such distances (Boesch and Boesch, 1984; McGrew, 1974, 1977). This suggests more foresight and curational behavior than is exhibited by modern apes.

2. It would appear that meat and marrow acquired through either scavenging or hunting were becoming much more important for subsistence than they are for chimpanzees. Stone tool cut-marks have been observed on bones, recovered from Oldowan sites, of animals ranging from small mammals to megafauna; the average sized animal represented at FLK Zinj in Bed I of Olduvai is in the weight range of modern wildebeests (Bunn and Kroll, 1986). As we have noted before, unlike chimpanzee food resources whose procurement involves the use of tools (immobile termite or ant nests or nutting trees), animals obtained by hunting or scavenging would not be as predictable, requiring the adoption of a foraging pattern that would emphasize the carrying of stone or, as Potts (1984) has suggested, the caching of rock resources.

3. Concentrations of tools and technological by-products appear to be much higher at early hominid sites (often numbering in the hundreds or even thousands) than in any chimpanzee tool-using localities reported thus far. Some of these concentrations may well represent a palimpsest of hominid activities (e.g. reoccupation of the same locality at different times), but even so, the cumulative density of collected hominid materials appears to be greater than that observed in any extant non-human primates. We would also suggest that if the component of Oldowan technology made from decomposable,

organic material had also been preserved (such as simple containers, probes, digging tools, etc.), the amount of material culture exhibited at these sites would have been considerably greater. In the best of cases, where observational studies of modern wild chimpanzees have been conducted in habitats where all the organic materials they use can be seen and recorded, there are no analogous concentrations of material culture. This suggests a more intensive use of tools by early hominids than is observed among chimpanzees.

4. We would also argue that it may be beyond the cognitive capabilities of chimpanzees to modify stones in an Oldowan manner. The motor skills of the chimpanzee may be adequate for such operations, despite the alleged lack of a "precision grip" among apes, as trained chimpanzees have no problem in buttoning and unbuttoning their shirts or even tying their shoelaces (a series of operations that is beyond some kindergarten students). However, the seeking out of acute angles or overhangs on cobbles and cores, as well as the judgement of the correct angle and force of impact required to effectively flake stone, may well be beyond the capabilities of chimpanzees even in the best cases of Pavlovian classical conditioning (see below).

Acheulean industries

As Wynn (1989) has pointed out, the operational skills required for making later Acheulean artifact forms appear to be much more sophisticated than those inferred for Oldowan hominids. These include the ability to envisage such geometric relations as symmetry of planform and cross-section, and the ability to create a straight or regular edge. Besides manifesting a clearer sense of spatial geometry, the technical sophistication of these forms is such that even modern humans learning to make stone tools often require many months of apprenticeship to reach the requisite level of finesse. This often includes careful preparation of biface edges to steepen the angle for soft hammer flaking (e.g. with bone, antler, wood, or soft stone).

Just as striking is the recurrent and strong stylization seen in many Acheulean assemblages. It seems clear that the cultural norms for manufacturing special shapes of handaxes (lanceolate, ovate, ficron, tranchet tip, etc.) were much more standardized in later *Homo erectus*/archaic *Homo sapiens* than in earlier hominid forms, a trend which becomes even more exaggerated over time. For many prehistorians, this suggests higher levels of cognition and communication; however, compared with the rate of progress of technologies in stone and other materials within the last 40 000 years, major changes in later Acheulean technology appear to have been remarkably slow. This would appear to indicate major cognitive, communicative, or cultural differences between late *Homo erectus*/archaic *Homo sapiens* and the anatomically modern *Homo sapiens sapiens* of the Upper Palaeolithic.

The spread of stone industries at different evolutionary stages

One clue to the communicative skills of *Homo erectus* may be seen in what appears to have happened to hominid technologies during their supposed spread out of Africa between 1.5 and 1.0 million years ago. Sites such as Ubeidiya in southwestern Asia suggest that Acheulean hominids were populating the Levant by approximately 1.2 million years ago and had arrived in East Asia by between 0.7 and 1.0 million years ago. It is possible that the earliest populations to inhabit Eurasia had moved out of Africa before the Acheulean technological tradition was well established, but it is interesting to note that most of the earliest stone age sites in Eurasia are devoid of handaxes and cleavers. It seems that the earliest migrants to both Europe and Asia (outside the Levant) do not have Acheulean elements in their stone technologies.

Sites in Europe that are believed to be over 700 000 years old (usually in palaeomagnetically reversed sediments) include Isernia La Pineta, Italy; Prezletice, Czechoslovakia; Stranska Skala, Czechoslovakia; Karlich, Germany; Vallonet and Soleihac, France (Klein, 1989). In each of these cases, handaxes and cleavers are normally absent from the artifact inventories.

The emerging pattern would indicate, then, that during the earlier migrations of hominid populations out of Africa via the Middle East into previously uninhabited areas, the tradition of manufacturing large bifacial artifacts characteristic of the Acheulean disappeared. In our view this may have been due to the fact that as hominid populations traversed tracts of landscape where large raw materials suitable for making handaxes or cleavers were absent or not readily located, the tradition of making Acheulean forms would have died out after one generation. The communicative and cognitive processes of these hominids may not have been strong enough to maintain the Acheulean tradition.

During the next several hundred thousand years, the "Mode II" technologies with large bifacial forms finally and successfully spread throughout much of Europe and western Asia. This could have occurred with the spread of a new wave of Acheulean people (some of them perhaps early archaic *Homo sapiens*) moving into Europe and encountering established populations with "Mode I" technologies. With settled populations already in place with a fully-developed mental map of the environment and its resources, the concept of making these new "Mode II" forms could spread easily without hindrance (as they had originally in Africa). After around 500 000 years ago, then, we see the

Acheulean successfully "arrive" in Europe as evidenced at numerous sites (e.g. Torralba and Ambrona, Spain; Abbeville and St. Acheul in France; Swanscombe and Hoxne in England; and Torre in Pietra in Italy).

The movement to uninhabited or very sparsely inhabited regions could then have produced a "technological bottleneck" for the first wave of *Homo erectus* to spread from Africa: their standard tool forms were abandoned due to lack of suitable raw materials (large flakes, blocks and cobbles), and the techniques for making them were lost quickly over time. The apparent lack of Acheulean technologies in eastern Asia during the Early and Middle Pleistocene has traditionally been explained by anthropologists in a number of ways: in terms, for example, of the geographical barriers of the Himalayas and tropical forests creating a "cultural backwater" in eastern Asia; of adaptation to new environments where handaxes and cleavers were not as useful, and of the use of bamboo as a major cutting tool. The general lack of Acheulean technologies in Eastern Asia may as well have been due to the "idea" and know-how of this technology having been lost in transit due to breaks in the supply of readily available raw material and the inability of *Homo erectus*'s cognitive and communicative abilities to carry the tradition through these gaps over several generations. Thus, when suitable stone resources were eventually located in Asian localities inhabited later, the Acheulean did not "reemerge", and was never reinvented.

The linguistic and cognitive limitations of *Homo erectus* and related species may have made their cultural traditions particularly vulnerable to such lapses and losses. Such limitations may in fact have also contributed to the remarkable persistence of the Acheulean over vast stretches of time and space, when raw materials permitted: the concept of the artifact may have been profoundly fixed and dependent upon its physical manifestation and form.

In contrast to the apparent difficulty encountered by *Homo erectus* populations in successfully transplanting their technology to new environments in previously uninhabited regions, *Homo sapiens* seems to have had no such problem. There are at least two cases of hominid tool-making populations colonizing previously uninhabited continents during the past 100 000 years, which is the time period that appears to have witnessed the emergence of anatomically modern humans (as well as archaic forms of *Homo sapiens* until about 32 000 years ago). These are the peopling of Australia and the peopling of the Americas.

In our present state of knowledge, based upon radiocarbon dates and studies of the mitochondrial DNA of anatomically modern human populations, the earliest occupants of Australia arrived around 40 000 years ago or earlier, almost certainly from Southeast Asia. Even when sea levels were at their lowest during the late Pleistocene, it appears that at least 80 kilometers of water had to be crossed in one stretch between the Sunda shelf and Sahul (Australia, New Guinea, and Tasmania), as well as several other shorter distances of water between islands. Whether or not the very earliest Australian inhabitants were anatomically modern is still not known. However, at least by the time of settlements of c. 30 000 to 35 000 B.P., the inhabitants were modern humans; skulls from a similar time range in Southeast Asia from the late Pleistocene, such as Zhoukoudian, Ziyang, Liujiang in China, Niah Cave in Borneo, and Wajak in Java, have also yielded anatomically modern hominid fossils.

The early Australian stone industries strongly resemble those of Southeast Asia in the Late Pleistocene: large core forms (including "horsehoof" cores), some blade elements, and some microlithic elements. The pattern of rapid technological innovation associated with *Homo sapiens sapiens* elsewhere can be seen during the next millenia in the form of microlithic technologies, the spear thrower, and ground stone axes. Thus, the migration to Australia, and the subsequent dispersion within the continent, was accomplished without the cultural and technological breaks apparent in the earlier *Homo erectus* migrations.

An analogous situation in a later time period, and unequivocally associated with anatomically modern populations, is the peopling of the Americas during the past 15 000 years or so. A striking feature of stone industries in North America between 12 000 and 10 000 years ago is the nearly ubiquitous fluted projectile point (first "Clovis points", and then a millennium later, "Folsom points", "Cumberland points", etc.).

These well-made and highly distinctive artifact types are found from Alaska to Mexico and from California to the East Coast. Radiocarbon dates so far are so tightly clustered that they do not suggest a clear route of diffusion of these forms, nor their place of origin, which might have been on the Beringia subcontinent now submerged by postglacial sea level rise (Toth, 1991). What this evidence does suggest is a very rapid migration of populations with a highly idiosyncratic lithic technology that did not change significantly as they spread geographically throughout vast areas of North America.

Summary

The different patterns of cultural diffusion during population migrations in the case of *Homo erectus* and the Acheulean, and in the case of the late Pleistocene spread of modern peoples into Australia and North America, may show something very important not only about the culture of these two hominid forms but also about their communication systems. Even if biface technology was learned primarily through demonstration and non-verbal communication, it may well have depended for its long-term survival on verbalized knowledge and traditions to carry it through times and places in which the necessary raw material was unavailable or its location undiscovered. *Homo erectus* may well have been deficient in the ability to verbalize the traditions of Acheulean technology sufficiently to pass such accumulated knowledge on to succeeding generations in a verbal rather than a demonstrative way. On the other hand, *Homo sapiens sapiens* has proven to be a highly efficient transmitter of ideas across great spans of time and space, largely through verbal communication.

Technological hallmarks of anatomically modern humans

It is worth asking the question: are there aspects of the Palaeolithic technologies of anatomically modern humans (*Homo sapiens sapiens*) that are sufficiently distinctive from those of earlier archaic *Homo sapiens* forms to be considered diagnostic of our extant subspecies? Although it is difficult in the present state of knowledge to ascertain which of the elements listed below could only be accomplished with anatomically modern cognitive and communicative skills, as opposed to accumulated cultural knowledge, we nevertheless find that the following features have as yet been detected only in association with modern humans:

1. Prismatic blade cores produced by indirect percussion;
2. A well-developed bone/antler/ivory technology;
3. Geometric microliths for composite tools (the "Howieson's Poort" large microlithic component of South Africa may be an exception, but may be associated with anatomically modern humans as well);
4. Mechanical devices such as the spear thrower, and bow and arrow technology;
5. Heat-treatment of raw materials to increase their flaking properties;
6. Pressure-flaked projectile points and knives (e.g. Doian of Somalia; Solutrean of Spain/France);
7. Ground stone artifacts (e.g. in Australia and New Guinea);
8. Ceramic vessels (e.g. Jomon, Japan);
9. Needles for sewing;
10. Representational mobiliary and parietal art.

We would argue that several of the above features strongly suggest, or would seem to require, verbal communication and, moreover, that the entire list indicates a cultural inventory being rapidly enriched with new ideas, discoveries and inventions which may presuppose a well-developed language ability. Some of the new technologies, such as prismatic blades or geometric microliths, may be seen to some degree as translations of methods and techniques used in earlier tools, such as Levallois cores, which sometimes show elaborate preparation and a rather complicated series of technological acts. Moreover, the composite tools of the microlithic industries may be said to have been foreshadowed by hafting, which presumably arose with earlier flaked stone tool industries. Other innovations, such as the proliferation and elaboration of the use of bone, antler, and ivory; the advent and very early refinement of the pressure flaking technique; the early experiments with ceramics; and the advent of groundstone technologies bespeak a burst of new, diverse ideas and experimentation with new materials and novel techniques.

The discovery of heat treatment of flint and siliceous materials to improve their texture and flaking qualities may subtly indicate a new threshold in the process of human invention: it shows the progression from an observation of an external phenomenon (i.e., one in which the hominid is not actively participating but which is taking place among external agents or processes), to the discovery of the principle behind it, to the design of means to control or harness that process. The invention of mechanical devices such as the spear-thrower and the bow and arrow argue even more strongly for a very rational manipulation of physical processes, which would again indicate significant cognitive if not communicative abilities. Last but not least, the florescence of artistic expression, and the mastery of techniques in a myriad of media and forms, show beyond doubt a concern with conceptualization and expression that goes far beyond the materialistic. All of these features mark a radical departure by Upper Palaeolithic times from the cultural and technological patterns detected in the earlier archaeological record. They argue strongly, though not irrefutably, for the attainment of modern cognitive and, presumably, communicative levels by this time.

Suggested avenues for future inquiry

Tool-making abilities of apes

One interesting avenue for future investigation would be to try to teach apes to make stone tools by example, a possibility demonstrated in a pioneering experiment by Wright (1972) with an orangutan in the Bristol Zoo. Of great interest would be to find out whether these intelligent primates can master some of the basic principles of stone fracture known to *Homo habilis* and his successors, e.g. striking acute edges and striking the core at the proper angle, and also how such learning occurs. This could give some insight into the cognitive threshold of early tool-making hominids. As a direct consequence of the conference which has given rise to this volume, we have recently initiated a collaborative project with Sue Savage-Rumbaugh, Duane Rumbaugh and Rose Sevcik of the Language Research Laboratory in Atlanta, to study the stone tool-making and tool-using abilities of Kanzi, a pygmy chimpanzee, whose linguistic achievements are described by Savage-Rumbaugh and Rumbaugh in Chapter 3 of this book.

Studies of tool-making in verbal and non-verbal contexts

Another potentially fruitful set of experiments would involve teaching modern humans to make various types of stone tools in two different ways: 1. with verbal instructions giving detailed explanations of the techniques and strategies of artifact production, along with visual demonstrations by a competent flintknapper; 2. with only visual, non-verbal instruction by a flintknapper. In such experiments one might document possible limitations in developing tool-making skills for different technologies when no verbal communication is used and, conversely, the potential enhancement of learning with the aid of language. Although the learning pattern in pre-anatomically modern hominids may have been different from that observed in *Homo sapiens sapiens*, such an approach might prove to be very informative for inferring levels of communicative sophistication from levels of technology.

Finally, to carry out such an experimental investigation into the consequences for tool-making abilities of verbal and non-verbal instruction, one should consider the possible differences in learning patterns with immature and mature subjects. As is known from studies of foreign language acquisition, young adolescents can learn foreign languages more efficiently than adults, presumably because hormonal changes modify the way the brain learns new communicative systems after puberty. It may therefore be important to examine immature subjects as well as adults in such an investigation.

Studies of modern human pathological conditions

Although fraught with potential ethical problems, it might be possible to construct, in a therapeutical and safe manner, an experimental program that would examine possible relationships between deficiencies in language ability (due to birth defects, disease, or injuries) and proficiency (or lack of it) in flaked stone tool-making abilities. Although such subjects are in no way "proto-human" beings, a study of their abilities could yield insights into possible relationships between tool-making, cognition, and language.

Studies of brain activity

Another possible way of gaining an understanding of the relationship of technology to language might be to map the areas of the human brain that are stimulated by a range of different activities: for example while making various sorts of stone tools, while speaking, and while listening to someone else speak. Such studies might show similarities in neural activity that might suggest interconnections of possible evolutionary significance.

Recent neuroscientific advances in positron emission tomography (PET) have allowed researchers to start examining which areas of the cerebral cortex are stimulated during the exercise of various cognitive functions by measuring regional changes in blood flow (Petersen *et al.*, 1988; Posner *et al.*, 1988). The methodological problem at present is that the subject must remain very still for the procedure to take effect, so that applying such analysis to kinetic activities such as stone tool-making may be, at present, beyond the capabilities of the research design. Even going through the thought processes involved in tool-making would probably stimulate so many motor control areas, as well as areas involved in cognitive operations, that, at least in the present state of knowledge, such investigations would probably yield ambiguous results (Petersen, pers. commun.)

Another, more fundamental problem is that there is no guarantee that the thought processes and cortical areas involved in stone tool-making by *Homo sapiens sapiens* are similar to those for *Homo habilis*, *Homo erectus*, or archaic forms of *Homo sapiens*. Nonetheless, such investigations might prove to be very interesting and potentially valuable.

New archaeological directions

In recent years, prehistorians have been trying to break out of the typological-cum-technological criteria used to assess cognitive levels of early hominids and to examine other possible, and perhaps more reliable, indicators of proto-human mental abilities. These potential indicators include: 1. Hunting patterns (e.g. scavenging versus hunting, or the overall level of efficiency as indicated by the age, sex, types of species and species diversity of archaeologically repre-sented fauna); and 2. evident land-use patterns as shown by site distributions, resource procurement, etc. For example, Binford (1983) has stressed that anatomically modern humans were different from more archaic hominid populations in their ability to organize their operations in their environment, predict the availability of resources, and cache or curate aspects of their technological repertoire.

In their analysis of patterns at Mousterian sites (presumed to be associated primarily with Neanderthals) and at Upper Palaeolithic sites (presumed to be associated primarily with anatomically modern humans) in the Negev Desert, Marks and Freidel (1977) suggest that Mousterian people possessed a "radiating" foraging pattern with a central base camp and outlying resources, whereas Upper Palaeolithic populations practised more of a "circulating" system whereby people moved from camp to camp in their foraging rounds, each camp serving a variety of functions. They go on to postulate that the blade technologies of the Upper Palaeolithic were an important part of this circulating system, since they provided a more efficient way of utilizing raw stone material than the Middle Palaeolithic/Mousterian flake technologies, reducing the frequency with which raw materials had to be replenished at quarry sites. It may be too soon to say whether such differences in foraging patterns are in any way universal, but future investigations should shed light on this question.

Conclusions

The relationship between technology, cognition, and language will almost certainly remain a controversial issue for many years to come; at present, all of the methodological approaches to this problem are indirect, since neither hominid cognitive processes nor linguistic patterns fossilize in a direct manner. By-products of such processes can be recognized to some degree in the record of material culture (technology, art, etc.) and can provide at least some minimal indicators of prehistoric hominid capabilities. Comparative studies of the mental capabilities of modern humans and non-human primates may also shed some light on relationships between technology, cognition, and language.

It is our contention that other aspects of the palaeolithic archaeological record may ultimately provide more reliable indicators of cognitive and linguistic capabilities: in particular, understanding the organizational and land-use patterns of hominid groups through time in a given geographical area. A feature that may distinguish anatomically modern humans from earlier hominid forms is the ability to predict when resources will become available and to schedule foraging activities accurately in anticipation of their availability. Thus, a critical indicator of cognitive advance might consist in evidence of planning, or thinking many steps ahead of the present situation, and of curating or caching elements of material culture that will be required for these future activities.

Investigations into such questions, if carried out at a large number of sites and site complexes in different regions and time periods, could greatly enhance our understanding of the cognitive complexity entailed in hominid foraging patterns throughout much of the Pleistocene. Thus in the future, archaeologists may gain a great deal of insight from detailed examination of such elements as the geographical distribution of archaeological sites, evidence for seasonality at specific sites, and patterns of transport of raw materials from their geological sources. Such studies may, in time, reveal patterns that would characterize different stages of hominid evolution and might suggest different levels of cognition, thus providing some indication of potential linguistic ability, regardless of the level of complexity or sophistication of lithic technology.

Acknowledgements

Special thanks to Dean Falk and Steven Peterson for supplying information regarding recent experimental research into mapping activity areas of the human brain. The perspective and content of this paper has benefited greatly from discussions with a number of fellow researchers, including Dean Falk, Thomas Wynn, J. Desmond Clark, as well as a number of our colleagues and students at Indiana University.

References

Binford, L.R. (1983). *In Pursuit of the Past*. New York: Thames and Hudson.
Boesch, C. & Boesch, H. (1981). Sex differences in the use of natural hammers by wild chimpanzees: a preliminary report. *Journal of Human Evolution*, 10, 585–93.
Boesch, C. & Boesch, H. (1984). Mental map in wild chimpanzees: an analysis of hammer transports for nut cracking. *Primates*, 25, 160–70.
Bunn, H.T. & Kroll, E.M. (1986). Systematic butchery by Plio/Pleistocene hominids at Olduvai Gorge, Tanzania. *Current Anthropology*, 5, 431–52.
Calvin, W.H. (1983). *The Throwing Madonna: Essays on the Brain*. New York: McGraw Hill.
Collins, R.L. (1963). On the inheritance of handedness: I. Laterality in inbred mice. *The Journal of Heredity*, 59, 9–12.
Dennell, R.W., Rendell, H. & Hailwood, E. (1988). Late Pliocene artefacts from Northern Pakistan. *Current Anthropology*, 29, 495–8.
Falk, D. (1980). Language, handedness, and primate brains: did the australopithecines sign? *American Anthropologist*, 82, 72–8.
Hannah, A.C. & McGrew, W.C. (1987). Chimpanzees using stones to crack open oil palm nuts in Liberia. *Primates*, 28(1), 31–46.
Harris, J.W.K. (1983). Cultural beginnings: Plio-Pleistocene archaeological occurrences from the Afar, Ethiopia. *The African Archaeological Review*, 1, 3–31.
Hay, R.L. (1976). *Geology of Olduvai Gorge: A Study of Sedimentation in a Semiarid Basin*. Berkeley: University of California Press.
Isaac, G. Ll. (1984). The archaeology of human origins: studies of the Lower Pleistocene formations west of Lake Natron, Tanzania. In *Background to*

Evolution in Africa, ed. W.W. Bishop & J.D. Clark, pp. 229–57. Chicago: University of Chicago Press.

Isaac, G. Ll. & Harris, J.W.K. (1978). Archaeology. In *Koobi Fora Research Project*, vol. 1, ed. M.G. Leakey and R.E.F. Leakey, pp. 64–85. Oxford: Clarendon Press.

Klein, R.G. (1989). *The Human Career*. Chicago: University of Chicago Press.

Kortlandt, A. (1986). The use of stone tools by wild-living chimpanzees and earliest hominids. *Journal of Human Evolution*, 15, 77–132.

Leakey, M.D. (1971). *Olduvai Gorge: Excavation in Beds I and II, 1960–1963*. Cambridge: Cambridge University Press.

McGrew, W.C. (1974). Tool use by wild chimpanzees in feeding upon wild driver ants. *Journal of Human Evolution*, 3, 501–8.

McGrew, W.C. (1977). Socialization and object manipulation of wild chimpanzees. In *Primate Biosocial Development*, ed. S. Chevalier-Skolnikoff & F.E. Poirier, pp. 261–81. New York: Garden Publishers.

McGrew, W.C. (1979). Evolutionary implications of sex differences in chimpanzee predation and tool use. In *The Great Apes*, ed. D.A. Hamburg & E.R. McCown, pp. 441–64. Menlo Park: Benjamin Cummings.

Marks, A.E. & Freidel, D.A. (1977). Prehistoric settlement patterns in the Avdat/ Aqev area. In *Prehistory and Paleoenvironments in the Central Negev, Israel*, vol. 2, ed. A.E. Marks, pp. 131–58. Dallas: Department of Anthropology, Southern Methodist University.

Nishida, T. (1973). The ant-gathering behavior by the use of tools among wild chimpanzees of the Mahali Mountains. *Journal of Human Evolution*, 2, 357–70.

Nishida, T. & Hiraiwa, M. (1982). Natural history of a tool-using behavior by wild chimpanzees in feeding upon wood-boring ants. *Journal of Human Evolution*, 11, 73–99.

Petersen, S.E., Fox, P.T., Posner, J.I., Mintun, M. & Raichle, M.E. (1988). Positron emission tomographic studies of the cortical anatomy of single-word processing. *Nature*, 331, 585–9.

Posner, M.I., Petersen, S.E., Fox, P.T., & Raichle, M.E. (1988). Localization of cognitive operations in the human brain. *Science*, 240, 1627–31.

Potts, R.B. (1984). Home bases and early hominids. *American Scientist*, 72, 338–47.

Schick, K.D. (1987). Modeling the formation of Early Stone Age artifact concentrations. *Journal of Human Evolution*, 16, 789–807.

Sugiyama, Y. & Koman, J. (1979). Tool-using and making in wild chimpanzees at Bossou, Guinea. *Primates*, 20, 513–24.

Susman, R.L. (1988). Hand of *Paranthropus robustus* from Member 1, Swartkrans: fossil evidence for tool behavior. *Science*, 240, 781–4.

Toth, N. (1985a). Archaeological evidence for preferential right-handedness in the lower and middle Pleistocene, and its possible implications. *Journal of Human Evolution*, 14, 607–14.

Toth, N. (1985b). The Oldowan reassessed: a close look at early stone artifacts. *Journal of Archaeological Science*, 12, 101–20.

Toth, N. (1987). Behavioral inferences from Early Stone Age artifact assemblages: an experimental model. *Journal of Human Evolution*, 16, 763–87.

Toth, N. (1991). The material record. In *The First Americans: Search and Research*, ed. T.D. Dillehay and D.J. Meltzer, pp. 53–76. Boca Raton, FL: CRC Press.

Toth, N. (in prep.). The functional significance of spheroids: results of an experimental study.

Toth, N. & Schick, K. (1986). The first million years: the archaeology of proto-human culture. *Advances in Archaeological Method and Theory*, 9, 1–96.

Toth, N. & Woods, M. (1989). Molluscan shell knives and experimental cut-marks on bones. *Journal of Field Archaeology*, 16, 250–5.

Warren, J.M. (1980). Handedness and laterality in humans and other animals. *Physiological Psychology*, 8, 351–9.

Washburn, S.L. (1960). Tools and human evolution. *Scientific American*, 203, 63–75.

Wright, R.V.S. (1972). Imitative learning of a flaked stone technology – the case of an orangutan. *Mankind*, 8, 296–306.

Wynn, T. (1989). *The Evolution of Spatial Competence*. Illinois Studies in Anthropology No. 17. Chicago: University of Illinois Press.

16

Tools and language in human evolution

IAIN DAVIDSON AND WILLIAM NOBLE

Introduction

Humans and chimpanzees shared a common ancestor some time before 3.5 million years ago, and still share a huge proportion of their genetic material. Yet their behaviour, and the flexibility of their behaviour, are fundamentally different, despite the best efforts of ethologists and psychologists to explore the similarities. At some time in the course of human evolution these differences emerged, and we might hope that the archaeological record of humans and their ancestors could shed direct light on this process of emergence. One difference between the two species is language, naturally present in humans and not in apes. We propose that all human ancestors without language should be considered as closer to chimpanzees than to modern humans in their behaviour. Two events in the record of the prehistoric evolution of human behaviour can be said to be the first that unambiguously entail the existence of language: the colonisation of Australia, before 40000 years ago, by people crossing the sea to an unknown shore; and the appearance of sculptures and bas reliefs with coded symbols in different parts of Europe before 32000 years ago.

In several recent papers (Davidson, 1988; 1989; 1991; Davidson & Noble, 1989; Noble & Davidson, 1989; 1991; unpubl.) we have explored the theoretical, methodological and empirical grounds for arguments about the timing and implications of the emergence of language. We found existing arguments to be unsatisfactory on all of these grounds and offered a hypothesis that the origin of language occurred relatively late in human evolution. This hypothesis is informed by an ecological theory of language and perception derived in part from Ludwig Wittgenstein and James Gibson; and it is illustrated with a speculative scenario of how reference, displacement and symbol production might have arisen. In developing this account, we emphasise that language is characterised by those three features: reference, displacement and productivity. They are properties which afford the deployment of signs in arbitrarily derived and conventionally confined ways for the exchange of meanings. We have endeavoured to make our account consistent with the other evidence for the evolutionary emergence of the distinctive characters of modern humans.

Randall White (1989, 81) has commented on the changes visible in the European Upper Palaeolithic:

> In some form or other behind all of these developments must stand language. To provide the base for further social and technical changes, images must be shared and communicated. Language may well have existed before the Upper Palaeolithic, perhaps in a concrete form closely tied to specific natural objects. The revolutionary developments of that period, however, presuppose innovations in the manipulation and sharing of images. Although certain types of neural "hardware" were no doubt a prerequisite for these innovations, the inception of image-based representation should not be seen as the crossing of a neurological threshold. On the contrary, it was more probably a cultural transition based on the establishment of shared conventions of representation.

We disagree with details of this account, specifically whether images can ever be created *as* images without language; what a "concrete" form of language might be; and whether anything describable as a "cultural convention" can exist without language. But we would agree that "neural hardware" (and other "wetware") are necessary though not sufficient conditions for language. We do not believe that endocranial structures identified by palaeoneurologists, nor other skeletal structures such as the hyoid, settle the question of the emergence of behaviours that are associated with them in the present time. In an evolutionary argument, structures must emerge before some of the functions they can be used for.

How long before the Upper Palaeolithic, how and where language emerged, is subject to heated debate. The issue will probably remain controversial. The identification of language as a condition for the two events, of the colonisation of Australia and the appearance of coded signs, results from our knowledge of some essential characteristics of language: reference; displacement; and symbol production. In searching for the earliest appearance of such characteristics, separately or together, we may inspect the archaeological evidence of fossil skeletons; stone, wood or bone tools; fire; shelter; or, non-utilitarian objects such as ochre, shaped tooth fragments, or collected fossils. We believe that the real issues are those that surround the interpretation of the traditional materials of archaeology: tools, fire and shelter. In this chapter we will concentrate on the evidence of the tools. In these alone can we trace a series of changes during the Lower and Middle Pleistocene that may be germane to the origins and development of hominid communication. We will suggest that the evidence from the record of stone and bone tools can be interpreted to support a late origin for language, and that this interpretation is consistent with other evidence about the evolution of human behaviour.

Francois Bordes and the Institute Conference

In 1974 Francois Bordes attended a conference of the Australian Institute of Aboriginal Studies amid controversy about the factors determining variation in Australian Aboriginal stone artefact assemblages (see e.g., J.P. White, 1977). Also present was Irari Hipuya, from the Highlands of Papua New Guinea, who had been an informant for Peter White's ethnoarchaeological studies of stone artefact manufacture (White & Thomas, 1972; White, Modjeska & Hipuya, 1977; White & Dibble, 1986). One evening Bordes was invited to manufacture a "handaxe"[1] from Australian stone raw material, to lay to rest the possibility that the raw material was a factor that could account for the lack of typological riches in Australian stone assemblages. So Bordes did this, using the full toolkit of a modern experimental stone knapper. The occasion was one of great good humour and, we suppose, history. But Irari Hipuya, seated on the ground close to Bordes, does not belong to the culture of archaeologists, and he did not, so far as we know, understand the position that handaxes have in that culture. He watched with what seemed to be growing bewilderment until finally, with no prompting, he took two of the blocks of rock and began to make flakes by striking one on the other in a well known technique. He then sorted the flakes that resulted into three groups – two groups of usable tools and one group of fragments of no use to him. By the end of the occasion, Bordes had a "handaxe", and Irari Hipuya had two piles of useful tools. There were, in addition, many flakes produced by Bordes, and at least two cores[2] produced by Irari Hipuya. None of the observers (including Davidson) looked at either the flakes produced by Bordes, or the core(s) produced by Irari Hipuya. But for Bordes the flakes were clearly unimportant, although he had to produce them to make a handaxe. For Irari Hipuya, the cores were ultimately unimportant, although he had to obey the mechanical rules of flaking cores in order to produce flakes.

We tell the story because it illustrates a major problem in attempts to understand the significance of prehistoric stone tools. Bordes was exhibiting what we call the "finished artefact fallacy", which has bedevilled the traditional study of stone artefacts, and introduced a false understanding of what it is that artefacts tell us about hominid abilities. The fallacy is the belief that the final form of flaked stone artefacts as found by archaeologists was the intended shape of a "tool". As a result the seeming complexity of hominid stone tools from 2 million years ago onwards, as contrasted with the simplicity of the use of stones by modern chimpanzees (e.g. Boesch & Boesch, 1981; Kortlandt, 1986), is a major element in any argument for an early origin for language.

Gowlett (1984a; 1984b), for example, attempted to relate to language the sequence of operations necessary for the production of cores (choppers, chopping-tools and handaxes). In doing so he implied that the only artefact produced by such a sequence was the final one. Of course this is not the case: many of the flakes produced in the sequence are potentially of use as tools (at least for a creature that can use them to cut or scrape). Gowlett's error is in supposing that the links in the sequence of operations were all part of a single chain. He argues as if he believes that core tools were produced at Olduvai or Koobi Fora 2 million years ago to the same pre-planned end as Bordes produced a handaxe in Canberra in 1974. An alternative scenario is that flakes were removed from cores adventitiously, at different times and in different places. Thus the loops of Gowlett's chain may have been produced separately and only become *links* by means of present-day analysis. And only through our imagining that the chain existed for the prehistoric stone knapper is language a necessary part of the productions of early hominids.

The same fallacy attends Holloway's (1969) treatment of stone tool manufacture in which he portrays the process as parallel to the production of speech. We have analysed Holloway's arguments in detail elsewhere (Noble & Davidson, 1991).

The standard story of palaeolithic tools

Stone artefacts are made either by flaking or grinding, or by a combination of these processes. The grinding of stone tools seems to be present from the date of the earliest occupation of Greater Australia (Jones, 1989; Groube *et al.*, 1986) before 40 000 years ago. Not only is this the earliest evidence in the world, but, as we have already argued, it must postdate the emergence of language. In looking at the early evidence for hominid stone tools, therefore, we confine our attention to flaked stones.

Stone may be flaked by exploiting properties of fine-grained rocks that also have a regular (isotropic) structure. A stone thrown up by a truck can remove a flake from the interior of a car's windscreen approximately in the shape of a cone. If the stone strikes perpendicular to the glass, with the right force, the shape will be perfectly conical. The removal of a flake from a piece of rock (which becomes a core after the flake is removed) follows the same mechanical principles (see, for example, Cottrell & Kamminga, 1987). In this case, however, the force is not applied in the middle of the rock, and a flake is detached from its side. The success in detaching a flake, and the form of the flake detached will be determined by the angle of the edge of the core, and the angle of application as well as the amount of the force. By and large, all stone artefact manufacture has had to take place under these constraints (see Crabtree, 1972). For this reason it is not necessarily the case that all of the patterning that can be observed in stone artefacts is the result of an awareness of that pattern on the part of the prehistoric manufacturers. Even today most knappers learn by imitation and experiment, much as in other craft activities. Yet it is usually presupposed, in the literature on stone tools, that the intentions of producers were represented linguistically. Part of the finished artefact fallacy is that categories recognised by archaeologists were forms created, successfully, by self-conscious prior intent. Bradley & Sampson (1986, 29–30) reconstruct this process in their approach to replication of the products of Acheulean knappers:

> They began by making a master decision about their shaping objective, then chose a raw material blank calculated to provide the best possible chance of reaching that objective. They then carried out an alternating sequence of flaking decision followed by hammer blow followed by decision and so on. Each flaking decision generated another blow. A sequence of flakes was produced, each reflecting a preceding intention. The endproduct of the process was determined not only by traditional tool-design habits, but also by the blank size and shape and by the knappers' ability to match each blow to the preceding intention. They could stop the process at any stage leading to the endproduct, leaving partly completed work or because of knapping failure or for external reasons.

Many of these "steps" derive from the self-conscious decision making necessary for the experimental replication of desired endproducts. But that does not imply similar decision making or self-consciousness by the prehistoric knappers. Once it is conceded that some of the products archaeologists observe are "partly completed" or "failures", the whole question opens up. None of the classification schemes admits the possibility of "partly completed" or "failed" products, yet they are an evident outcome of experimental replication. It seems less certain, then, that we can make unqualified judgements that there were *intended* endproducts.

It is our suggestion instead that tool making earlier than the Upper Palaeolithic can be (more parsimoniously) understood in terms of a small number of learned motor actions, and that the categories elaborated by archaeologists actually mask the fewer and simpler regularities which are all we may legitimately infer.

Oldowan

The earliest stone tools are generally isolated and their patterning (if any) is not well understood. The Omo tools dated to about 2 million years ago (e.g Chavaillon, 1976; Merrick & Merrick, 1976) are most convincingly identified by the fact that the quartz flakes must have been carried (presumably by hominids) into the sedimentary context where they were discovered. The earliest stone industry, for which patterning at more than one site has been defined, is called Oldowan, after Olduvai Gorge where it was first described. According to Fagan (1989, 112) it is characterised by the following features:

> * Use of pebbles as raw material
> * Cores with edges flaked from both sides ...
> * Some of the cores possibly fashioned into deliberate core tools (were these choppers?)
> * Both heavy- and light-duty tool forms, some modified into crude scrapers
> There is a tendency to describe the Oldowan as a very simple technology. It is true that there are few formal Oldowan tool types, but the artefacts show a skilled appreciation of basic stone-flaking techniques and flaking sequences that were *envisaged in the mind's eye.* (our emphasis)

Fagan (1989, 115–16) goes on to point out that Toth's (1985) work "led him to argue that much of the variety in Oldowan artefacts was, in fact, the result of flake production", influenced by raw material supply. Willoughby (1985) has also demonstrated that there are features of these assemblages which show patterning resulting from access to raw material. In particular this accounts for the differences between the sites of Koobi Fora that have few "core tools" and "spheroids" and contemporary sites at Olduvai Gorge where there are many.

The importance of this nexus between access to the raw material supply and the form of artefacts is emphasised when it is realised that there is abundant archaeological evidence for hominids carrying stones around the landscape. This is shown by research in which flakes from a single flaked core have been put back together, only to reveal that certain flakes and cores are missing (e.g. Toth, 1982, 91). In addition some sites consisting of bones marked by cuts from stone tools have no flaked stone – implying that the tools were taken away by the hominids (Isaac, personal communication, 1984). Access to raw materials, therefore, may be supposed to have varied as the hominids moved (carrying stones), and, therefore, the forms left at the sites of activities may be expected to vary.

Toth's experimental replications (1985, 107; see Figure 16.1) seem to demonstrate that there is significant continuity in forms, a point also made by Villa (1983, 101; see Figure 16.2).

Nevertheless, when different core forms were used as a starting point "the mean shape of flakes from different cores were surprisingly distinctive with no overlap" (Toth, 1982, 202). It seems plausible that, for these earliest industries, many of the types that archaeologists have seen are not a reflection of intentional manufacturing strategies, but fortuitous products of initial actions. Experiments involving the observation of novice knappers (Toth, 1982, 247–8)

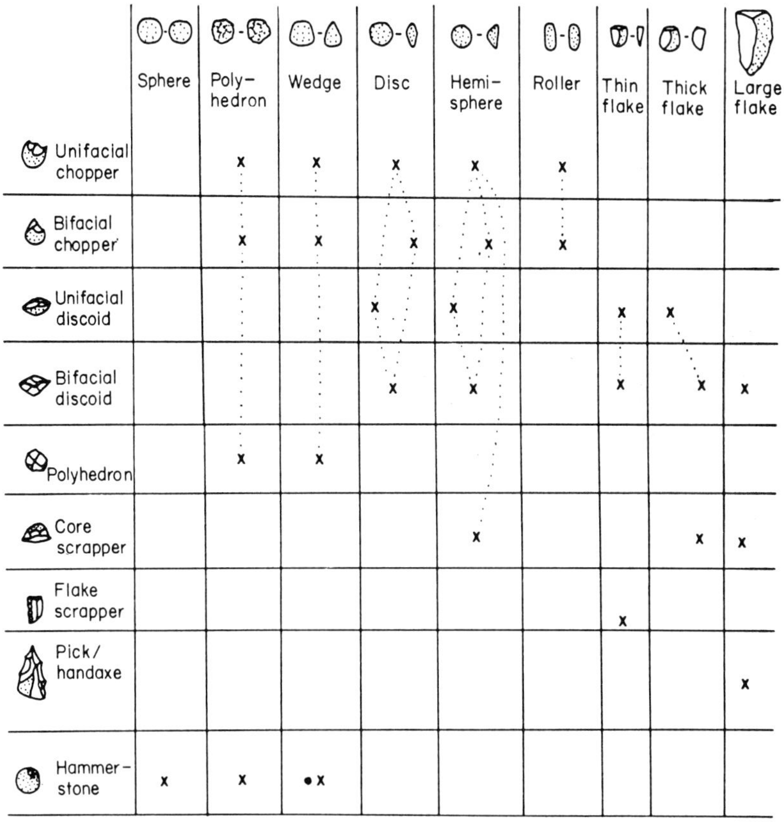

	Sphere	Poly‑hedron	Wedge	Disc	Hemi‑sphere	Roller	Thin flake	Thick flake	Large flake
Unifacial chopper		x	x	x	x	x			
Bifacial chopper		x	x	x	x	x			
Unifacial discoid				x	x		x	x	
Bifacial discoid				x	x		x	x	x
Polyhedron		x	x						
Core scrapper					x			x	x
Flake scrapper								x	
Pick/handaxe									x
Hammer‑stone	x	x	•x						

Fig. 16.1. Chart, from Toth, 1985, showing influence of initial form in determining probable endproducts. This chart also shows the continuous nature of the variability that is partitioned by archaeological classifications.

showed that these students produce many forms similar to Oldowan ones, particularly when bifacial flaking is done. Nonetheless, it may be that the rotation of a core to produce bifacial flaking was a significant novelty in hominid motor actions.

Acheulean

Fagan's (1989, 137–8) description of the Acheulean industry is:

> In Africa, Europe, and some parts of Asia, *Homo erectus* is associated with a distinctive toolkit that includes not only a variety of flake tools and sometimes choppers but bifacially flaked hand axes. . . . [T]hey are sometimes artifacts of great refinement and beauty . . ., so much so that people ignore all the other tools *Homo erectus* made.
>
> The hand axe . . . has been found . . . in all shapes and sizes . . . evidently made

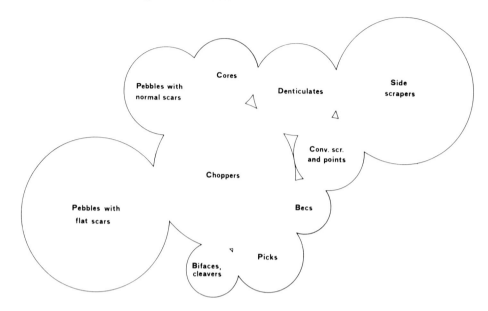

Fig. 16.2. Chart, from Villa, 1983, showing the morphological and technical relationships among components of the Terra Amata assemblage. The chart indicates the continuous nature of the variability that is partitioned by archaeological classifications.

with considerable care. ... *The maker had to envisage the shape of the artifact,* which was to be produced from a mere lump of stone, then fashion it not with opportunistic blows but with carefully directed hammer blows. (Our emphasis)

Bradley & Sampson (1986, 29) have listed four assumptions in the typological analysis of Acheulean assemblages:

> First ... that elaborately flaked toolshape – particularly biface shape – reflects the traditional design habits of an Acheulean group ... Secondly, ... that the group remained untroubled by such design restrictions when making less elaborately flaked tools on flakes, cores or blocks because these fulfilled the needs of the moment and were of no lasting value to their makers. ... [T]hird ... that tool shapes ... represent evolutionary stages of stylistic refinement ... Students of the Acheulean complex have occasionally slipped into a fourth assumption that such marker fossils are indeed biological organisms subject to the laws of Darwin and his successors.

The symmetry, "standardisation" and "refinement" of handaxes is said by some to indicate an aesthetic (Edwards, 1978) or mathematical (Gowlett 1984a, 183–7) sense. In reality handaxes themselves are both more diverse from region to region than the standard story would have it (Roe, 1968) and yet, on summary statistics for individual sites, are often astonishingly standardised ("stylistically refined"). Gowlett (1984a) argues for standardised formal design

Table 16.1. *Summary ratios for African and European handaxes*

Columns show mean, standard deviation and coefficient of variation for ratios of thickness to breadth and breadth to length of handaxes.

	Thickness: breadth			Breadth: length		
	mean	SD	CV	mean	SD	CV
Africa						
Kilombe EH	0.47	0.12	25.53	0.62	0.08	12.90
Kilombe AH + AC	0.46	0.10	21.74	0.61	0.07	11.48
Kilombe AC	0.48	0.09	18.75	0.60	0.07	11.67
Kilombe AH	0.43	0.10	23.26	0.62	0.08	12.90
Kariandusi	0.53	0.11	20.75	0.58	0.07	12.07
Olorgesailie	0.48	0.12	25.00	0.58	0.09	15.52
Europe – Caves						
Pech de l'Aze 4f	0.43		23.20	0.78		11.50
Combe Grenal 58	0.54		22.20	0.70		14.30
Combe Grenal 59	0.52		26.90	0.70		15.70
Combe Grenal 60	0.46		28.30	0.76		18.40
Pech de l'Aze II	0.53		22.60	0.69		10.10
Pech de l'Aze 9	0.56		23.20	0.69		15.90
Europe – Open						
Pendus	0.62		19.30	0.71		15.50
Dau	0.44		27.30	0.73		13.70
Bouheben	0.54		25.90	0.68		26.50
Cantalouette	0.51		23.50	0.75		21.30

Notes:
SD = standard deviation; CV = coefficient of variation.
Source: African data from Gowlett (1978); European data from Rolland (1986).

among the makers of handaxes at Kilombe. In an earlier paper (Gowlett, 1978, 352–3) he showed similarities between the summary statistics for four separate assemblages at Kilombe, one from Kariandusi, and another from Olorgesailie. The numbers for a variety of ratios expressing different aspects of the shape of handaxes showed near identity among these several assemblages (see Table 16.1).

Also shown in Table 16.1 are the comparable statistics for some European assemblages (from Rolland, 1986, 137). The similarities are no longer astonishing, simply because they are now familiar. For Gowlett (1984a, 185) the Kilombe data suggest that "*Homo erectus* of 700000 years ago had a geometrically accurate sense of proportion, and could impose this on stone in the external world". Earlier he suggested (Gowlett, 1978, 353) that the similarities between sites 200 km apart might be because the makers "belong to the same cultural microcosm". This would be difficult to claim for the likeness between the African and European examples. Dibble (1989, 421–4) has drawn attention to related issues and explains some of the similarities by the practices and definitions used by archaeologists:

> what causes the high correlation is that there are no pieces as wide or wider than they are long. Many of the rounder and thicker artefacts would most probably be called cores instead of bifaces, and by definition, a biface is always longer than it is wide.

McGrew, Tutin & Baldwin (1979) have demonstrated local regularity, together with regional variety, in the dimensions of termiting sticks made by chimpanzees. This seems to us an apt parallel to bring to bear on hominid stone tools. From such a position, the regularity of handaxes can be seen, not as the result of design, but rather as the unintended by-product of a repertoire of flaking habits – that produce flakes – limited by the form of the raw material. The "handaxes" "made" on flakes in Africa and Southern Europe (Rolland, 1986), and on flint nodules in other parts of Europe, and their shared features of size and shape, were feasibly constrained by limits on hominid abilities to produce large flakes as blanks, and by limits on their abilities to manipulate and flake those blanks (or the naturally occurring nodules). Such initial contingencies are sufficient to account for what archaeologists call "traditional design habits". In the course of replication of artefact forms from the Caddington site, Bradley & Sampson (1986, 35) noted that "while making other sorts of biface implements, we found that a Caddington handaxe shape inevitably emerged at some stage of reduction" (see Figure 16.3). The fallacy of the finished artefact is not only about whether the forms were tools, but also about whether there was a sense in which they were finished.

So what *was* the artefact (if indeed there was any "artefact" recognised by its makers)? Binford (1972) showed that the patterns of covariation of occurrence of handaxes and small scrapers made on flakes suggested that they resulted from different patterns of behaviour. One possibility is that "handaxes" were carried as cores and used as tools, and that flakes were also used as tools, but that the contexts of use and abandonment varied. The test of this must come from conjoining Acheulean knapping floors to show whether or not flakes produced in the manufacturing process were removed, and use–wear analyses to show whether such flakes were used. A suggestion of the likely results is given by Potts (1989, 481). He has recently shown at Olorgesailie that there are bones of an elephant, accompanied by flakes that derive from bifacial cores. This demonstrates that the flakes taken from "handaxes" were used as tools, which suggests that "handaxes" were cores (or also cores) and not the "intended finished artefact".

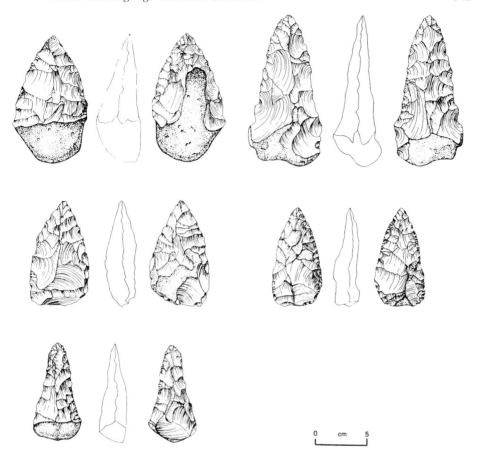

Fig. 16.3. Replicated "handaxe" forms. From Bradley & Sampson, 1986.

Levallois technique

Stone industries, from some time before 100000 years ago, are generally regarded as having included techniques of preparing the core. Of these the Levallois technique is the most discussed. Fagan (1989, 159) describes it thus:

> ... Prepared cores were carefully flaked to enable the toolmaker to strike off large flakes of predetermined size.
>
> The Levallois technique meant that the stoneworker would shape a lump of flint into an inverted bun-shaped core (often compared to an inverted tortoise shell). The flat upper surface would be struck at one end, the resulting flake forming the only product from the core. Another form was the disk core, a prepared core from which several flakes of predetermined size and shape were removed.

On the face of it the conventional textbook view of the Levallois technique seems strange. It is said to be a technique of core preparation that had as an

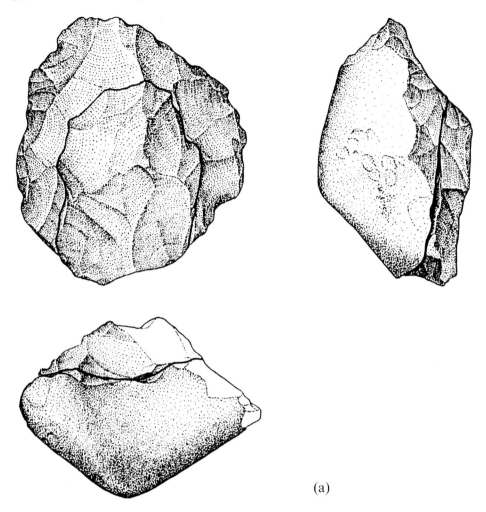

(a)

Fig. 16.4. Levallois core and conjoined flake, depicted together (*a*) and separately (*b*). From Schafer, 1990.

objective the removal of a single large flake at the final stage of production. We need to examine whether there are characteristics of such flakes that are different from those of the flakes produced throughout the "preparatory" flaking that led not only to our ability to identify them but to a choice by the hominids to use them. In addition, the presence of cores from which the final flake has not been struck, and of conjoined pairs of core and flake suggests that not all of the expectations are met (see Figure 16.4 from Schafer 1990). If the core was prepared for the removal of a single final flake, we would expect that flake either to be used at the site or to be taken away, not abandoned in conjoinable proximity to the core (see Villa, 1983, 201). Much recent work has

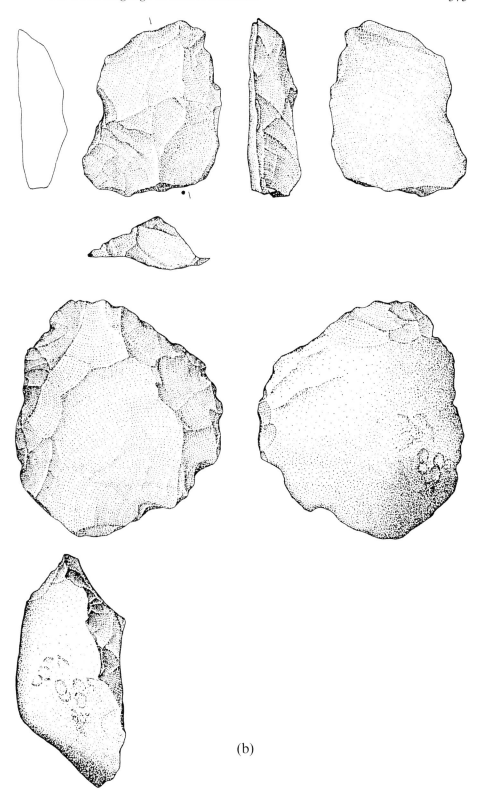

(b)

shown that the flakes identified as Levallois can be produced by a variety of techniques, and that the cores were used to produce many "prepared" flakes (see, e.g., Boëda, 1988; Marks, 1988).

We are currently investigating the possibility that there is a simpler explanation of the Levallois technique. In this view, Levallois cores were used for the production of many flakes, but at some stage, recognised by the "classic" Levallois core, the angles of the margins were such that few further flakes could be produced. At this stage, a large flake was sometimes removed which, if it reached the margins of the core would rejuvenate the edge angles so that more flakes could be removed. If it did not reach the margins then core and flake would be abandoned. In this scenario, the "classic" tortoise core and flake represent failures, not predetermined products. The time depth of intentionality is reduced to decisions about the next flake, and not to decisions about the final form of a core envisaged from the first blow on an unstruck nodule.

The Mousterian

Fagan (1989, 158) describes the conventional picture of the Mousterian industries, the industries associated with Neanderthals, as follows:

> Mousterian and other Middle Palaeolithic tools were made for the most part on flakes, the most characteristic artifacts being points and scrapers ... Some of these were composite tools, artifacts made of more than one component – for example, a point, a shaft, and the binding that secured the head to the shaft, making a spear. The edges of both points and scrapers were sharpened by fine trimming, the removal of small, steplike chips from the edge of the implement. ... The complexity of Mousterian technology is striking. The French sites have yielded a great diversity of Mousterian artifacts and toolkits. Some levels include handaxes; others, notched flakes ...

The distinctive flakes of the Mousterian were produced from disc shaped flat cores. But discoids are a major class of stone artefacts in Oldowan and Acheulean industries, first appearing in the earliest levels at Olduvai Gorge. The illustrations of these (see Figure 16.5) show strong similarities to the illustrations of cores produced by the Levallois technique, or from the Mousterian industry. Examination (by Davidson) of specimens in the Clark collection at the University of California at Berkeley confirmed this. Gowlett (1986, 251), writing about the same material, has drawn attention to this issue from a perspective completely opposite to ours:

> Mary Leakey (1971: 31...[see Figure 16.5]) specifically notes that some of the discoids, especially a series of small quartz examples, are "remarkable for the refinement of workmanship". They are bifacially worked around the whole circumference, radially flaked, and reasonably regular in shape. We could, *tongue in cheek*, regard them as the earliest Mousterian disc-cores, or they may be completed core tools. (our emphasis)

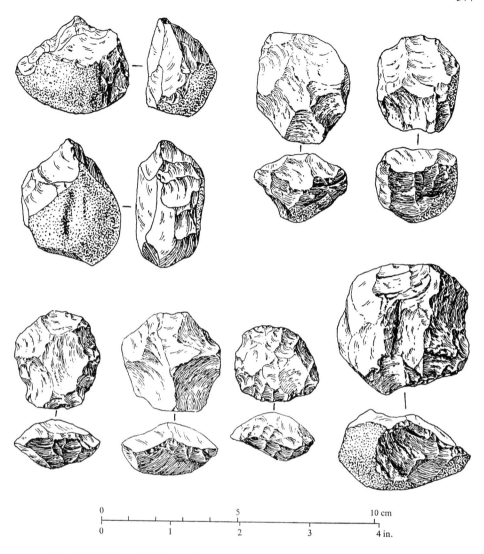

Fig. 16.5. Discoids from the lowest levels at Olduvai Gorge. From Leakey, 1971.

The conventional narrative (cf. Landau, 1984) is forcing Gowlett's interpretation of the stones. Because a developmental sequence is presumed, he must adopt a joking attitude to instances that contradict it. In this case the production of "prepared cores" seems to have been one of the earliest forms of stone working in the hominid experience, yet the conventional narrative has this as a phenomenon emerging after a million or so years of gradual elaboration of technological sophistication. There is, in reality, no reliable developmental sequence. Instead there is, in the conventional narrative, a paradigm, in the sense of a "shared commitment of a scientific group" (Kuhn, 1977, 294), such that early exemplars of forms taken as properly belonging to a later point in the sequence, are treated as anomalies. One function of the present chapter is to emphasise the anomalies in an effort to overthrow the paradigm.

In our model of the evolution of human behaviour we have consistently assumed that aspects of behaviour which at one time were a minor part of the range of behaviours available to hominids could later become much more important. Stone tools are no exception. It is reasonable to suggest that the technique indicated by the flaking of discoids at the lowest levels of Olduvai Gorge became more important as the dominant means of producing flakes in the Mousterian. We doubt whether it is necessary to invoke deliberate decision making to account for this. The determination of analysts to demonstrate "preparation" of the core in the Levallois technique and the Mousterian does not imply a similar determination on the part of the makers of those objects to produce those results.

Much of the variability described for the Mousterian is based on differences in relative frequencies of different artefact types as defined and elaborated by Francois Bordes (1961). The discussion of the meaning of this variability is one of the most famous debates of palaeolithic archaeology (see e.g. Binford, 1973; Bordes, 1973; Mellars, 1973). The fundamental issue in the debate has always been whether the typological units among which the variation is observed are comparable entities. Obviously, in positing the "finished artefact fallacy", we are not convinced that this is so, and instead prefer the interpretation of Mousterian variability being developed by Nicolas Rolland (Rolland, 1981; Rolland & Dibble, 1990) and Harold Dibble (e.g. 1987; see Figure 16.6). In this picture, the variety of forms of flaked stone artefacts can be accounted for by the sequence of use and reduction of a relatively small number of simple flake forms. The extent of reduction and the variety of forms produced can be seen as determined largely by the relation of the hominid activities to the availability of stone raw materials. In addition, Holdaway (1989) and others have cast doubt on some of the evidence for shafting of Mousterian points.

Rolland (1981) has shown that one of the major sources of variation in the content of assemblages in the Mousterian of Europe is the relative frequency of "implements" in the assemblage. Here, Rolland uses the word "implements" to mean artefacts with a shape modified by marginal flaking. This is another version of the finished artefact fallacy. Many of the unretouched flakes were also tools, but are commonly given less importance in discussions of assemblage composition. In fact, it can be shown that retouch is more abundant in assemblages with logistic requirements for tool maintenance because of

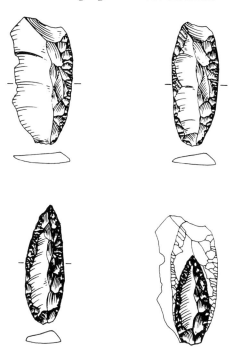

Fig. 16.6. Variability in scraper form produced by successive use and reshaping. From Dibble, 1987.

individual foraging mobility, as opposed to the logistics that permit tool replacement, because of easier access to raw material (e.g. Kuhn & Stiner, 1989; Rolland & Dibble, 1990). The concatenation of all these lines of analysis and argument suggests that much of the variability in assemblage composition is a sign of differing access to raw material in the context of the logistics of a mobile foraging strategy. We stress that this variability can exist as a result of behavioural contingencies and without "planning".

The shape of things to come

Glynn Isaac (1978, 145) once wrote (tongue in cheek?) that the "stone tools belonging to the last 100 000 years or so . . . seem rather fussy, hide-bound and unadventurous". He went on to point out that these

> regionally differentiated, rule-bound stone artifacts owe their distinctness from earlier material to crucial developments in aspects of human behaviour that have nothing to do with stone tools. It seems possible that they reflect dramatic increases in the information capacity of human communication systems, notably language.

Some regional differentiation seems to take place within the African Middle Stone Age (Allsworth-Jones, 1986a; Clark, 1988; Volman, 1984) corresponding in some respects to the European and Southwest Asian Mousterian. It is not clear, however, whether this regionalisation represents anything distinct from the sorts of regional variations in technique also evident in the Lower Palaeolithic. Nevertheless, significant novelties occur in several parts of Africa: for example, the Howieson's Poort industries, as at Klasies River Mouth in southern Africa (Singer & Wymer, 1982, 87–119), and the Aterian industry of the Sahara region, in northern Africa. Howieson's Poort industry possibly dates to the beginning of the Upper Pleistocene, but the Aterian is probably later (e.g., Davidson, 1974).

In both industries there are distinctive artefacts, confined to relatively small regions and narrow time periods, shaped in ways that cannot be related either to the technology of their production (as handaxes can) or to the modification of the working edge as a result of the constraints imposed by the technology of use (as scrapers or denticulates can). Thus Howieson's Poort industries have small tools shaped into crescent and other geometric forms (see Figure 16.7), collectively called microliths, formed by blunting the margins of blades, and the Aterian has leaf shaped points that are not just tiny handaxes, and that have, in addition, tangs which seem unrelated to the working edge. Nevertheless, the novelty at Howieson's Poort is not the start of a continuous tradition with these characters, although the traditional paradigm requires that it be so ("... it seems likely that the ideas involved in microlith manufacture were retained by some groups, rather than re-invented later" [Gowlett, 1984b, 115]). The prehistory of Australia demonstrates that reinvention of microlith was more than likely (White & O'Connell, 1982). The Howieson's Poort industry at Klasies River Mouth also contains large amounts of ochre (including at least one "crayon"), and a bone point made by scraping and polishing (Singer & Wymer, 1982, 116–17), although Volman (1984, 215) suggests the polishing could be through use. Such bone objects in which the form of the object is not determined by the mechanical constraints of the technique of manufacture are hardly known before the Upper Palaeolithic in Europe, or elsewhere.

Some of the most remarkable artefacts of the Mousterian are the leaf shaped points that occur in central Europe towards the end of the period (Allsworth-Jones, 1986b; Svoboda & Siman, 1989). These might actually be bifacial cores, in much the same manner as we have hypothesised that "handaxes" were. If that is so it will be another example of the wishful thinking of archaeologists, like their optimism about Mousterian burial, in seeing similarity between these forms, and later forms such as Solutrean leaf points.

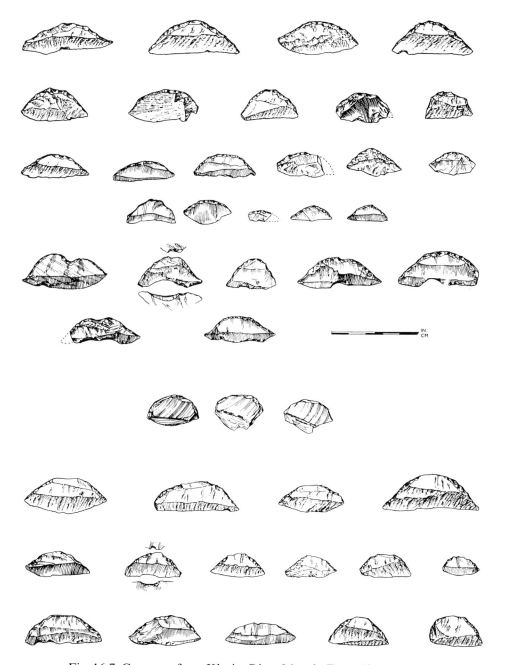

Fig. 16.7. Crescents from Klasies River Mouth. From Singer & Wymer, 1982.

The Upper Palaeolithic

The really big change seems to come, in Europe, with the Upper Palaeolithic. In general there are contrasts in spatial organisation of activities, stone industries, bone work, and ochre use between Middle and Upper Palaeolithic, even at sites with a sequence that covers the transition, such as Arcy-sur-Cure, in France (Catherine Farizy, personal communication). Fagan's (1989, 174) description of the Upper Palaeolithic is as follows:

> The classic stoneworking technology used for Upper Palaeolithic tools was based not only on percussion and the use of bone hammers but on punch-struck blades, using a hand-held or chest impelled punch to produce parallel-sided blades . . . [These] were made into a variety of tools, among them burins and scrapers, which were typical of all stages of the Upper Palaeolithic. Burins . . . were used for grooving wood, bone, and particularly antlers, which were made into spears and harpoon points.

Most importantly for the argument we are developing here, there are many objects shaped into forms by cutting and grinding rather than flaking. And these were bone, ivory or antler objects, not stone ones: bone projectile points made of antler (see e.g., Newcomer, 1977, 299–300), and the remarkable ivory sculptures of stylised and composite creatures, such as those from Vogelherd, Geissenklosterle and Hohlenstein-Stadel (Hahn 1986).

By the time of the Upper Palaeolithic in Europe, as we have stated before (Noble & Davidson 1989, 338), it was all over. Modern human beings were making tools to the shapes they decided upon, with little restriction imposed by the mechanical constraints of the technique of manufacture.

Discussion

The position we arrive at, in concluding this review, is that the stones and bones of the Upper Palaeolithic and the colonisation of Australia both exhibit the marks of planning that seem to entail a capacity for consciousness of what is being done, not just responsiveness to the unfolding contingencies of the here-and-now environment. It is language that makes possible this "reflective turn", the capacity to express what Dennett (1983, 346), using Grice's terminology, refers to as "second-order intentionality" (having "beliefs and desires" about one's "beliefs and desires").

In our view (Davidson, 1991; Davidson & Noble, 1989; Noble & Davidson, 1989; 1991; unpubl.), language arises from the discovery *that* meaning is conveyed by signs. It is evident that primates, among many other species, convey meanings through vocal and other signs (e.g., the vervet calls described by Seyfarth, Cheney & Marler, 1980; Cheney & Seyfarth, 1990). But these creatures show no unambiguous evidence of realizing that it is those signs which convey those meanings. Our speculation about how that discovery could have been made relies on a sequence in which increasing visually-guided control of movement of the forelimbs by upright hominids, evolving in part through improved capacity for one-handed aimed throwing (Calvin, 1982), led to the capacity to indicate silently (by manual pointing) significant environmental events, such as the presence of predators or prey. An extension of pointing, namely gestural indexing of characteristic features of different animals, would be made possible by the same forelimb controllability. The selective advantage of such noiseless manoeuvres lies in improved food getting (and avoidance of predators) together with concealment of one's own whereabouts.

The discovery of these signs as conveyors of meanings arose, we argue, when they were left as traces in a pliable medium, such as mud or sand. This move objectifies the sign, makes it able to be witnessed as a device linked to its referent through the morphological similarity of gesture, sign and signified. Continuing production of visible signs, and of any vocal gestures associated with them, would afford discovery of both their malleability and their property of being removable from immediate circumstances of reference. In this chain of argument it is evident that bodily gesture plays a pivotal role. While we acknowledge the insights gained from Hewes (e.g., 1973), we do not see this as a "gestural origin" theory. Rather, as in Marshack's (1989, 335) characterisation, it is a "gestural–mimetic–depictive model", in which each of these elements is a key to the discovery that results from their conjunction. The term "depiction", suggestive of deliberate intention to portray in fixed visible form, may be too strong to cover the inauguration of what may come, subsequently, to be more deliberately practised. It is the fixing, and making visible, of meaningful signs – their external representation – that is the key issue.

It needs to be stressed also that the creatures who made this discovery were not chimpanzees, but the ancestors of modern humans. Their evolution included an increasing capacity for visuo-motor control, associated in part with a (supposed) increasing capacity for aimed throwing, and in part with the (archaeologically evident) coordinations entailed in stone tool manufacture and use. As has been noted by others (e.g., Bradshaw & Nettleton, 1982; Lieberman, 1984; MacNeilage, Studdert-Kennedy & Lindblom, 1984) the choreographies embodied in these sorts of practices pre-adapt the practitioners for control of vocal articulation. Our argument implies that control of vocal articulation does not, on its own, deliver language. We offer, rather, an account of how the hominid capacity for articulatory control would be recruited in a behavioural environment which yielded the discovery of the symbolic potential of meaningful signs.

Acknowledgements

Iain Davidson acknowledges helpful discussion, during the formulation of these ideas, with Katalin Siman, Jordi Estevez, Ken Juell, Lewis Binford, Pamela Willoughby, Desmond Clark, Curtis Marean, Stephen Sutton and Luke Godwin. William Noble, likewise, acknowledges Jeff Coulter and David Olson. We both acknowledge fruitful exchanges with Whitney Davis. Our thanks also to Tim Ingold, whose skilful editing again obliged us to express ourselves more clearly. But do not blame any of them; blame us.

Notes

1 Handaxes are recognised as stone artefacts from which flakes have been removed. Flakes were generally removed from opposite sides of the original stone ("bifacially") and from both edges. In plan they are oval or pear shaped, symmetrical about the long axis of that shape. They are also symmetrical about the plane of the plan.
2 Cores are any stones used to produce flakes. They have the negative scars where flakes have been removed. Handaxes are, therefore, cores by definition.

References

Allsworth-Jones, P. (1986a). Middle Stone Age and Middle Palaeolithic: the evidence from Nigeria and Cameroun. In *Stone Age Prehistory*, ed. G.N. Bailey & P. Callow, pp. 153–68. Cambridge: Cambridge University Press.
Allsworth-Jones, P. (1986b). *The Szeletian and the Transition from Middle to Upper Palaeolithic in Central Europe*. Oxford: Clarendon Press.
Binford, L. R. (1972). Model building – paradigms and the current state of Paleolithic research. In *An Archaeological Perspective*, ed. L.R. Binford, pp. 244–94. New York: Seminar Press.
Binford, L. R. (1973). Interassemblage variability – the Mousterian and the "functional" argument. In *The Explanation of Culture Change: Models in Prehistory*, ed. C. Renfrew, pp. 227–54. London: Duckworth.
Boëda, E. (1988). Le concept Levallois et évaluation de son champ d'application. *L'Homme de Neandertal, vol 4. La Technique. Etudes et Recherches Archéologiques de l'Université de Liège*, 31, 13–26.
Boesch, C. & Boesch, H. (1981). Sex differences in the use of natural hammers by wild chimpanzees: a preliminary report. *Journal of Human Evolution*, 10, 585–93.
Bordes, F. (1961). *Typologie du Paléolithique Ancien et Moyen*. Bordeaux: Delmas.
Bordes, F. (1973). On the chronology and contemporaneity of different palaeolithic cultures in France. In *The Explanation of Culture Change: Models in Prehistory*, ed. C. Renfrew, pp. 217–26. London: Duckworth.

Bradley, B. & Sampson, C. G. (1986). Analysis by replication of two Acheulian artefact assemblages. In *Stone Age Prehistory*, ed. G.N. Bailey & P. Callow, pp. 29–45. Cambridge: Cambridge University Press.

Bradshaw, J. L. & Nettleton, N. C. (1982). Language lateralisation to the dominant hemisphere: Tool use, gesture, and language in hominid evolution. *Current Psychological Reviews*, 2, 171–92.

Calvin, W. H. (1982). Did throwing stones shape hominid brain evolution? *Ethology and Sociobiology*, 3, 115–24.

Chavaillon, J. (1976). Evidence for the technical practices of early Pleistocene hominids, Shungura Formation, Lower Omo valley, Ethiopia. In *Earliest Man and Environments in the Lake Rudolf Basin*, ed. Y. Coppens, F. C. Howell, G. Ll. Isaac & R. E. F. Leakey, pp. 565–73. Chicago: University of Chicago Press.

Cheney, D. L. & Seyfarth, R. M. (1990). *How Monkeys See the World*. Chicago: University of Chicago Press.

Clark, J. D. (1988). The Middle Stone Age of East Africa and the beginnings of regional identity. *Journal of World Prehistory*, 2, 235–305.

Cottrell, B. & Kamminga, J. (1987). The formation of flakes. *American Antiquity*, 52, 4, 675–708.

Crabtree, D. E. (1972). An introduction to the technology of stone tools. *Occasional Papers of the Idaho State Museum*, 38, 1–17.

Davidson, I. (1974). Radiocarbon dates for the Spanish Solutrean. *Antiquity*, 48, 63–5.

Davidson, I. (1988). Comment. *Rock Art Research*, 3, 100–1.

Davidson, I. (1989). Bilzingsleben and early marking. *Rock Art Research*, 7, 52–6.

Davidson, I. (1991). The archaeology of language origins: a review. *Antiquity*, 65, 39–48.

Davidson, I. & Noble, W. (1989). The archaeology of perception: Traces of depiction and language. *Current Anthropology*, 30, 125–55.

Dennett, D. C. (1983). Intentional systems in cognitive ethology: The "Panglossian" paradigm defended. *Behavioral and Brain Sciences*, 6, 343–90.

Dibble, H. L. (1987). Reduction sequences in the manufacture of Mousterian implements of France. In *The Pleistocene Old World: Regional Perspectives*, ed. O. Soffer, pp. 33–44. New York: Plenum Press.

Dibble, H. L. (1989). The implications of stone tool types for the presence of language during the Lower and Middle Palaeolithic. In *The Human Revolution: Behavioural and Biological Perspectives on the Origins of Modern Humans*, Vol. I, ed. P. Mellars & C. Stringer, pp. 415–31. Edinburgh: Edinburgh University Press.

Edwards, S. W. (1978). Nonutilitarian activities in the Lower Paleolithic: a look at the two kinds of evidence. *Current Anthropology*, 19, 135–39.

Fagan, B. (1989). *People of the Earth. An Introduction to World Prehistory*, (6th edn). Glenview, IL: Scott, Foresman and Co.

Gowlett, J. A. J. (1978). Kilombe – an Acheulian site complex in Kenya. In *Geological Background to Fossil Man*, ed. W. W. Bishop, pp. 337–60. Edinburgh: Scottish Academic Press.

Gowlett, J. A. J. (1984a). Mental abilities of early Man: a look at some hard evidence. In *Hominid Evolution and Community Ecology*, ed. R. Foley, pp. 167–92. London: Academic Press.

Gowlett, J. A. J. (1984b). *Ascent to Civilization*. London: Collins.

Gowlett, J. A. J. (1986). Culture and conceptualisation: the Oldowan–Acheulian gradient. In *Stone Age Prehistory*, ed. G. N. Bailey & P. Callow, pp. 243–60. Cambridge: Cambridge University Press.

Groube, L., Chappell, J., Muke, J. & Price, D. (1986). A 40 000-year-old human occupation site at Huon Peninsula, Papua New Guinea. *Nature*, 324, 453–5.

Hahn, J. (1986). *Kraft und Aggression. Die Botschaft der Eiszeitkunst im Aurignacien Süddeutschlands?* Tübingen: Verlag Archaeologica Venatoria, Institut für Urgeschichte der Universität Tübingen.

Hewes, G. (1973). Primate communication and the gestural origin of language. *Current Anthropology*, 14, 5–24.

Holdaway, S. (1989). Were there hafted projectile points in the Mousterian? *Journal of Field Archaeology*, 16, 79–85.

Holloway, R. L. (1969). Culture: a human domain. *Current Anthropology*, 10, 395–412.

Isaac, G. Ll. (1978). The first geologists – the archaeology of the original rock breakers. In *Geological Background to Fossil Man*, ed. W. W. Bishop, pp. 139–47. Edinburgh: Scottish Academic Press.

Jones, R. (1989). East of Wallace's line: issues and problems in the colonisation of the Australian continent. In *The Human Revolution: Behavioural and Biological Perspectives on the Origins of Modern Humans*, Vol. I, ed. P. Mellars & C. Stringer, pp. 743–82. Edinburgh: Edinburgh University Press.

Kortlandt, A. (1986). The use of stone tools by wild-living chimpanzees and earliest hominids. *Journal of Human Evolution*, 15, 77–132.

Kuhn, S. & Stiner, M. (1989). Bones and Stones: Foraging Practices, Technology, and Land Use in the Italian Mousterian. Paper presented at the symposium in Honor of Lewis R. Binford, The Organization of Land and Space Use, Technology, and Activities in Past and Present Societies. University of New Mexico, Albuquerque, November 1989.

Kuhn, T. S. (1977). *The Essential Tension*. Chicago: University of Chicago Press.

Landau, M. (1984). Human evolution as narrative. *American Scientist*, 72, 262–8.

Leakey, M. D. (1971). *Olduvai Gorge. Volume 3. Excavations in Beds I and II, 1960–1963*. Cambridge: Cambridge University Press.

Lieberman, P. (1984). *The Biology and Evolution of Language*. Cambridge, MA: Harvard University Press.

MacNeilage, P. F., Studdert-Kennedy, M. G. & Lindblom, B. (1984). Functional precursors to language and its lateralization. *American Journal of Physiology* 246, (Regulatory Integrative Comparative Physiology) 15, R912–14.

Marks, A. E. (1988). The curation of stone tools during the Upper Pleistocene. A view from the Central Negev, Israel. In *Upper Pleistocene Prehistory of Western Europe*, ed. H. L. Dibble & A. Montet-White, pp. 275–85. Philadelphia: University Museum, University of Pennsylvania.

Marshack, A. (1989). Comment. *Current Anthropology*, 30, 332–5.

McGrew, W. C., Tutin, C. E. G. & Baldwin, P. J. (1979). Chimpanzees, tools, and termites: cross-cultural comparisons of Senegal, Tanzania, and Rio Muni. *Man: The Journal of the Royal Anthropological Institute*, 14, 185–214.

Mellars, P. A. (1973). The character of the middle–upper palaeolithic transition in south-west France. In *The Explanation of Culture Change: Models in Prehistory*, ed. C. Renfrew, pp. 255–76. London: Duckworth.

Merrick, H. V. & Merrick, J. P. S. (1976). Archeological occurrences of earlier Pleistocene age, from the Shungura formation. In *Earliest Man and Environments in the Lake Rudolf Basin*, ed. Y. Coppens, F. C. Howell, G. Ll. Isaac & R. E. F. Leakey, pp. 574–84. Chicago: University of Chicago Press.

Newcomer, M. (1977). Experiments in upper palaeolithic bone work. *Methodologie Appliquée a l'Industrie de l'Os Préhistorique*, pp. 293–301. Paris: CNRS.

Noble, W. & Davidson, I. (1989). On depiction and language. *Current Anthropology*, 30, 337–42.

Noble, W. & Davidson, I. (1991). The evolutionary emergence of modern human behaviour: language and its archaeology. *Man: The Journal of the Royal Anthropological Institute*, 26, 602–32.

Noble, W. & Davidson, I. (unpubl.). Tracing the emergence of modern human behaviour: methodological pitfalls.

Potts, R. (1989). Olorgesailie: new excavations and findings in Early and Middle Pleistocene contexts, southern Kenya rift valley. *Journal of Human Evolution*, 18, 477–84.

Roe, D. (1968). British Lower and Middle Paleolithic handaxe groups. *Proceedings of the Prehistoric Society*, 34, 1–82.

Rolland, N. (1981). The interpretation of Middle Palaeolithic variability. *Man: The Journal of the Royal Anthropological Institute*, 16, 15–42.

Rolland, N. (1986). Recent findings from La Micoque and other sites in south-western and mediterranean France: their bearing on the "Tayacian" problem and Middle Palaeolithic emergence. In *Stone Age Prehistory*, ed. G. N. Bailey & P. Callow, pp. 121–51. Cambridge: Cambridge University Press.

Rolland, N. & Dibble, H. (1990). A new synthesis of Middle Paleolithic variability. *American Antiquity*, 55, 480–99.

Schafer, J. (1990). Conjoining of artefacts and consideration of raw material. Their application at the Middle Palaeolithic site of Schweinskopf-Karmelenberg. In *The Big Puzzle. Proceedings of the International Symposium on Refitting Stone Artefacts, Monrepos 1987*. Bonn: Holos.

Seyfarth, R. M., Cheney, D. L. & Marler, P. (1980). Monkey responses to three different alarm calls: evidence of predator classification and semantic communication. *Science*, 210, 801–3.

Singer, R. & Wymer, J. (1982). *The Middle Stone Age at Klasies River Mouth in South Africa*. Chicago: University of Chicago Press.

Svoboda, J. & Siman, K. (1989). The Middle–Upper Paleolithic transition in southeastern Central Europe (Czechoslovakia and Hungary). *Journal of World Prehistory*, 3, 283–322.

Toth, N. (1982). *The Stone Technologies of Early Hominids at Koobi Fora, Kenya: An Experimental Approach*. Ann Arbor: University Microfilms International.

Toth, N. (1985). The Oldowan reassessed: a close look at early stone artifacts. *Journal of Archaeological Science*, 12, 2, 101–20.

Villa, P. (1983). *Terra Amata and the Middle Pleistocene Archaeological Record of Southern France*. Berkeley: University of California Press.

Volman, T. P. (1984). Early prehistory of southern Africa. In *Southern African Prehistory and Paleoenvironments*, ed. R. G. Klein, pp. 169–220. Rotterdam: A. A. Balkema.

White, J. P. (1977). Crude, colourless and unenterprising? Prehistorians and their views on the stone age of Sunda and Sahul. In *Sunda & Sahul*, ed. J. Allen, J. Golson & R. Jones, pp. 13–30. Sydney: Academic Press.

White, J. P. & Dibble, H. (1986). Stone tools: small-scale variability. In *Stone Age Prehistory*, ed. G. N. Bailey & P. Callow, pp. 47–53. Cambridge: Cambridge University Press.

White, J. P., Modjeska, N. & Hipuya, I. (1977). Group definitions and mental templates. An ethnographic experiment. In *Stone Tools as Cultural Markers*, ed. R. V. S. Wright, pp. 380–90. Canberra: Australian Institute of Aboriginal Studies.

White, J. P. & O'Connell, J. (1982). *A Prehistory of Australia, New Guinea and Sahul*. Sydney: Academic Press.

White, J. P. & Thomas, D. H. (1972). What mean these stones? In *Models in Archaeology*, ed. D. L. Clarke, pp. 275–308. London: Methuen.

White, R. (1989). Visual thinking in the Ice Age. *Scientific American*, 261, 74–81.

Willoughby, P. R. (1985). Spheroids and battered stones in the African Early Stone Age. *World Archaeology*, 17, 44–60.

17

Layers of thinking in tool behavior

THOMAS WYNN

Introduction

There are two ways in which stone tools might inform us about language. First, language may affect how tool-use and tool-making are learned, which in turn might have visible consequences for the tools themselves or for tool assemblages. This is the line taken by Guilmet (1977), who argues for a relationship between verbal instruction and standardization, and Krantz (1980), who argues for a relationship between phonemic speech, innovation, and the rapid dissemination of technical ideas. While there are some minor logical problems with this approach, the major reason for its relative unpopularity is that it does not effectively treat the issue most scholars find central – the evolution of language in the narrow sense of a grammatical system.

Stone tools can inform us about language in this narrow sense only if stone tools and language share critical features. This is the approach taken by most scholars who have attempted to relate stone tools to language. The posited critical features include reflexivity (Kitahara-Frisch 1978), rule governed behavior (Isaac 1976), a principle of opposition (Foster 1975), and common sequencing neurology (Frost 1980; Hewes 1973), to name just a few of the more serious attempts. The search for a common critical feature or features is hampered by an old problem in evolutionary studies. How can we distinguish between similarities that are simply analogous and those that are in fact homologous? Only if tool behavior and language use the *same* critical features can we use the first as evidence for the second. Similarity alone is insufficient; we must be able to demonstrate that the similarity results from a common source. There are two ways to look for a common source. Most persuasive would be to demonstrate that tool behavior, or some identifiable characteristic of tool behavior, and language make use of the same neural structures or functions. This would not of course answer the question of evolutionary priority but it would be provocative. The second, and less persuasive, approach is through cognitive science by attempting to identify common patterns of thinking. This is the approach I will take in the following discussion.

I have three goals in this chapter. The first is to present a tentative scheme of layered cognition in tool behavior. The second is to argue that there is, in fact, no equivalent to linguistic syntax in tool behavior, thereby seriously weakening the potential of prehistoric tools to inform us about language. Third is to suggest that tools, while not much help in understanding the evolution of language in its narrow sense, may still yield insights into the more general question of the evolution of semiotic behavior.

Tool behavior and cognitive layers

It is useful to think of tool behavior as a layered system, in much the same way that Van Sommers (1984) has described drawing. The sequential action of tool behavior is organized on several different levels simultaneously. We can examine the patterns of thinking and principles of organization encountered at each level and also examine how each level influences or is tied to the others. I am using this idea of layering as a heuristic which helps us understand some of the complexity of tool behavior. It is a way to identify different patterns of thinking and a way to discuss their interrelationships. It is more than just a classification, however, because the layering heuristic implies that lower layers or their elements are subordinate to and organized by the higher layers. In other words, I am hypothesizing that the general structure of tool behavior has similarities to that of language, with its layering of phones and phonemes, syntax and semantics. However, this general similarity may be merely trivial and carry no necessary implications about the relationship between tools and language. For this we must examine the layers themselves. I suggest that there are at least three layers of thinking in tool behavior: a biomechanical layer; a layer that constructs most of the action sequences; and a problem solving layer.

Breaking up tool behavior into its layered components yields some important benefits. One of the problems with studying the evolution of tool behavior, especially the transition from ape-like to human-like, is that "tool-use" and "tool-making" lump together several different components, not all of which can have evolved hand in hand (Wynn 1990). We can ask how each component – each layer in the current scheme – has evolved and also how the entire package has changed, as for example with the addition of new components. This adds some needed definition to the study of tool evolution. It also allows us to explore the comparison to those other well-known layered behaviors, semiotics and language.

In the discussion that follows I will describe each layer of tool behavior, discuss for each some questions of evolution and also discuss possible comparisons to individual layers in language.

Biomechanics

The lowest level of organization is that of biomechanics, which consists of the constraints imposed by the anatomy and physiology of the tool-users. The variety of biomechanical constraints on tool behavior is quite large. "Simple" constraints include the amount of force that can be delivered, speed of repetition, scale of precision, and so on; all will affect the kind of tool an organism can use or make. Biomechanical constraints have both anatomical and neurological components. The combined mechanical action of anatomical units will allow a range of possible movements. This range is often limited; horses do not have a very effective backhand, for example. There are also neurological constraints on what can be done, as in the case of handedness in modern humans. The anatomy of the left hand is no different from that of the right, yet almost all people prefer one over the other. The preference lies in the brain, not in limb anatomy.

It is with regard to biomechanical constraints that anatomical evolution has a direct relevance for the evolution of tool behavior. The now classic arguments for the relation between precision grip and stone tool manufacture come immediately to mind. The anatomy of fingers, hands, arms and shoulders constrain what can be done in stone. Moreover, this anatomy can be recognized in fossils, as Marzke and Shackley (1986) have recently demonstrated. Biomechanical constraints also affect the need for tools. An oft-cited example is that of hominids who required tools to break into bone diaphyses. This example has an interesting, but rarely mentioned, corollary. Chimpanzees may use few stone tools because they are biomechanically equipped with relatively large incisors and canines. If hominids had been similarly equipped (or better equipped), then stone tools may not have been necessary. Anatomical constraints and possibilities affect the tool behavior of all organisms, not just of hominids.

Biomechanics may also be of some relevance to discussions of the evolution of language. The debate about Neanderthal vocal structure is a fine example (Lieberman 1984). It is also possible, however, that biomechanical constraints on tool behavior may have implications for language. It appears now that we can recognize handedness in tool manufacture from fairly early on (Toth 1989). Handedness has long been recognized as being tied to cortical asymmetry, as has language. Indeed, several arguments for language evolution have been based on the presumed handedness of early ancestors (Frost 1980). If brain asymmetry is a tight neurological link between handedness and language in its narrow definition, then we would have an easy solution to the problem of language origins: *Homo habilis* appears to have been preferentially right handed, it must have had language. Such an argument would only work, however, if the link between handedness and language were *very* specific. If asymmetry is tied to language only through general semiotic ability or analytical ability, then handedness would inform us only about these more general features of cognition, not about language itself. Given the state of our knowledge about asymmetry we cannot eliminate the more general interpretation and, unfortunately, cannot accept this easy solution. We must therefore examine some other characteristics of tool cognition.

Sequence construction

Tool behavior is sequential. It consists of motor actions strung together into episodes, usually terminated by a recognizable result – the completed task or artifact. As such it is similar to speech, which also consists of elements strung together into utterances. The sequential nature of tool behavior and language tempts us, quite reasonably, to suppose that the respective sequences are constructed in the same or at least in similar ways. Given the conservative nature of evolution, where there are few radically new developments, this hypothesis makes some intuitive sense.

Unfortunately, it is rather difficult to make a disciplined comparison between language and tool behavior. Over the last 80 years social science has made a concentrated assault on language, the result of which is a sophisticated body of theory on how utterances are constructed. Nothing comparable exists for tool behavior. True, social science has shown serious interest in technology as an element in social and cultural systems, but this emphasis is almost invariably on technology as an indivisible entity acting within the larger system (Pfaffenberger 1988; Wynn 1990). There is almost no concern with how tools are made and used and there are no well-developed theories of how sequences of tool-use are constructed. When we come to compare tool behavior to language, it is very easy to allow powerful linguistic theories to encompass tool behavior, much as they have swamped semiotic theory in general. Without much reflection, it is easy to coin "techmemes" and other bastard concepts, and conclude that sequence production in the two domains is essentially the same.

However, what we do know about tool behavior suggests that it is rather different from language in the way it constructs sequences, a difference that has important consequences for the use of archaeological data on tools to study language evolution.

Modern ethnography almost entirely ignores tool behavior (Pfaffenberger 1988; Lemonnier 1986). As a consequence it is very difficult to find rigorous descriptions of how tools are made and used and, equally important, how these behaviors are learned. In the last decade there has been some movement to remedy this situation, within cognitive anthropology at least. Cognitive anthropology is rather different in its methods from cognitive psychology, making comparison between the two fields sometimes difficult. Anthropology relies on ethnographic descriptions, rather than limited controlled experiments. The richness of such descriptions is their strength, but it also makes it more difficult to point to the kind of simple cognitive processes that can be isolated in controlled experiments. Two studies are of particular interest because they directly address the questions of technique and learning. They also suggest something interesting, though not surprising, about how sequences are constructed in tool behavior.

John Gatewood (1985) has studied the acquisition of technical skills required on a commercial trawler. Gatewood hired on as a complete novice, with no prior knowledge of the equipment or procedure, and documented his experience in learning this new technology. He then corroborated his results by interviewing other novices. The methodological difference between this kind of study and controlled testing should be immediately obvious. Despite this difference, Gatewood has identified, just as validly I believe, important characteristics of thinking in day-to-day circumstances. His major conclusion was that the technical skills of purse seining are learned in a relatively primitive fashion. The novice does not learn by simply internalizing a body of knowledge presented to him by others. He is, in fact, given little or no information at all.

> They did not bother naming parts of the hardware, nor did they explain the purpose of each task I was assigned. Rather, they just told me to do several things in a linguistic form similar to: "Put that [point] through there [point]" (Gatewood 1985:204).

The novice first learns the tasks by serial memorization and has no clear idea of how each action relates to others or, indeed, how they combine to accomplish the result.

> Typically, this understanding involves a simple memorization after the fashion, "First I do job_1, then I do job_2, then I do job_3, . . ., then I am finished." In other words, this first level of understanding is in the form of a string of beads (Gatewood 1985:206).

Gatewood found that such string-of-beads organization was used by all novices, at least initially. Later, after beginning to master some of the tasks, novices began to assemble smaller tasks into larger complexes of action, which could then be conceived as hierarchies of routines and sub-routines. However, the tasks were not learned as hierarchies of action. These were the result of later reflection. Initially a novice learns by chaining actions into longer and longer sequences by memorization. The result is, in Gatewood's apt phrase, like a string of beads.

Gatewood has identified a pattern of thinking and also its principle of organization. Action is sequential but the sequence does not initially have any hierarchical structure. It is learned by memorization, by chaining one action to another. This is relatively primitive compared to some kinds of thinking but appears to be sufficient for most kinds of tool sequences. Other well-known characteristics of tool behavior corroborate Gatewood's assessment.

Because tool sequences are organized like strings of beads and learned by observation and memorization, apprenticeship is essential to the learning of tool-use and tool-making. This is clear, I think, from Gatewood's example, but numerous examples can doubtless be found from the reader's own experience and also from the ethnographic record.

> Tom [a silversmith] said that the Navajo learn by watching and then doing, following exactly as possible what they have seen their teachers do (Adair 1944:75).

> ... Marie [a Navajo weaver] doesn't "tell" when teaching. She "shows". The Navajo word for "teach" means "show" and is absolutely literal (Reichard 1934:21).

> Copying, and trial and error, rather than explicit teaching, are certainly the methods by which Duna men learn about flaked stone (White *et al.* 1977:381).

Every individual learns a tool sequence by constructing his or her own string of beads – by repetition and rote memorization. This is a result of the very simple ordering principle that organizes this level. Chaining one act to the next by means of spatial and/or temporal contiguity is not a powerful organizing principle. One must build long sequences by accretion onto previously memorized sequences. One can *talk* about sub-routines and hierarchies within the sequence, but one does not *learn* them this way.

This form of sequence construction is not limited to the narrow domain of tool behavior. It is commonly encountered in any human behavior requiring precise motor coordination. Instrumental musicians, for example, use much the same technique in learning complex passages of music. It is also the essence of most sport. The "motor memory" of the sport psychologist reminds one of Gatewood's strings of beads. Tactics in sports are not, of course, sequential, but the units of action, which are the pieces employed in the tactical game, must be learned by repetition. Just as one cannot hope to beat a world class fencer by reading a book about fencing, one cannot learn to build a fine violin by reading an instruction manual. One must practice, repeating basic actions and sequences until they have been learned at a very primitive cognitive level.

Sequences constructed by contiguity also appear in the pre-linguistic com-munication of infants. In a study of the symbolic behavior of young children, Bates (Bates, 1979) argues that symbolic ability involves the development of three capacities: imitation, whole-part relations, and communicative intent. Of special relevance to my argument is her discussion of imitation. Bates sees imitation as the reproduction of perceived contiguities, that is, actions that are perceived as connected and then reproduced, even when the action is not understood by the actor. It is the contiguity that is at the core of the behavior. She uses the example of an infant who picks up and holds a telephone to its ear like an adult, even though it has no comprehension of phone conversation. The child

> ... need not analyze the bond between vehicle and referent, i.e., the relation-ship that makes the means work. Instead, he can simply exploit the observed contiguity between means and ends. With continued experience in using such solutions, the child may *later* come to understand how things work (Bates 1979:346).

Gatewood's description of learning salmon fishing is remarkably similar to this account of imitation. He and Bates have identified what appears to be the same kind of sequential organization by observing very different realms of behavior. Salmon fishing, an adult technology, is learned in the same manner that infants learn sequences of action. This chaining appears to be very primitive. It is, in fact, one of the characteristics of Piaget's stage of sensori-motor intelligence. "In other words, sensori-motor intelligence acts like a slow-motion film, in which all the pictures are seen in succession but without fusion, and so without the continuous vision for understanding the whole"(Piaget 1960:121). Once again this resembles Gatewood's description of his initial learning of a task. It is important to note that sensorimotor intelligence is not entirely non-hierarchical; the actions of sensorimotor intelligence can often be analyzed in terms of routines and sub-routines. I want to emphasize again that other levels (and stages) of intelligence bear on tool behavior; I am not arguing that all tool behavior is a matter of sensorimotor intelligence. But I am arguing that the principles we use to construct and master the sequences of action in tool behavior are simple and characteristic of much of sensorimotor intelli-gence. They may be longer and require more memory than the initial schemes of infancy but they are organized in the same way.

This kind of thinking does not have much potential for internal evolution. Chaining biomechanical elements together by means of spatial or temporal contiguity is a relatively simple cognitive task. We can identify it in human tool behavior, the behavior of infants, and the behavior of chimpanzees. A chimpanzee juvenile learns termiting by observing its mother and imitating her actions; it constructs its own strings of beads through spatial and temporal contiguity. The style of thinking is the same as that used by humans in learning tool behavior. What differs is the length of the sequences. Constructing sequences does employ neural storage for the motor memory, which is a kind of long term memory. If memory capacity increases, then the constructed sequences can be longer and, arguably, the resulting behavior more complex. Selection for memory would, therefore, have an impact on tool behavior.

I do not think that this is a very profound insight into tool behaviour. While Gatewood's terminology of "strings of beads" is especially appropriate to his ethnographic description, this kind of sequence building is recognizable in sport and infant play and communication, as well as in adult tool-use and tool-making. Lieberman (1984) has recognized it as a component of speech as well, where the syllable sequences of words are learned and executed through the "automatized" motor memory of the structures of the human mouth and throat. On the other hand, the hierarchical organizations identified by Green-field (1991) as an active organizational principle in the object manipulation of infants are little in evidence. However, as Gatewood and Bates both point out, the novice and the child both eventually conceive of their sequential productions in terms of routines and sub-routines. There appears to be some hierarchical structure inherent in the sequences. What is unclear is the degree to which this is an active organizational feature in adult tool behaviour, rather than an organization attributed to the sequences by later reflection. Unfortunately, the methodological differences between anthropology and psychology are a problem on this point. The relevance that controlled experiments on infants' object manipulation may have to adult tool behavior has not been assessed. Even if some hierarchical principle *is* applied by novices, its effect is secondary. The central cognitive principle appears to be stringing actions together by perception of contiguity.

This kind of thinking bears only a superficial resemblance to syntax. True, both generate sequences. However, syntax has far more productive power. While syllable sequences within words require rote, motor memory, sentences clearly do not. Syntax produces complex sequences with far less effort than is required for the motor sequences we have discussed here.

Constellations of knowledge and problem solving

The construction of sequences in tool behavior is comparatively easy to observe and document, making the above interpretation at least reasonable. However, it is clear that sequence construction is only one element of tool behavior, albeit a crucial one. Tool behavior, in mammals at least, also entails problem solving, the ability to adjust behavior to a specific task at hand, and, for this, rote sequences are not enough. Unfortunately, the kinds of behavior employed to solve problems with tools and the cognitive underpinnings of these behaviors have not been extensively studied. Preliminary studies suggest that it may be useful to think of problem solving in terms of "constellations of knowledge". The clearest treatment of this kind of technological thinking is in the Kellers' studies of blacksmithing.

A constellation consists of all of the elements appropriate to a task (Dougherty and Keller 1982). They are brought together in the mind of the tool user by the specific characteristics of that task.

> Each microepisode requires a unique association of process, materials, implements, and desired end point, embedded within the larger constellation of making a fleur-de-lis [the task at hand in this example] (Dougherty and Keller 1982:767).

Constellations are not like string-of-beads sequences because they are not learned by rote memorization. Rather, they come into existence at the time of use. Their constituent elements are determined by the task at hand, especially by visual images the artisan has of the goal.

> What has led to the particular image is a combinatorial arrangement of selected bits of prior socially given knowledge stored in Charles Keller's [the smith] head, in the heads of other smiths and museum staff, in the literature, and in exemplars (Keller and Keller 1991:5).

Just as important to this kind of thinking is the feedback between the image of the task and the actual events of the process. The goal must be constantly altered, however slightly, in the light of developments in the procedure, which in turn affect other elements of the constellation. It is a dynamic relationship that could not be accomplished by rote memory.

The elements that come together in a constellation are quite varied. They include "... aesthetic, stylistic, functional, procedural, [and] financial ... standards in conjunction with conceptions of self and other, and material conditions for work." (Keller and Keller 1991:9). The rote motor sequences of Gatewood's strings of beads constitute only one kind of element. Thus, in a sense, constellations are a "higher" level of organization that incorporates sequences. The motley of elements that come together in a constellation make this kind of thinking difficult to characterize, a problem that is exacerbated by the fairly rapid conversion of constellations into recipes.

Constellations are often repeated over and over again, and as a task is repeated the original dynamic interplay of elements and goal is replaced by an almost unconscious recipe. "When a production has become routine much of the detail of the task becomes 'taken-for-granted' and is difficult or impossible to articulate" (Keller and Keller 1991:3). Much of tool behavior consists of such recipes; most tool-use is, after all, a mundane kind of behavior that is practised on a regular, often daily, basis. After years of repeating a task, many of its elements have been so routinized that it takes a serious effort for the artisan to identify and separate them. Nevertheless, constellations differ from rote sequences in their manner of construction and use. Even routine tasks require some attention to and adjustments of the constituent elements (think of driving to work, for example), though the dynamic feedback between goal and procedure is much less marked than that found in constellations generated for a novel task.

When anyone, trained artisan or not, confronts a situation requiring tools, he or she must assemble a constellation of knowledge that brings together a diversity of appropriate elements. What makes an element appropriate? This question bears on a number of anthropologically interesting matters, including matters relating to cultural evolution. At a minimum we must consider three aspects of "appropriateness": function, idiosyncracy, and tradition.

Of course, the task needs to be accomplished. The constellation must "work", and working is a kind of appropriateness. The elements brought together in a constellation must get the job done. However, it would be a mistake to think that function somehow determines the characteristics of any constellation of technological knowledge. The task-at-hand establishes a broad set of constraints on the possible solutions. A very large number of procedures, tools, styles, and so on will actually get the job done. The task does not determine which; indeed it cannot because the task-at-hand and the constellation are in no way directly tied to one another. They are connected only in the mind of the artisan who must choose what to do. "The form of designed things is decided by choice or else by chance; but it is never actually entailed by anything whatsoever"(Pye 1964:9). Pye, writing about furniture design, seriously challenges the notion that function determines form. This is a theme taken up by the archaeologist Sackett (1982) and the ethnographer Lemonnier (1986). The array of tools and procedures that can successfully achieve a task-at-hand is, if not infinite, at least immense. When an artisan assembles a constellation of knowledge he is choosing, often quite arbitrarily. Certainly, characteristics such as economy – how much time and effort the artisan is willing to invest – come into play, but these are matters of the personal history of the artisan, not of the task *per se*.

We saw in the discussion of sequence construction that tool behavior is learned largely through apprenticeship, which is essentially non-verbal. The apprentice learns by practice and failure. Both the Kellers and Gatewood note an important consequence of apprenticeship – each actor constructs his own constellations. They are not shared. When Gatewood inquired about the cognitive segments (Gatewood's term for constellations) employed by other seiners, he found that there were marked differences in the content, even though the same words were used as labels (for example, "pursing" or "lifting rings" would call up an idiosyncratic constellation of knowledge for each seiner). The constellations are not shared primarily because each has been constructed by correspondences recognized by each person as he learned the task. Once again other ethnographic descriptions corroborate the idiosyncratic nature of constellations. Adair (1944), in his study of Navajo silversmiths, found that ". . . the way smiths learned was by finishing each piece" (p. 75). Regardless of even gross errors, the apprentice was made to complete every piece begun. If major constellations of tool behavior must be constructed by each actor, and if such constellations are not shared, then such a practice not only makes sense, it is essential. The artisan brings idiosyncratic ways of doing things, aesthetics, and so on to any task. As such, much that is appropriate is idiosyncratic.

Tradition and community standards probably constrain the range of appropriate constellations more than even personal history. The artisan draws largely from the pool of solutions known by his own community. These tend to be very constraining. Lemonnier, in his study of Papuan material culture, observed that the Anga know how to make a deadlier kind of arrow that is produced regularly by other groups, but nevertheless do not do so. The wood used and the game hunted is the same; this Anga group simply did not choose to make that kind of arrow (Lemonnier 1986). In addition to exerting constraints on what an artisan is willing to produce, community standards are very conservative, to the point of violating what would seem obvious "functional" considerations. A well-known archaeological example is that of Chalcolithic copper axes, which for centuries were made to resemble the ground stone axes that preceded them. Community standards are, from the perspective of an individual artisan, the range of appropriate forms (and acceptable deviations) from which he can choose. In their simplest guise, they consist of all of the known possible solutions to a task-at-hand. But often, as in the case of the Anga, community standards are only a subset of the known possible solutions. It is in this respect that we can speak of knowledge about tools being shared. The bits and pieces of knowledge that make up community standards are presumably shared by all community artisans. However, they form only one part of the considerations brought to bear when an artisan assembles a constellation to achieve the task-at-hand. They constrain the solution but cannot themselves achieve a specific task. The artisan combines them with idiosyncratic knowledge, action sequences acquired by rote, considerations of economy, and so on.

The above is a description, and a rather clumsy one, of what goes on in an episode of tool behavior. It is not itself a model of thinking but it does allow consideration of its cognitive underpinnings.

One process that appears to be involved in assembling constellations is association. I mean this in a general sense, not in a more technical sense, for example, of classical conditioning. When the artisan envisions the goal (the completed task-at-hand), certain salient features of the desired end correspond to images, muscle tensions, equipment, and so on. These correspondences come together into the constellation of knowledge. The key organizational mechanism here appears to be feature correspondence (rather than the temporal and spatial contiguity of rote memorization). Certain features of the task call up elements of the artisan's repertoire. Interestingly, this is largely a non-linguistic process. Both the Kellers and Gatewood found it impossible to elicit consistent labels for the behaviors they studied. "Rather, one experiences visual imagery and muscular tensions appropriate to certain actions, but can only grope for words. ..." (Gatewood 1985:206). The constellations are clearly not constructed through lexical categories. The use of feature correspondence is a more powerful organizing tool than the simple rote chaining we saw in sequence production. It is not, however, a very complex form of reasoning. Indeed, as we saw in the discussion of recipes, it can rapidly sink below the level of immediate consciousness.

The use of feature correspondence is the common property of all constellations. Constellations are also plans of action, however, and plans of action can have varying degrees of complexity, which can be examined for their cognitive requirements. Modern tool behavior incorporates two kinds of planning – trial and error and contingency planning. The former is simpler in its cognitive requirements. The artisan envisions a goal, which calls up appropriate notions of tools, motor sequences, and so on. He then proceeds until failure, revises his constellation, and begins again. The dynamic interplay between goal and constellation takes place in action. From a Piagetian perspective this kind of planning lacks the element of "reversibility", the ability to perform the action in thought, to imagine the coming failure, and to return to the starting point *in thought*, before actually committing the error. Reversibility is a characteristic of thinking used in contingency planning, where failures are anticipated and alternative procedures prepared ahead of time. In reality, modern tool behavior incorporates both; despite the use of reversibility, the artisan can rarely envision all of the possible problems and solutions of a new task. The important point here is that constellations can incorporate complex plans that require complex cognitive skills like reversibility. However, they need not. Indeed, all constellations share only the association of elements by feature correspondence, which is not itself a complex cognitive skill.

There are several ways in which constellations can be said to evolve, and it is this which makes the notion of constellation useful in studies of human evolution. The basic feature of constellations, association of elements, does not itself evolve; presumably all organisms that learn to use tools create constellations in the same basic way. One need only examine chimpanzee tool episodes to see the construction of constellations and the use of recipes. However, some of the other cognitive underpinnings can evolve. For example, the organizational features applied to plans of action can evolve in complexity. This is something that I have attempted to demonstrate from the analysis of the geometry of stone tools (Wynn 1989). There is a point in the development of stone tools when the use of reversibility became necessary; the goals could not have been achieved by trial and error alone (Wynn 1979).

Constellations have also clearly evolved in terms of the number and variety of elements incorporated into each individual constellation. It is here that we encounter technical evolution or progress as it is commonly conceived. The tasks themselves may become more complex (for whatever reason), requiring more elaborate constellations of knowledge. Tool behavior may enter into more and different behavioral domains. If a tool comes to mark social status, as many modern tools clearly do, then these elements enter into the constellation assembled by the artisan. Tools may carry semiotic load. They can be intentional symbols or, more commonly, unintended indices. The latter results from the constraints imposed by community standards. If a community will accept only one possible solution, this solution can come to mark an artisan as a member of that community.

Focusing on constellations is a useful way to look at tool-use and tool-making, but does not tell us anything very new about human behavior. The idea of a constellation of knowledge is especially appropriate in its applicability to new situations. This is how artisans and non-artisans alike think when they solve problems with tools. It encompasses mundane, everyday behaviour, as well as skilled craftmanship. Indeed, we can understand constellations as an interesting kind of problem solving, but one that is not fundamentally different from other goal directed behavior. Language, on the other hand, appears to have some very unique features. As a consequence tool behavior, even if it can be used as evidence for general cognitive processes, cannot be used as evidence for language.

I would like to support this assertion by examining one contrast between the constellations generated for tool behavior and the speech acts generated by language. Language is shared in a much more thorough and specific way than tool behavior. Native speakers of a language can be said to share that language; all have acquired the same basic set of grammatical rules (or at least very similar sets) and all share at least a basic vocabulary. How these are acquired has been a matter of considerable investigation. Chomsky's argument for an innate grammar may be losing ground to the ideas of a partly innate Language Acquisition Device (LAD) that guides a child to learn the rules of his language and a Language Acquisition Support System (LASS), also partly innate, that structures the way adults speak to children at various stages of learning (Bruner 1985). Whether one favors the narrow Chomskian position or the less entailed LAD and LASS, it does appear that something is innate in the acquisition of language. The consequence of this (and a very useful consequence from an evolutionary perspective) is that individuals share their language in a very detailed way.

The same cannot be said for tool behavior. While one could perhaps argue that humans have a predisposition to manipulate objects, this is only very general and, moreover, probably shared with chimpanzees. There appears to be no Technology Acquisition Device. Many people never acquire competence in tools comparable to competence in language. There do appear to be Technology Acquisition Support Systems, but they are not general and not innate. Indeed, apprenticeship appears to be an entirely cultural system whose role is to force novices to construct their own technologies, only some of whose elements are shared with others.

While the constellations of knowledge employed in tool behavior may be uninformative about language, they may still be able to tell us about the more general phenomena of semiotics. Tools are not words and there may be no syntax in tool behavior, but tools can and do act as signs and symbols. Moreover, the semiotic role of tools has apparently evolved.

The most common semiotic role of a tool is as an index. An index signals its referent by physical contiguity (Casson 1981) or direct association. The choices an artisan makes when assembling a particular constellation of knowledge can come to be associated with one another and with the artisan, especially when constellations become recipes. A particular technological solution can act as an index of its maker (a Stradivarius violin is an example). The range of community standards from which an artisan chooses can act as indices for that community (Anasazi pots, for example). This indexical quality of tool behavior is largely an unintended consequence of the use of constellations of knowledge. Nevertheless, it is a consequence that can be exploited as a source of information, a kind of semiotic.

Occasionally, the symbolic value of a tool becomes the predominant consideration in its use and manufacture. In these cases the mechanical considerations often fade and the sign can no longer actually be used as a tool. However, the iconic connection – the "toolness" – of the sign remains fundamental to the meaning. The Christian cross is an example. For the Romans it was a simple tool. Early Christians attached symbolic value to it because it was an icon for the passion of Christ. Today an artisan making a crucifix does not attend to its possible use as an instrument of execution, but does attend to its role as a symbol.

The interest in language evolution has tended to overshadow interest in general semiotic evolution. There have been some interesting studies of the role of symbols, loosely defined, in the European Upper Palaeolithic (Gamble 1982) and in the earlier Middle Palaeolithic (Maringer 1960; Chase and Dibble 1987) but most of these investigate semiotics in a rather rudimentary way. They look for the presence or absence of symbolic behavior but rarely try to characterize the nature of the symbolic system. There does seem to be potential for studying the evolution of semiotic behavior in a more sophisticated way (see Davidson and Noble 1989, for an interesting discussion of iconicity). The indexical role of tools has apparently evolved in a manner that cannot be grasped in terms of a simple opposition between presence and absence. For example, we know that Acheulean artisans, beginning about 1.5 million years ago, incorporated some notion of standard shape into their constellations of knowledge. Since the shape appears not to have been constrained closely by function, it must have been a matter of choice constrained by community standards of some sort, a choice that has a potential indexical role. Can we say that handaxes signal something? Unfortunately, we do not yet have an answer. Bifaces were manufactured according to this pattern, more or less (a troubling caveat), over three continents for hundreds of thousands of years. This situation has no modern analogs and its implications as regards a possible indexical role for the biface are far from clear. My purpose here is not to solve a long standing enigma, but to point to an untried direction of inquiry. Instead of formulating unconvincing arguments about handaxes and syntax, perhaps archaeologists should pursue the more promising approach of general semiotics.

Conclusion

The kinds of thinking applied in tool behavior do not appear to have any remarkable or unique features. Reliance on rote "strings of beads" and the use of "constellations of knowledge" to solve practical problems are general kinds of thinking that can be recognized in other domains of behavior. Nowhere can we recognize patterns of thinking that appear to be peculiar to tools; to use the jargon of cognitive science, there appear to be no "domain specific" features. Arguably, language does employ such domain specific features. Lieberman (1984) argues persuasively that humans employ neural pathways that are devoted to the rapid production and discrimination of speech. Chomsky (e.g. 1980) argues for an innate universal grammar and Bruner (1985) argues for a language acquisition device. While disagreeing on much else, all would appear to agree that something is special about language. The kind of thinking used with tools, on the other hand, appears to be very general. Greenfield (1991) has argued that infants use the same kinds of thinking in object manipulation and in early speech (something she terms a "supramodal hierarchical processor"), but that after the age of three, the two kinds of thinking take different developmental paths. If we accept for the moment that object manipulation is relevant to tool-use (by no means an established point), then it appears that the ethnographic evidence discussed in this chapter supports Greenfield's hypothesis. Tool behavior employs cognitive organizations (and neural circuitry) that develop along lines that are separate from those required for language, and which appear to be more generally available.

This conclusion has implications for understanding human evolution. First, it highlights the evolutionary continuity between apes and humans. "Strings of beads" and "constellations of knowledge" can be used to characterize the behavior of chimpanzees as well as humans. Chimpanzee solutions appear to be less demanding in terms of both memory (having shorter sequences and fewer elements in constellations) and sophistication of the plans of action (there is an apparent lack of contingency planning, for example), but the style of thinking is the same. This is only to be expected given the general nature of the thinking used in tool behavior. Second, evolution of a more human-like tool behavior could have long preceded the appearance of language. Long term memory capacity and problem solving ability, the core abilities in tool behavior, could well have evolved without the appearance of the domain specific features necessary for language. The relatively simple and general kinds of thinking used in tool behavior suggest that it may well have been an older adaptation. However, the tools themselves cannot answer this question for us. They can inform us about a few general cognitive abilities only; the evidence for language must come from elsewhere.

References

Adair, J. (1944). *The Navajo and Pueblo Silversmith*. Norman: University of Oklahoma Press.

Bates, E. (1979). *The Emergence of Symbols*. New York: Academic Press.

Bruner, J. (1985). Vygotsky: a historical and conceptual perspective. In *Culture, Communication and Cognition: Vygotskian Perspectives*, ed. J. Wertsch, pp. 371–91. Cambridge: Cambridge University Press.

Casson, E. W. (1981). *Language, Culture, and Cognition*. New York: Macmillan.

Chase, P. & Dibble, H. (1987). Middle Paleolithic symbolism: a review of current evidence and interpretations. *Journal of Anthropological Archaeology*, 6, 263–96.

Chomsky, N. (1980). Rules and representations. *Behavioral and Brain Sciences*, 3, 1–61.

Davidson, I. & Noble, W. (1989). The archaeology of perception: traces of depiction and language. *Current Anthropology*, 30, 125–56.

Dougherty, J. W. D. & Keller, C. M. (1982). Taskonomy: a practical approach to knowledge structures. *American Ethnologist*, 5, 763–74.

Foster, M. L. (1975). Symbolic sets. Paper presented in the symposium "Toward an ideational dimension in archaeology." Meetings of the Society for American Archaeology, Dallas.

Frost, G. T. (1980). Tool behavior and the origins of laterality. *Journal of Human Evolution*, 9, 447–59.

Gamble, C. (1982). Interaction and alliance in palaeolithic society. *Man*, 17, 92–107.

Gatewood, J. (1985). Actions speak louder than words. In *Directions in Cognitive Anthropology*, ed. J. Dougherty, pp. 199–220. Urbana: University of Illinois Press.

Greenfield, P. M. (1991). Language, tools and the brain. The ontogeny and phylogeny of hierarchically organized sequential behaviour. *Behavioral and Brain Sciences*, 14, 531–95.

Guilmet, G. M. (1977). The evolution of tool-using and tool-making behavior. *Man*, 12, 33–47.

Hewes, G. W. (1973). An explicit formulation of the relation between tool-using and early human language emergence. *Visible Language*, 7(2), 101–27.

Isaac, G. L. (1976). Stages of cultural elaboration in the Pleistocene: Possible archaeological indicators of the development of language capabilities. In *Origins and Evolution of Language and Speech*, ed. S. Harnad, H. Steklis & J. Lancaster, pp. 275–88. Annals of the New York Academy of Science, vol. 280.

Keller, J. & Keller, C. (1991). Thinking and acting with iron. *The Beckman Institute Cognitive Science Technical Reports*, CS–91–08. (University of Illinois at Urbana-Champaign).

Kitahara-Frisch, J. (1978). Stone tools as indicators of linguistic ability in early man. *Kagaku Kisoron Gakkai Annals*, 5, 101–9.

Krantz, G. (1980). Sapienization and speech. *Current Anthropology*, 21, 773–92.

Lemonnier, P. (1986). The study of material culture today: toward an anthropology of technical systems. *Journal of Anthropological Anthropology*, 5, 147–86.

Lieberman, P. (1984). *The Biology and Evolution of Language*. Cambridge, MA: Harvard University Press.

Maringer, J. (1960). *The Gods of Prehistoric Man*. New York: Knopf.

Marzke, M. & Shackley, M. (1986). Hominid hand use in the Pliocene and Pleistocene: Evidence from experimental archaeology and comparative morphology. *Journal of Human Evolution*, 15, 439–60.

Pfaffenberger, B. (1988). Fetished objects and humanized nature: towards an anthropology of technology. *Man*, 23, 236–52.

Piaget, J. (1960). *The Psychology of Intelligence*. M. Piercy & D. Berlyne trans. Totawa: Littlefield, Adams and Co.

Pye, D. (1964). *The Nature of Design*. London: Studio Vista.

Reichard, G. (1934). *Spider Woman: A Story of Navajo Weavers and Chanters*. Glorietta, NM: Rio Grande Press.

Sackett, J. (1982). Approaches to style in lithic archaeology. *Journal of Anthropological Archaeology*, 1, 59–112.

Toth, N. (1989). The prehistoric roots of a human concept of symmetry. Paper delivered at the symposium "Symmetry of Structure", organized by the Hungarian Academy of Sciences, Budapest.

Van Sommers, P. (1984). *Drawing and Cognition: Descriptive and Experimental Studies of Graphic Production Processes*. Cambridge: Cambridge University Press.

White, J., Modjeska, N. and Hipuya, I. (1977). Group definitions and mental templates. In *Stone Tools as Cultural Markers*, ed. R. V. S. Wright, pp. 380–90. New Jersey: Humanities Press.

Wynn, T. (1979). The intelligence of later Acheulean hominids. *Man*, 14, 379–91.

Wynn, T. (1989). *The Evolution of Spatial Competence*. Urbana: University of Illinois Press.

Wynn, T. (1990). Natural history and the superorganic in the study of tool behavior. In *Interpretation and Explanation in the Study of Behavior: Comparative Perspectives*, vol 2, ed. M. Bekoff & D. Jamieson, pp. 98–117. Boulder, CO: Westview Press.

18

The complementation theory of language and tool use

PETER C. REYNOLDS

Interpretations of similarity and difference

For decades, anthropologists have been torn between two opposed interpretations of primate tool-using skills. In one interpretation, human tool use is seen as a more evolved version of a simian system, characterized by quantitative differences between humans and apes (Beck, 1980; Blurton-Jones, 1972). In this perspective, hominids developed a greater reliance on tools for subsistence, which in turn led to the development of a greater variety of tool types, more differentiation of function, and increased skill in their manufacturing – all quantitative changes from a simian baseline. This interpretation is supported by both archaeological and primatological evidence. Stone tool-making skills do appear to become more precise over time, showing evidence of finer workmanship and more attention to form (Isaac, 1976); and field and laboratory studies indicate that nearly all species of anthropoid primates can hurl missiles, smash objects with hammers, and use sticks as extensions of the arm (Parker, 1974).

The other interpretation of human tool use, while recognizing these facts, emphasizes the differences in the finished products made by humans and apes (Bordes, 1971; Holloway, 1969; Leroi-Gourhan, 1943, 1957). The artifacts manufactured by humans are typically made of distinct parts, often useless in themselves, which fit together to make a functional whole; and this new entity is then used as a base for subsequent operations. In baking a cake, for example, sugar, flour, baking powder, flavoring, etc. are all mixed together into a batter, which then functions as a unit in subsequent steps such as baking, cooling, and the application of frosting. In the archaeological record as well, hominid tools begin as a differentiated system of tools, with the different types presumably made for different purposes, and carried for miles to be used again. Moreover, they exhibit the intentional shaping of matter and specialization by form and material that seem closer to a hafted spear than to a termite stick produced by a chimpanzee. After all, why do home-reared apes not bake cakes, or produce comparable part/whole constructions, even when exposed to human tutors who engage in such technical activities as a normal part of everyday life? This is the question that the second interpretation tries to answer.

These two major approaches to tool use are based on different types of observation, and each has a range of applicability, but neither can explain the observations on which its opposite is based. This is readily apparent if one sees both types of evidence at first hand. Before I began my research into the structure of human tool use, I spent almost every day for two years observing social behavior in groups of macaque monkeys, both *Macaca fascicularis* and *M. mulatta* (Oakley & Reynolds, 1976); and I saw a wide range of behaviors involving physical objects, including poking with sticks, making noise by banging with sticks on metal containers, and tug-of-war games with a length of rope. I also carried out some research at the San Diego Zoo, presenting gorillas with pieces of fruit in a transparent latch box, which required them to undo the various bolts and snaps in order to retrieve the food–actions which they readily performed. At the Yerkes Primate Center in Atlanta, I filmed immature gorillas, orangutans, and chimpanzees living together in a large outdoor playground, and I witnessed numerous examples of stick throwing, tire swinging, ball rolling, and rope coiling. I also spent some time associating socially with a signing gorilla, which drank out of baby bottles, stacked children's blocks, and used a potty chair. So I have a first-hand appreciation of nonhuman primate object-using skills, and some partiality for continuity theories (Reynolds, 1982, 1983); but when I first began videotaping three- and four-year old children in a preschool, I was shocked by what I saw: five years of observations on nonhuman primates did not prepare me for a human level of object use and tool-using skills.

It was *not* that the children built complex, multipart constructions using sophisticated motor patterns without precedent in apes. To the contrary, the motor component is very similar to the basic primate repertory of running, jumping, swinging, gripping, throwing, poking, hitting, biting, tugging, bouncing, twirling, rolling, pushing, and so on (Parker, 1974). Although there are some documented differences between human and nonhuman primate object-using skills, the fact that the play of children and monkeys can be described so unambiguously with ordinary English words attests to their similarity. Indeed, when the differences in locomotion type and hand morphology are taken into account, the video images of low-level action patterns of humans and nonhuman primates are close to identical. Nor are there profound differences in the complexity of the objects produced by preschool children. They either play with objects provided by adults, such as plastic buckets, swings, and wheelbarrows, or they *pretend* to make things without actually modifying matter much at all. In the course of my observations at the preschool, recorded on videotape for two hours a day for six months, the children *pretended* to make cakes, babies, cavalry forts, Indian teepees, towns, airplanes, automobiles, roads, and houses, but the transformations of matter actually produced by this behavior were generally unphotogenic piles of sand, lumps of mud, and tin cans arranged in a circle. From a technological perspective, these children did not make anything that would attract the attention of a classical archaeologist, for a lump of mud that is a 'pretend cake' is often indistinguishable from an ordinary lump of mud. Thus, a thorough-going behaviorist who compares human action patterns in a preschool to action patterns in a chimpanzee colony is right to insist that there are no profound discrepancies between the kinds of objects produced by apes and by children – for none are apparent on the video.

However, anthropologists learned a long time ago to attend to the audio as well as the video, and a pretend cake can be discriminated from a lump of mud because the children *tell* you it is a cake. I remember a little girl who came up to me with a pebble displayed on the palm of her hand. 'Look at my spider', she said. Now I suppose a real scientist would have corrected her: 'That's not a spider; it's a stone.' But anthropologists, for better or for worse, are trained to accept the context of discourse constructed by their subjects. So I asked: 'Aren't you afraid of getting bitten?' 'No', she said, 'It's a *friendly* spider.' This simple dialog, which assumes a disparity between the morphology of the referent (a stone) and the meaning attributed to it (a spider), reveals a profoundly significant aspect of tool-using behavior in humans: the fact that it is combined with symbolic thought. Moreover, it demonstrates the limitations of conventional primatological methods, for there is no way that a non-English speaker, much less a behaviorist, could penetrate into this symbolic world, which is invisible on the video.

Pretend tool use complicates the discussion of tool-using skills, but it does not resolve the conflict between the two major approaches to human evolution. A theorist committed to ape–human continuity might rejoin: 'We have known all along that animals have symbolic capabilities that are beyond behaviorist methods to demonstrate; humans only appear unique because we already share the contexts of symbolic discourse.' A theorist committed to hominid uniqueness, when examining the same data, might respond: 'We have known all along that the symbolic faculty in humans, particularly language, creates a new level of cognitive functioning, and that the behavioral similarities pointed out by primatologists are largely irrelevant to understanding these capabilities.' The continuity and discontinuity aproaches can be used to bring animals into the club, or to keep them out, but they do not greatly contribute to our understanding of what to look for or where to look for it. Anthropology needs a framework that encompasses both cultural and primatological observations, both the audio and the video, the jungle and the blackboard jungle.

The great tool-use fallacy

The development of a comprehensive theory bridging both anthropological and primatological observations has been hampered by the failure to recognize the social dimension of tool-using skills. Although the social aspect of tool use *does* appear on the video, its significance has gone largely unrecognized because 'tool using' is too narrowly defined by primatologists and archaeologists alike. In industrial societies, 'tool use' is defined as something that is done with the hands by an individual working alone (Ingold, this volume); and in popular books in Western culture, such as Jean Auel's novels about the Paleolithic (e.g., Auel, 1980), the heroine discovers fire, invents the bow and arrow, tames the horse, and so on. Although this view of human history is dismissed by anthropologists as simplistic, it is not as far from canonical scientific theory as many archaeologists would like to believe. Indeed, evolutionary notions of tool use are permeated with Eurocentric assumptions that cannot be reconciled with the activities of children in a preschool, much less survive cross-cultural comparison (Reynolds, 1991).

The concept of the stone tool as the product of an artisan working alone is an artifact of archaeology itself. The archaeological method distorts the normal tool-making process, first, by consigning tools to museum collections where they are divorced from the social contexts in which they were made and used, and secondly, by recreating in the laboratory a purely technical notion of tool-making, in which the archaeologist, alone with a hammerstone and a piece of flint, attempts to reproduce the tools found in museum drawers. To confirm this supposition, I simply rented the movies that archaeologists show to their classes; and almost without exception, stone tool-making was presented as a relationship between artisan, tools, and raw materials, demonstrated for the camera by a single individual working alone (see Johnson, 1978, for this literature). The distortions introduced by archaeological methods are easy to see when filmed demonstrations of flint-knapping are contrasted to films of Australian Aborigines, who until recently depended on this technology for a living.

Cooperative construction, complementary action

Human technology involves the cooperative construction of artifacts, and even stone tools are normally made by groups, not individuals. As an anthropologist working in Australia, I had an opportunity to see stone tool-making for myself when two Australian Aborigines were flown into Canberra from the Northern Territory by the Institute of Aboriginal Studies in order to participate in a filming session on traditional crafts.[1] The two elderly men, Billy Dempsey and Slippery Morton, had learned the craft from their forefathers, and they walked into the institute still dressed in the dusty blue denim of the outback. Cameras had been set up in a meeting room in the basement, and they obligingly sat down on the floor and began to manufacture examples of Aboriginal knives from large chunks of quartzite provided by the anthropologists. Since the traditional knives of northern Australia are equipped with handles molded from the resin of spinifex grass seeds, which must be shaped while hot, Dempsey and Morton first built a fire in a metal trash can lid in the middle of the improvised video studio. Then they set to work while cameras recorded their activities.

In reviewing a videotape copy of this footage, I was struck by the complementarity in the actions of the two men. With little fanfare, each individual would anticipate what the other was about to do and facilitate it by performing the complementary action. In the preparation of the thermoplastic to be used as the handles for the stone knives, one of the men emptied the spinifex seed from a modern plastic bag into a traditional wooden dish, about a meter long and shaped like an elongated oval, and then smoothed out the seed to an even depth. Although one man already was gripping the tray at opposite ends with both hands, the other grasped the tray as well, and the two of them together lowered it onto the fire to heat. Also, in tending the fire, one man handed a stick to the other, who in turn raked the coals. Then the other picked up a stick and joined in, the two of them stirring the coals at the same time. As the seeds began to heat, they switched their attention to the dish, stirring the contents with their sticks simultaneously. Then as the fire began to wane, one of them lifted the dish so the other could stir up the blaze underneath by putting more fuel on the fire. The next scene begins with cooperative stirring of the thermoplastic which has formed in the dish. One of the men takes the warm, viscous liquid and molds it into a ball about the size and shape of a baking potato. The material is then rolled back and forth between the two hands, and rotated end over end as it moves. The ball of the thumb is used to shape it by pressure, and its surface is smoothed by the joint action of the two thumbs squeezing inward. Then the two men divide the labor into two tasks, with one of them molding the thermoplastic into roughly shaped knife handles, while the other gives them their final form and attaches them to the stone knife blades that had been prepared earlier.

From an examination of even a few minutes of videotape, it is clear that stone tool-making is a cooperative process, not an example of an isolated craftsman working alone, and that even in this so-called primitive craft *the basic principles of a manufacturing system are present:* task specialization, symbolic coordination, social cooperation, role complementarity, collective goals, the logical sequencing of operations, and the assembly of separately manufactured parts. In evolutionary theories of technology, however, Aborigines are portrayed as jacks-of-all-trades with crude stone tools; and even in anthropology their technology is designated by such condescending terms as *simple* and *unspecialized* – creating the impression that it lacks task specialization, role complementarity, technical expertise, and the logical organization of technical skills. Irrespective of their stated intentions, these conventional characterizations reinforce the popular prejudice that Aborigines are technical illiterates lacking technology in any meaningful sense of the term; but as I show elsewhere (Reynolds, 1991), the seemingly profound discrepancies between 'primitive' and 'advanced' technologies are created by the ideologies of industrialism itself.

The essence of human technical activity is anticipation of the action of the other person and performance of an action complementary to it, such that the two people together produce physical results that could not be produced by the two actions done in series by one person. I call this process *heterotechnic* cooperation ('different crafts') to emphasize the complementarity of social roles. Heterotechnic cooperation may be contrasted with *symmetric* cooperation, in which all the participants do the same thing at the same time in order to facilitate a common goal, as when both men stir the spinifex together or a pack of wolves runs down a caribou. Thus, human technology is not just 'tool use,' and not just 'cooperative' tool use, but tool use combined with a social organization for heterotechnic cooperation. This heterotechnic aspect of human tool use, characterized by complementary technical roles among the participants, is manifested in all human societies by a distinct form of social organization that I call the *face-to-face task group* – a social structure defined by *the shared intention to transform matter and energy through the cooperative and complementary use of tools and tool-using skills by a group of people in face-to-face contact.*

Although human beings can and often do work alone, even solitary labor is almost always a social activity because it is typically directed towards social ends, requires materials conveyed through social exchange, is typically one step of a larger, cooperative endeavor, and uses skills developed as a member of society. Moreover, solitary labor is the exception in the human species rather than the rule, and in all societies work is typically performed by face-to-face task groups of people cooperating to accomplish a common goal. Even in the manufacture of stone tools, which can in fact be done by a single individual working alone (as indeed has been proven by the archaeologist turned flint-knapper), people such as the Australian Aborigines who depend on this technology for their livelihood nonetheless produce stone tools as a collective enterprise, as for example when one person molds the knife handles, while another attaches them to the blades.

The social structure of subassemblies

The social cooperation that can be seen in the videotapes of Australian Aboriginal tool-making described above has been confirmed under more natural circumstances through anthropological fieldwork. In 1965, a group of Australian Aborigines, who apparently had never before been contacted by Westerners, met an oil exploration party in the Central Desert, which took them by Land Rover to the Warburton Ranges Mission. However, the Aborigines did not like life at the mission, and in September 1966 they walked back to the bush, taking up foraging once again, and bringing with them two anthropologists, Richard and Betsy Gould, who described their experiences in a book (Gould, 1969) and recorded a typical day in the Aborigines' lives on 16 mm film.

The Goulds' film shows many examples of the specialized task groups that characterize human technologies wherever they are found. As the film opens, the Aborigines wake up before dawn to begin their round of activities. The children are given wooden bowls and sent to fetch water (symmetric cooperation), while the women discuss making sandals from *taliwanti* bark, a local plant. One woman, Katapi, says she will look for ripe *ngaru* (fruit of the shrub *Solanum eremophiolum*) while the other women gather *taliwanti*. If she finds fruit she will send up a smoke signal; otherwise she will join them later. Katapi goes off alone, carrying her infant son in one arm, her firestick in the other, wooden tools stuck in her hair. The other women help each other to place wooden bowls full of water on top of their heads (heterotechnic cooperation). Balancing the water bowls on their heads, the women take up their fire-hardened digging sticks and head out of the camp, followed by all of their children and all six dogs. Mitapuyi and Walanga, the two adult men, go off in another direction to wait for emus in a blind they have recently constructed. They carry chewing plugs of tobacco, made from wild species of the true tobacco plant, *Nicotiana*, to while away the time in the blind. After waiting a number of hours, Mitapuyi misses the emu because his spear breaks as it leaves the spear-thrower. The emu hunt unsuccessful, the men head home in the late morning, hunting lizards on the way. They follow the trail of a large lizard, a goanna, to its underground nest. While one man stamps on the ground to trap it by caving in its burrow, the other uses a stick to dig out the entrance – another example of complementary roles. Mitapuyi kills it with the stick and tucks it into his belt. Then the men head home with their catch.

Meanwhile, the women collect the *taliwanti* plants, loading them into the water bowls. After observing Katapi's smoke signal, the women and children follow her tracks to a patch of ripe *ngaru*. It is growing with another plant that also produces edible fruit, called *kamurarpa*. The women use two different techniques to collect the two different kinds of fruit. Then using sharp, flat sticks, which they carry in their hair, they split open the *ngaru* fruit and remove the seeds, which are thrown away, filling the bowls with the edible husks (symmetric cooperation). About 35 pounds of fruit are collected in an hour. The women and children walk briskly back to camp around noon, hurrying to get home before the afternoon wind begins to sear the sandhills. The women spend the afternoon grinding fruits with grind stones on a flat stone slab. They divide themselves into two groups, as Katapi works on one kind of fruit, while the other three women work the other (two tasks, one symmetric group). The *kamurarpa* fruits are made into a paste and rolled in balls by Katapi, while the *ngaru* is parched over the fire. Meanwhile, the men are preparing a ritual artifact.

Thus, the Australian Aborigines in Gould's description, in the course of a typical day, formed specialized task groups to get water, load water containers, hunt emu, hunt lizards, gather *taliwanti* bark, gather fruit, prepare a ritual board, and process fruit in two different ways. Some of these tasks, specifically the hunting and ritual, are men's tasks, while others are women's tasks, still others are jobs for children. Although these small-scale societies are frequently described as lacking any technical specialization but a 'sexual division of labor', the entire population can be regarded as a pool from which specialized groups are recruited, each one dedicated to a particular type of technical task. Sometimes, all the members of a demographic category do function as a single task group, as when all the children collect water together, all of the women gather fruit together, or all of the men hunt together; but just as often they divide into subgroups, each group charged with a particular task, as when Katapi goes off to find a specific kind of fruit. Moreover, within each task group, people perform complementary roles, as when women help each other to lift the water bowls or the men help each other to catch a goanna. Although in small-scale societies the family and the domestic group form the labor pool from which task groups are recruited, the task group itself is a technically specialized organization of labor. Consequently, it is misleading to classify societies like Australian Aborigines as having 'unspecialized' technologies where everyone is just a jack-of-all-trades. In actual fact, they are based on the learning of many different technical roles and on participation in many different kinds of task groups (Hamilton, 1980).

I was able to confirm these observations among another group of hunters and gatherers when two social anthropologists, Kirk and Karen Endicott, invited me to videotape at their field site on the Malay peninsula. The Endicotts were working with a group called the Batek (not to be confused with the better known Batak of Indonesia), a forest-dwelling Semang people who can fabricate almost their entire material culture using only steel knives which they obtain through trade (Karen Endicott, 1984; Kirk Endicott, 1983; K. & K. Endicott, 1984). My videotapes show that the Batek, like the Australian Aborigines, have both a multitude of task groups and role complementarity within these groups. The integration of specific tasks with complementary roles can be easily seen in the harvesting of rattan. Rattan is prepared from the trunks of certain species of climbing palms that grow wild in the Malaysian rain forest, and the collection of this material for the market plays an important role in the trade economy. Batek men collect it in the forest and transport it down the river for sale to Malay villagers, who, in turn, ship it to Singapore where it is made into furniture. Green rattan is heavy and not very buoyant, and because of this it is tied into bundles and loaded onto rafts made of hollow bamboo. The rafts are pushed through rapids, lowered over cataracts, and finally steered down the quieter stretches of the river to trading towns. Although the operations involved are simple and straightforward, they require a human level of technical role complementarity. Someone must lift the rattan bundles while another ties them up; the rafts must be lifted in synchrony to negotiate the rapids; and maneuvering on the river requires a man in the stern to provide power and another in the bow to steer.

The principles of technical specialization definitive of a human level of technology can also be observed in the play groups of Batek children. On one occasion, the children arranged river cobbles in a rough grid pattern on a sand bar in the middle of a stream (illustrated in Reynolds, in press). The rows of stones were said to represent streets and 'shophouses' – a common type of building in Malaysian towns that has a store in front and living quarters in back. In all likelihood, these children had never seen a Malaysian town but had only heard about them from parents or older friends, since Batek children do not usually go on trading trips. In this respect, the pretend content is very similar to the sand box play that I filmed in an Australian preschool, where many pretend play items such as 'Indian teepees' had probably never been seen by the children except on television. In both societies, many of the constructions of pretend play are representations of essentially mythical objects.

The Batek children built their 'shophouses' on the sandbar with the same kind of cooperative labor and task specialization that can be seen in the preschools of industrial society. Four or five children laid out the stones on the sandbar to represent houses and smoothed the sand with bamboo scrapers to represent streets. Two other children gathered suitable stones from a riffle further along the stream and loaded them into toy boats made from longitudinal sections of hollow bamboo. Then they pushed the boats over the water to the sand bar, where the stones were unloaded onto ramps, then rolled up the slope by others to the construction site.

In human societies, task groups are coordinated with one another, such that my task group and your task group are both instrumental to a superordinate technical task that requires the successful completion of both of our subtasks. In Papua New Guinea I videotaped a traditional feast, an elaborate undertaking which requires the integration of the products of multiple task groups into a single, multifaceted event. The preparation of the fire pit, for example, involved a number of technical steps each performed by a different group: the digging of the pit itself, the collection of firewood, the building of a fire to heat the rocks, the lining of the pit with hot rocks, the collection of banana leaves, the use of banana leaves as an insulation layer, and so forth. Although any one of these tasks could probably be done by a single individual working alone, and certainly by a small group of people performing them in sequence, in actual fact the work is done by multiple, specialized task groups in which the output of one group is the input to another. Moreover, relations in these face-to-face task groups are empirically distinct from those of the kinship system and the so-called division of labor. Even in societies where almost all groups are recruited on the basis of marriage and descent, as in precolonial Australia and in rural Papua New Guinea, these social categories do not substitute for specialized task groups but define the subpopulations from which the latter are recruited.

The case of the missing task group

The social structure of subassemblies is conspicuously missing from anthropological theories of human evolution. Although anthropologists have long insisted on technology as one of the critical factors in the emergence of hominids, contemporary evolutionary theory, by confounding technology with tool use, has failed to recognize the profound differences in social context that can be easily demonstrated from films of humans and apes; and, at the same time, it has reinforced the unilinear evolutionary notion that technical skills in small-scale societies are organized in a radically different way from our own. Although each society defines in its own way the tasks that are considered appropriate to each demographic category, the rules by which labor is recruited, and the system of exchange by which products are moved from one group to another, in all societies there is interposed, between the political economy and the hand-held tool, a distinct level of technical social organization – the face-to-face task group with heterotechnic cooperation.

For example, in Raymond Firth's classic monograph (1965), *Primitive Polynesian Economy*, based on intensive field studies in Tikopia, the task groups are cameo characters, never mentioned in the detailed explication of kinship groups and social exchange that dominates the book, but they are easy to see in the case studies, which are apparently taken verbatim from the ethnographer's field notes. Thus, Firth's 'Case B' – the removal of bark from paper-mulberry trunks for conversion into cloth – illustrates the role complementarity, task specialization, and cooperative activity definitive of the face-to-face task group: 'girl cut slit down trunk, and she and young men each pried bark from trunk with wooden levers. Lad assisting by cutting fresh levers as others broke. Several children impressed to help by holding trunk, pulling bark free, and rinsing bark in sea water. This group typical of many domestic semi-skilled tasks.' As the founder of modern field anthropology, Bronislaw Malinowski, recognized in his studies of the Trobriand Islanders, back in the 1920s, *recruitment* to task groups is based on age and sex categories in small-scale societies, but *each cooperative task has specialized roles within it that are filled on a temporary basis,* the participants switching on and off as needed (Malinowski, 1922, pp. 158–9).

But when archaeologists and social historians describe technical processes, they often seem to forget everything they ever knew about social relations and go at it like authors of cookbooks, listing sequences of steps without ever actually describing who does what when in real-life situations. This literary approach to technical skills is very misleading, for even simple technical processes that could in theory be performed in sequence by a single group or even by a single person working alone may in fact be subdivided into a host of subtasks, each one performed by a different task group. In fact, on the basis of the video of Aborigines making stone knives, I would say it is impossible to judge the 'simplicity' or 'complexity' of a task simply by cataloging the physical processes involved or by examining the finished product. So too, in the case of the Melanesian feast, any one of these men could probably do any of the tasks himself, from start to finish, *but in fact they are never done by individuals working alone* – so tool-using skills in humans cannot be described simply as a list of action patterns and physical processes. The social structure of subassemblies is an integral part of the technical process.

Moreover, manual tool-using skills and the face-to-face coordination of cooperative action are not 'primitive' forms of social organization that disappear as technology advances. After completing my fieldwork in the Pacific, I returned to the United States; and as an unemployable anthropologist, I took a job as a software instructor in Silicon Valley for a company that manufactured microcomputers (Reynolds, 1987). Even though social theorists had led me to expect a technical organization that could not be more different from the village technology I had videotaped in Papua New Guinea and Malaysia, I was far more impressed by the enormous continuity in the social structure of manual skill. Although one often reads in the popular press that 'computers are built by other computers', conjuring up a Frankenstein image of technics out of control, the pre-production models of computers – the ones tested by design engineers, demonstrated to prospective investors, and sent to government laboratories for radio emissions certification – are hand-built by skilled craftsmen with benches full of hand tools. The components, of course, are manufactured, the circuit boards and the chips, the disk drive and the keyboard, but each of these subassemblies must nonetheless be physically connected to each other by someone who can wield a soldering iron with the delicacy of a dentist's drill. The tools are laid out on the bench in a spatial map of the tasks to be performed; and small groups of technicians and engineers work together in face-to-face task groups, anticipating the intention of others and contributing the complementary action.

The face-to-face task group using hand tools to shape physical objects is not restricted to the computer industry but is essential to industrial technology in general. Even in the automobile industry, the paragon of mass production, new models of cars begin as hand-crafted prototypes which are tested in wind tunnels, driven around tracks, and smashed into barriers; only then are the parts manufactured on a mass scale, using assembly line techniques (Lacey, 1986). The small, face-to-face group was also critical in the development of modern automotive technology. When Henry Ford accepted the challenge to enter his horseless carriage in an automobile race in 1901, he recruited a team that included Charles B. King's former helper, Oliver Barthel, and the electrician from the Edison plant, Edward 'Spider' Huff. Huff was credited officially with the car's induction coil, and he also worked with Barthel on the device that turned out to be the Ford racer's most lasting contribution to automotive history: a special coil which they took to Barthel's dentist, Dr. W. E. Sandborn, who made a porcelain insulating case for it – the forerunner of the modern spark plug'(Lacey, 1986, p. 53).

Most theories of economics, however, focus on the mass production stage of modern industry while ignoring the product development stage – thereby reinforcing the premise of a profound disparity between 'primitive' and 'advanced' technologies. But field anthropology indicates that the heterotechnic task group is a building block of human technology, found in all societies, irrespective of the character of the larger system of power and exchange.

The phylogeny of subassemblies

The tool-using skills of contemporary human beings are embedded in a social process of cooperative production that presupposes role complementarity, the social exchange of tools and materials, and the construction of composite artifacts; but two decades of intensive primatological fieldwork confirm that nonhuman primate tool use is essentially an individualistic activity (Goodall, 1968; Hayes & Nissen, 1971; Premack, 1976; Teleki, 1974). A chimpanzee may be inspired to probe for termites by watching another, and a home-reared ape can learn to open the refrigerator by observing its human mother, but thenceforth it probes for termites or opens doors as an individual working alone. Although adumbrations of social coordination do occur in chimpanzees (Menzel, 1971), and the tutoring of apes in sign language demonstrates their capacities for denoting environmental events, in no species of nonhuman primate is heterotechnic tool use a normal part of the behavioral repertoire. Significantly, over half a century ago, on the basis of extensive experimentation with wild-caught chimpanzees in Africa, the psychologist Wolfgang Köhler concluded that humans and chimpanzees take a very different approach to inanimate objects (Köhler, 1959). Human tool use, he said, is a *social* activity based on *complementary* relations between the hands, whereas chimpanzee tool use is the incorporation of objects into whole-body locomotive skills. Elsewhere, I have published the results of a detailed frame-by-frame examination of films of chimpanzee ladder-building taken by Emil Menzel, Jr. (Menzel, 1972), which is one of the most complex examples of ape tool use known to primatology; and this research substantiates Köhler's conclusions (Reynolds, in press). In modern terminology, humans and chimpanzees have very different cognitive styles. Chimpanzees build complex constructions using gravity to hold the parts together, whereas humans not only make gravity-based constructions but also fabricate fasteners and joins.

There is no good word in English for 'a distinct object, not fastened to a surface or other object, that can be rotated as a unit through all dimensions of space without falling apart', but this concept is critical for understanding species differences in primate tool-using skills. A hammer, for example, can be rotated in space but a stack of coins will fall apart, and a hammer, unlike a doorknob, is free to be moved in all directions. Objects that can be moved in all directions without falling apart I call *liths*, from the Greek word for 'stone,' since the moveable rock not only embodies these distinctions but is the prototypical human tool. Liths are defined by their integrity and portability, not by the materials out of which they are made or by the simplicity of their construction. The following are all liths:

> a stone found on the beach
> a baseball
> a carpenter's claw hammer
> a brief case
> a table
> a pocket calculator

The following are not liths:

> a glass of water
> a stack of plates
> an appliance plugged into an electrical outlet
> spilled milk

A lith composed of liths joined together is called a *polylith*, and a polylith, like a lith, can be rotated through all dimensions of space without falling apart, but a *polypod* can only maintain its structural integrity when correctly oriented in a gravitational field. In short, polypods are held together by gravity, but polyliths are held together by *interlith joins*.

Polypods, it must be emphasized, are not 'simpler' than polyliths: the famous Lion Gate at Mycenae, made without mortar, is technically a polypod (although it was certainly built with polylithic techniques), and the balancing and support made possible by gravitational skills is as important to craftsmanship as bimanual coordination. In Papua New Guinea, for example, almost every evolutionary principle enshrined in anatomy textbooks about the functional differences between the hind limbs and the forelimbs is violated dozens of times a day. Although the Melanesians make well-crafted artifacts and manipulate objects with their two hands, the body itself is frequently an integral part of external constructions. Where a Westerner would use a nail or a knot, a Melanesian will often use his body as a fastener to hold subassemblies together.

For example, when trekking with two of my colleagues through the Ramu River area, on the north coast of Papua New Guinea, carrying our gear in knapsacks and pack frames, we came to a lagoon too broad and deep to wade across. There were two dugout canoes on the bank, but they were children's canoes, too small to ferry both an adult and a pack without tipping over. Our guide floated both canoes, loaded a pack in each of them, and then placed a board across the gunwales. He sat down on it, centering himself between the two vessels with a foot in each, and ferried the luggage across the lagoon, holding the canoes together by pressing his feet against the inner side of each boat and paddling with his arms. This skill, which a Melanesian would take for granted, violates all of the principles of functional specialization of the limbs which anatomists use to explain the evolutionary emergence of hominid tool use:

> The forelimbs are used for locomotion.
> Locomotion is implemented through object use (the boat and the paddle).
> The hind limbs are actively involved in the construction of an external object (the two-boat combination).
> Support of the body is implemented by an external object (the board across the gunwales) configured initially by the hands.

I also have a vivid recollection of a heavily burdened white-haired old woman climbing over a slat fence. She had a string bag filled with firewood slung across her back and a tobacco pipe clenched between her teeth. Using her hands, she pulled herself up on the fence, which was about the height of her chest, and clambered over the top. What makes this incident especially memorable is that she was also balancing two objects on her head while she climbed: a wood or ceramic vessel with a burning log about three feet long laid across its rim. Her body, in other words, was a complex construction requiring the simultaneous and intentional monitoring of a host of nested polypod constructions: she had to support the pot on her head by balancing motions, balance the log on the pot, and adjust to the displaced center of gravity caused by the heavy bag full of firewood – all while moving these assemblies up and over the fence in delicate synchrony. In industrial society the sedentary nature of crafts and the automation of physical work obscure the pivotal role of polypods in human tool-using skills, but in places that until recently lacked wheeled vehicles and beasts of burden, they are essential to the subsistence economy. In short, there is nothing primitive or unskilled about polypods; but, in humans, they are typically integrated with skills for the construction of polyliths, and these in turn presuppose interlith joins. It is the ability to fabricate interlith joins that I consider to be largely missing in apes.

The phylogeny of complementary roles

The distinction between polypods and polyliths is critical to the theory of primate tool use, for as I have argued elsewhere, there is abundant evidence that apes can construct polypods when left to their own devices, but there is no unequivocal evidence that they can construct polyliths without intervention by humans in the construction process (Reynolds, in press). Certainly, apes can combine two objects by inserting one inside the other, but a polylith results only when what I call the *insertion join* has already been fabricated by humans. Köhler's chimpanzee Sultan made a long stick out of two short ones whose ends had been hollowed out by the experimenter to facilitate joining them together, and Temerlin's (1975, pp. 93, 101) home-reared chimpanzee learned to attach a garden hose to the faucet and to use a screwdriver after seeing it used once. Apes can also facilitate insertion joins by chewing the ends of sticks to make them easier to insert into holes. But can apes undertake the intentional fabrication of insertion joins instrumental to polyliths? The evidence for other types of joins is also equivocal. Maple (1980) illustrates orang-utan 'knots' tied in pieces of rope and vine, but was the final shape of the string intentional or accidental? Was it intentionally shaped in order to use it as a fastener? Can apes, by themselves, construct joins integral to a polylith? These are questions that primatology needs to answer.

Certainly, primatology has shown that the basic social structures of subassemblies, observable in the tool use of any human population anywhere, are simply missing in nonhuman primate societies. One chimpanzee does *not* dig the hole in the termite nest while another prepares the stick. One chimpanzee does *not* provide a termite stick for another to use. One chimpanzee does *not* hold one end of the stick while the other strips the bark – unless it is planning on grabbing it away. A chimpanzee does not even hold a ladder in order that another may climb it – even though it may look that way to a casual observer (Reynolds, in press). Some transient instances of this sort of thing can be created in the laboratory, but in no nonhuman primate society is heterotechnic cooperation or polylithic construction a feature of tool-using behavior.[2] Indeed, there is no unequivocal evidence from even laboratory research that apes can make polylithic constructions without intervention of a human at some stage of the construction process (Reynolds, in press). But in virtually every human technology known to anthropology, polylithic constructions, such as baskets, textiles, decorated artifacts, sewn clothing, and hafted tools, are integral to the tool kit. In short, polyliths and heterotechnic cooperation are empirically associated, for they invariably occur together in humans but neither occurs in apes.

The *theory of complementation* asserts that the two most distinctive aspects of human tool use, namely, the heterotechnic task group and the construction of polyliths out of differentiated parts, are co-evolutionary developments, emerging together in a synergistic fashion.[3] Tool use in humans, as we have seen, is always firmly embedded in the social structure of subassemblies, whereas cooperative tool use in nonhuman primates, if it occurs naturally at all, is symmetric cooperation, in which all the participants in a group do the same thing, either serially or simultaneously. Consequently, any meaningful theory of hominid tool use must address both its social and its physical dimensions; and the complementation theory is a model of human evolution in which the ability mentally to represent reciprocal and complementary social relations is given equal status with the cognitive skills of causal inference and logical deduction, and with physical processes. That is, the theory asserts that a human level of tool use is characterized by face-to-face task groups exhibiting complementary roles instrumental to a common technical goal, and that *the construction of polyliths depends on this social organization.*

I submit that these developments cannot be explained by cognitive theories that simply postulate a capacity for interlith joins in hominids. Theories of cognition, like theories of intelligence, typically take the isolated craftsman as their point of departure, postulating mental operations in the mind of the individual tool-maker. Moreover the mental operations postulated by cognitive theories are themselves modeled on instrumental skills, such as logical deduction, causal inference, and mental mapping; and if they deal with social relations at all, they typically treat them as secondary or derivative. Heterotechnic cooperation, however, requires the *integration* of individual minds into a larger, cooperative framework. In the theory of complementation, chimpanzees do not lack polyliths because they are unable to perceive part–whole relationships or because they lack the manual skill to join one part to another. Rather, the theory asserts that chimpanzees lack polyliths because they lack the social organization that normally mediates such physical relationships.

Consequently, the theory of complementation maintains that the distinctive features of hominid tool use develop from the *social organization* of nonhuman primates, not from their instrumental skills and object manipulation, even though human technology presupposes such tool-using abilities. The likely evolutionary precursor for a human type of technical and economic organization, as I have argued elsewhere, is grooming in nonhuman primates, for this class of behavior is characterized by face-to-face groups with exchange dynamics and reciprocal, complementary roles (Reynolds, 1981, pp. 171–208). Another possibility, popularized by Bateson, is play behavior, which is characterized by reversible role relations in which the participants take turns in chasing and being chased (Reynolds, 1976). From this point of view, hominids diverged from apes when the social relations of grooming or play began to organize technical skills.

The theory of complementation can be tested through the use of both archaeological and primatological methods. According to the theory, in the simian adaptation, tool-using skills would be incorporated into the normal categories of nonhuman primate social behavior, whereas in the hominid adaptation, the social structure of grooming and/or play, with its reciprocal, complementary roles, would come to organize subsistence activity and associated technical skills. If this interpretation is correct, then one would expect arrangements of objects supported by gravity to co-occur with simian fossils, since they occur in contemporary apes, whereas the presence of polyliths would be definitive of *Homo*.

The complementation theory implies that the shaping of single objects or supporting them in gravitational fields preceded the assembly of polyliths, so prehuman hominids may have had a bizarre technology without parallel today. They would have been able to construct spatial arrangements of objects by placing one object on top of another, as chimpanzees stack boxes to reach a banana, but they would have been unable to make polyliths with differentiated parts. Their technology would have had walls but no tents, knife blades but no knife handles, fire but no fire bows, pottery but no baskets, animal skins but no textiles, fire-hardened spears but no hafted spear heads. When Mary Leakey (1971) wondered whether the spherical stones unearthed at Olduvai Gorge were hand-thrown missiles or parts of bolas, she was asking one of the fundamental theoretical questions in the anthropology of tool use, and it can only be answered by archaeology. At the base of the Paleolithic are there any of the following?

> knots
> pegs
> mortise and tenon joints
> woven textiles or baskets
> sewn garments
> hafted blades or projectile points
> shaping of one object to fit another
> composite objects held together with fasteners

All of the artifacts on this list are either joins or presuppose the intentional construction of interlith joins, so if the theory of complementation is correct, their presence in the archaeological record would indicate reversible, complementary roles in the social organization of tool use.

The linguistic implications

The theory of complementation also provides an avenue of exploration for the evolution of language. Because the theory maintains that all human tool use is embedded in a social process of heterotechnic cooperation, there is no longer a fundamental discrepancy between 'solitary' tool-using skills and 'social' communication. To the contrary, the face-to-face task group presupposes a sophisticated process of communication which allows a single task to be parceled out among two or more participants, each of whom performs an action complementary to or supportive of the other. Moreover, as linguists have so often pointed out, human language also presupposes what I am calling polyliths: an entity constructed out of parts which then functions as a unit through a range of subsequent transformations. In the following examples, the phrase 'the owner of the Cadillac' functions as a unit in three different linguistic roles, even though the phrase itself is constructed out of parts:

> The owner of the Cadillac was cited for speeding.
> The owner of the Cadillac drove it back to the shop.
> The owner-of-the-Cadillac theme is a cornerstone of the advertising campaign.

Also, as sociolinguists have emphasized, the participants in a conversation, through reciprocal and mutual actions directed towards the output of previous linguistic operations, jointly construct a culturally meaningful product, analogous to the cooperative constructions produced by task groups:

> 'Look at my spider.'
> 'Aren't you afraid of getting bitten?'
> 'No, it's a *friendly* spider.'

Conversely, the linguistic constructions of apes, as manifested by sign language studies, indicate, as one would expect, that they have the same problem constructing linguistic analogs to polyliths as they do in constructing physical polyliths. Linguists have noted the absence of 'grammar' and of 'function words' in ape sign-language performances, but these deficiencies in instrumentally acquired communicative skills come as no surprise given that the animals lack analogous abilities in their tool-using skills (Reynolds, 1986). The theory of complementation, however, suggests a more specific diagnosis that could be tested by fine-grained psycholinguistic studies. In a series of papers, I have shown how to to build linguistic-type models of nonverbal action, using frame-by-frame analysis of video footage (Reynolds, 1980, 1982, 1983, in press), so that it is both possible and practical to develop structuralist models that encompass the behavior of different species and different modalities of action. If the complementation theory is correct, it should be possible to show, using such methods, that apes can construct composite signs analogous to polypods, namely signs arranged by contiguity, while having difficulty with shorter and seemingly simpler combinations that require the construction of joins. Human language is to the heterotechnic task group as simian signing is to simian tool use.

The theory of complementation asserts that language and heterotechnic cooperation are co-evolutionary phenomena, that human tool use is an intrinsically social process, and that the development of fully-fledged grammatical relations in language is correlated with the construction of polyliths and joins. This concept of human evolution is profoundly at odds with the utilitarian science of the past two decades, in which everything is to be reduced to calories, climate, and competition; but the new approach has the virtue of taking seriously a century of social anthropology as well as the 'unproductive' play and grooming of nonhuman primates themselves.

Acknowledgments

I want to thank Billy Dempsey and Slippery Morton for their demonstration of stone tool-making; Peter Ucko and the Australian Institute of Aboriginal Studies for making available the video documenting their performance; the parents, teachers, and children of the University Preschool in Canberra for the opportunity to videotape play groups of children; Kirk and Karen Endicott for introducing me to the Batek people; the National Institute of Mental Health and Australian National University for funding my fieldwork in the Pacific; Roger Keesing and Derek Freeman for the support of the Anthropology Department of the Research School of Pacific Studies; The Harry Frank Guggenheim Foundation for funding comparative video analysis; Jackie Johnson for line drawings of constructions and tool-using skills; Nancy Lutkehaus for an opportunity to videotape on Manam Island, PNG; Sacha Josephides for the opportunity to videotape at Boroi Village, PNG; Emil Menzel, Jr., for providing me with a copy of his 16mm film footage on chimpanzee ladder-building; Joanne Rampelberg and Sunny Baker for making possible my entry into the 'Falcon' Computer Company; Myrdene Anderson and Sydel Silverman for making possible this presentation; and the editors of this volume, Kathleen Gibson and Tim Ingold, for their valuable editorial comments.

Notes

1 For a description of traditional Aboriginal artifacts, see McCarthy (1976) and Roth (1910).

2 Benjamin B. Beck's paper (1973) is often cited to demonstrate cooperative tool use in nonhuman primates, but the behavior in question is marginal or non-existent in the normal way of life of this species, as shown by the work of Kummer (1968).

3 In the theory of complementation, heterotechnic task groups and polylithic constructions are complementary and co-evolutionary developments, and I use the term *complementation* in preference to *complementarity* because the latter implies mutually exclusive events, such as the theory of complementarity in physics, which asserts that light is both a wave or a particle – but never the same thing.

References

Auel, J. M. (1980). *The Clan of the Cave Bear*. New York: Crown Publishers.

Beck, B. B. (1973). Cooperative tool use by captive hamadryas baboons. *Science*, 182, 594–97.

Beck, B. B. (1980). *Animal Tool Behavior: The Use and Manufacture of Tools by Animals*. New York: Garland STPM Press.

Blurton-Jones, N., ed. (1972.). *Ethological Studies of Child Behaviour*. Cambridge: Cambridge University Press.

Bordes, F. (1971). Physical evolution and technological evolution in man: a parallelism. *World Archaeology*, 43, 1–5.

Endicott, K. (1983). The effects of slave raiding on the aborigines of the Malay peninsula. In *Slavery, Bondage, and Dependency in Southeast Asia*, ed. A. Reid. St. Lucia, Queensland: University of Queensland Press.

Endicott, K. & Endicott, K. L. (1984). The economy of the Batek of Malaysia: annual and historical perspectives. *Research in Economic Anthropology*, 6, 29–52.

Endicott, K. L. (1984). The Batek De' of Malaysia. *Cultural Survival Quarterly*, 8, (2), 6–10.

Firth, R. (1965). *Primitive Polynesian Economy*, 2nd edn. London: Routledge & Kegan Paul.

Goodall, J. (1968). The behaviour of free-living chimpanzees in the Gombe Stream Reserve. *Animal Behaviour Monographs*, 1, 161–311.

Gould, R. A. (1969). *Yiwara: Foragers of the Australian Desert*. New York: Charles Scribner's Sons.

Hamilton, A. (1980). Dual social systems: technology, labour and women's secret rites in the eastern Western Desert of Australia. *Oceania*, 51, 4–19.

Hayes, K. J. & Nissen, C. H. (1971). Higher mental functions of a home-raised chimpanzee. In *Behavior of Nonhuman Primates: Modern Research Trends, Vol. 4*, ed. A. M. Schrier & F. Stollnitz, pp. 59–115. New York: Academic Press.

Holloway, R. (1969). Culture: a *human* domain. *Current Anthropology*, 10, 395–412.

Isaac, G. (1976). The activities of early African hominids: a review of archaeological evidence from a time span two and a half to one million years ago. In *Human Origins: Louis Leakey and the East African Evidence*, ed. G. Isaac & E. R. McCown. Menlo Park, CA: Benjamin.

Johnson, L. L. (1978). A history of flint-knapping experimentation. *Current Anthropology*, 19, 337–72.

Köhler, W. (1959). *The Mentality of Apes*. New York: Vintage Books.

Kummer, H. (1968). *The Social Organization of Hamadryas Baboons: A Field Study*. Basel: Karger.

Lacey, R. (1986). *Ford: The Men and the Machine*. Boston and Toronto: Little, Brown.

Leakey, M. D. (1971). *Olduvai Gorge, Vol. 3: Excavations in Beds I and II, 1960–1963*. Cambridge: Cambridge University Press.

Leroi-Gourhan, A. (1943). *L'Homme et la Matière*. Paris: Editions Albin Michel.

Leroi-Gourhan, A. (1957). *Prehistoric Man*. New York: Philosophical Library.

Malinowski, B. (1922). *Argonauts of the Western Pacific: An Account of Native Enterprise and Adventure in the Archipelagoes of Melanesian New Guinea*. London: Routledge & Kegan Paul.

Maple, T. L.(1980). *Orang-Utan Behavior*. New York: Van Nostrand.

McCarthy, F. D. (1976). *Australian Aboriginal Stone Implements*, 2nd edn., revised. Sydney: The Australian Museum.

Menzel, E., Jr. (1971). Communication about the environment in a group of young chimpanzees. *Folia Primatologica*, 52, 220–32.

Menzel, E., Jr. (1972). Spontaneous invention of ladders in a group of young chimpanzees. *Folia Primatologica*, 17, 87–106.

Oakley, F. B., & Reynolds, P. C. (1976). Differing responses to social play deprivation in two species of macaque. In *The Anthropological Study of Play: Problems and Prospects*, ed. D. F. Lancy & B. A. Tindall, pp. 179–188. Cornwall, NY: Leisure Press.

Parker, C. E. (1974). The antecedents of man the manipulator. *Journal of Human Evolution*, 3, 493–500.

Premack, D. (1976). *Intelligence in Ape and Man*. Hillsdale, NJ: Erlbaum.

Reynolds, P. C. (1976). Play, language, and human evolution. In *Play: Its Evolution and Development*, ed. J. Bruner, A. Jolly, & K. Silva, pp. 621–35. Baltimore: Penguin Books.

Reynolds, P. C. (1980). The programmatic description of simple technologies. *Journal of Human Movement Studies*, 6, 38–74.

Reynolds, P. C. (1981). *On the Evolution of Human Behavior: The Argument from Animals to Man*. Berkeley & Los Angeles: University of California Press.

Reynolds, P. C. (1982). The primate constructional system: the theory and description of instrumental object use in humans and chimpanzes. In *The Analysis of Action: Recent Theoretical and Empirical Advances*, ed. M. von Cranach & R. Harré, pp. 343–85. Cambridge: Cambridge University Press.

Reynolds, P. C. (1983). Ape constructional ability and the origin of linguistic structure. In *Glossogenetics: The Origin and Evolution of Language*, ed. E. de Grolier, pp. 185–200. New York & Paris: Harwood Academic Publishers.

Reynolds, P. C. (1986). Language: anthropoid ape. In *The Dictionary of Ethology and Animal Learning*, ed. R. Harré & R. Lamb, pp. 81–3. Oxford: Blackwell.

Reynolds, P. C. (1987). Imposing corporate culture. *Psychology Today*, March, pp. 33–8.

Reynolds, P. C. (1991). *Stealing Fire: The Atomic Bomb as Symbolic Body*. Palo Alto, CA: Iconic Anthropology Press.

Reynolds, P. C. (in press). Structural differences in intentional action between humans and chimpanzees – and their implications for theories of handedness and bipedalism. In *On Semiotic Modeling*, ed. M. Anderson & F. Merrell. Berlin: Walter de Gruyter.

Roth, W. E. (1910), *North Queensland Ethnography, Bulletin No.14, Records of the Australian Museums*, Vol. 8, No. 1.

Teleki, G. (1974). Chimpanzee subsistence technology: materials and skills. *Journal of Human Evolution*, 3, 575–94.

Temerlin, M. K. (1975). *Lucy: Growing Up Human*. Palo Alto, CA: Science and Behavior Books.

19

Tool-use, sociality and intelligence

TIM INGOLD

The evolution of intelligence

In a paper that has proved to be highly influential, Humphrey (1976) argued that the principal selective pressures behind the evolution of the 'higher intellectual faculties' of humans and other primates lay not in the technical dealings of these animals with objects in their physical environments, but in their *social* dealings with one another in the context of life in close-knit communities of conspecifics. Recently dubbed 'the Machiavellian intelligence hypothesis', Humphrey's idea has inspired a great deal of recent work on the complex dynamics of relationship in social groups of monkeys and apes. Introducing a collection of papers on the subject, Whiten and Byrne triumphantly declare that 'the idea of social intelligence is one whose time has come' (1988:1), although it has had many antecedents. Clearly, this work is intended to convey a message about the evolution of those cognitive and intellectual faculties that are apparently unique to humans, and of which language is commonly supposed to be the prime example. Yet to a social anthropologist like myself, what is most striking about it is that the model of human sociality that sets the pattern for depictions of non-human primate life seems to owe more to the researchers' own experience of their dealings with colleagues in academic conferences and senior common rooms than to the experience of everyday life in small and intimate groups. The society of primates is, paradigmatically, a 'collegiate community' (Humphrey 1976: 310)!

According to the traditional account of the evolution of human intelligence, which goes back to Darwin (1874:195–224) and Wallace (1870:302–21), the pressures that favoured intellectual advance lay in man's interactions with components of the physical environment, such as predators and prey. The use and manufacture of tools played a central role in this argument: men who were 'more sagacious' (as Darwin liked to say), with bigger and better brains, could design more ingenious tools, thereby securing a reproductive advantage.

Intelligence-enhancing variations would consequently tend to be preserved in future generations, leading to yet further advances in the technical sphere, and so on through mutual reinforcement. This view of the evolution of intelligence continues to enjoy considerable support (Washburn 1960). For example, Parker and Gibson (1979) assert that feeding strategies, and above all 'extractive foraging with tools', were the primary determinants of the kind of intelligence possessed by the earliest hominids, and that its further development was an adaptive response to complex hunting involving aimed missile-throwing, stone-tool manufacture, animal butchery, food division and shelter construction.

Following Humphrey's lead, critics of this view argue that the intellectual capacities demonstrated by non-human primates in laboratory settings, where they are tested on complex tasks involving the manipulation of gadgets of various kinds, is far in excess of anything that they might need for dealing with their physical environments under 'natural' conditions. Since natural selection does not, generally, perfect an attribute beyond the level that is necessary to ensure that an animal 'gets by' in its normal circumstances, some other factor must be adduced to account for this apparently excess capacity. That factor is complex sociality. Given the relatively long period of dependency of the young, which is conducive to the transmission of behaviour by learning, and the consequent overlap of generations in the social group, the potential for intra-group conflict is as great as the need to maintain community solidarity. The management of relationships with other individuals in the group therefore entails considerable skill, for at every moment the animal has to anticipate not only the immediate effects of its own actions, but also the ways these actions might be perceived by others and how their outcomes may be qualified by actions initiated by these others on the basis of their own perceptions. If the intellectual capacities of non-human primates were enhanced through selection for 'social skills', then – so the argument goes – the same must have been true in the evolution of our own species. For the technical problems that confronted our ancestors, in their procurement of subsistence, were not particularly difficult. Most would have already been solved, and had only to be learned by each generation from its predecessors. But it is precisely the importance of learning, which takes the weight off technical problems of adaptation, that sets such a heavy load on social ones.

I do not, in this chapter, intend to put up a defence of the traditional argument against the claims of advocates of the 'Machiavellian intelligence hypothesis'. My object is rather to show that the debate turns on a dichotomy between the spheres of technical and social relations, and more fundamentally between nature and society, which is rooted in the premises of modern Western thought. Moreover this same tradition of thought, whose penchant for constructing dichotomies is one of its main defining characteristics, has given us a distinction between intellect (as a property of mind) and behaviour (as bodily execution), along with the idea that all purposive or intentional action is preceded by an intellectual act of cognition, involving the construction of representations, the consideration of alternatives, and the formulation of plans. I dispute this, and with it the very idea of intelligence that informs (often implicitly) both 'technical' and 'social' accounts of its origins and evolution.

We may of course describe as 'intelligent' an animal whose actions manifest a certain sensitivity and responsiveness to the nuances of its relationships with the components of its environment. But it is quite another thing to attribute that quality to the operation of a cognitive device, an 'intelligence', which is somehow *inside* the animal and which, from this privileged site, processes the data of perception and pulls the strings of action. Indeed it makes no more sense to speak of cognition as the functioning of such a device than it does to speak of locomotion as the product of an internal motor mechanism analogous to the engine of a car. Like locomotion, cognition is an accomplishment of the *whole animal*, it is not accomplished by a mechanism interior to the animal and for which it serves as a vehicle. There is therefore no such thing as an 'intelligence' apart from the animal itself, and no evolution of intelligence other than the evolution of animals with their own particular powers of perception and action. In the study of human evolution, we are concerned to understand the specific powers of human beings, and to produce an account of how and why they came to be formed.

Now one of those powers, certainly possessed by individuals of our own species and possibly by other species as well, is the capacity to turn attention inwards towards the self rather than outwards towards the world – to explore, that is, our own past experience and to construe future possibilities of being. That is the power of imagination. People *do* construct plans and representations, and *do* consider alternative scenarios. Such constructions and considerations, however, are brought forth through intelligent activity – in this case the kind of activity we usually call 'thinking', just as material artefacts are brought forth through the intelligent activity we call 'making'. They are *not*, then, necessary preludes to action but rather its products. Even when actions *are* preceded by plans (and this probably applies to but a small part of what we do), the latter serve as standards or guidelines in our dealings with the world rather than as programmes that 'call up' the action as an automatic behavioural output. For the action issues from the agent, not from the plan (see Ingold 1986a: 312–42, for discussion of this point), and the quality of intelligence is not contributed by the plan but is rather immanent in the action itself.

Both tool-use and speech are, for humans, forms of intelligent activity, and what I have said about such activity in general applies more specifically to them as well. That is to say, we should see them, in the first place, as the actions of people, and not as the pre-programmed behavioural outputs of a device located somewhere inside each individual brain, which computes in advance what the body has to do and then issues the necessary commands. This is not to deny that both speech and tool-use require a certain neural organization, it is only to insist that in its functioning this organization is always immersed within a total context of bodily activity in an environment.

Three further, preliminary points need to be made at this stage. First, speech and tool-use are implicated in inward-directed action ('thinking') as well as in outward-directed action ('doing'). The speech entailed in thinking is what Vygotsky (1962) called 'inner speech' – a process in which words, rather than expressing thoughts already formed, assist in bringing those thoughts into being. In inner speech, Vygotsky wrote, 'words die as they bring forth thought' (1962: 125, 149). But people use material tools as well as words as aids to thinking: this is an important area of tool-use that should not be forgotten. Secondly, when it comes to doing rather than thinking, speech is no more confined to relations with conspecifics than is the use of tools to relations with non-human agencies and entities in the environment. Humans frequently use tools on one another, just as they often speak to non-humans. The latter, often considered irrational by Western standards, tends to be relegated by observers to the category of 'magic', but this is only because it confounds the boundary that *we* draw (but which others do not) between nature and society, and the kinds of activities appropriate to each domain. Thirdly, both speech and tool-use are not only ways of acting in the world, they are also ways of getting to know it. The visualist emphasis in psychological studies of perception tends to blind us to the vital role of touch in perceptual activity, which is often mediated by the very same manual tools by which we act upon, and aim to modify, the environment. Speech, too, serves at one and the same time to act upon others and to reveal their own intentions towards ourselves.

The argument that follows is divided into four sections. In the first, I am concerned to understand the nature of technique as skilled practice. Rejecting the cognitivist account of technical action as the mechanical execution of plans or schemata already arrived at by an intellectual process of reason, I show that the exercise of technique does not depend upon the prior existence of a separate body of knowledge in the form of a technology. Likewise, speech does not depend upon the objective existence of 'language'. In the second section, I demonstrate why it is impossible to draw a rigid distinction between social and technical relations, but show how an institutionalized opposition has neverthe-less come to be established, in the West, between technology and society – an opposition that has exerted a powerful influence on the way we think about human capacities and their evolution. In the third section I adopt an alternative perspective, that of hunters and gatherers with an allegedly 'simple' technology, to argue that tools – like words – are used to mediate an active engagement with the environment rather than to assert control over it. Meaning, thus, is not imposed on the world but arises out of that engagement. My conclusion, in the final section, is that to understand the activities of tool-use and speaking, they must be seen not as the behavioural products of the operation of higher intellectual faculties, internal to their possessors, but rather as integral to the functioning of an entire system of perception and action constituted by the nexus of relationships set up by virtue of the immersion of the agent in his or her environment.

Technology and technique

For the purposes of my present argument, it is vital to distinguish between *technology* and *technique*. I take technique to refer to *skills*, regarded as the embodied capabilities of particular human subjects (see Layton 1974:3–4), and technology to mean a corpus of generalized, objective knowledge, insofar as it is capable of practical application. Both technique and technology must, of course, be distinguished from *technics*. Briefly, I would define technics as the area of overlap between 'tools' and 'artefacts'. A tool, in the most general sense, is an object that extends the capacity of an agent to operate within a given environment; an artefact is an object shaped to some pre-existent conception of form (see Ingold 1986b for further discussion of the definition of these terms). Not all tools are artefacts, and not all artefacts are tools. The stone pebble that I use as a paperweight is a tool but not an artefact; a birthday cake is an artefact but it is not a tool. It is important to remember, too, that there can be techniques without technics, and for that matter, without tools. As Marcel Mauss recognized in his celebrated paper on techniques of the body, it is a fundamental mistake to think that 'there is technique only where there is an instrument' (1979:104). One obvious example of technique without instrumental aid is dance, but there are many others.

Why is it, then, that in both specialized anthropological and popular Western discourse, it tends to be assumed that technical activity is *ipso facto* tool-using activity? Consider, for example, Ellen's definition of subsistence technique: 'a combination of material artefacts (tools and machines) and the knowledge required to make and use them' (1982:128). Here, technique is regarded not as a property of skilled subjects, but as an inventory of instrumental objects together with their operational requirements. This view, I believe, results from a conflation of the technical with the mechanical, a conflation that has also given rise to the modern concept of technology. For what this concept does, in effect, is to treat the workman as an *operator*, putting into effect a set of mechanical principles, embodied in the construction of the instruments he uses, and entirely indifferent to his own subjective aptitudes and sensibilities. In other words, productive work is divorced from human agency and assigned to the functioning of a device. Thus, technique appears to be 'given' in the operational principles of the tools themselves, quite independently of the experience of their users. If all technical activity is tool-using activity, it is because the technique is seen to reside, outside the user, in the tool, and to come 'packaged' – like the instruction manual for a piece of modern machinery – along with the tool itself.

My contention, to the contrary, is that technique is embedded in, and inseparable from, the experience of particular subjects in the shaping of particular things. In this respect, it stands in sharp contrast to technology, which consists in a knowledge of objective principles of mechanical functioning, whose validity is completely independent both of the subjective identity of its human carriers and of the specific contexts of its application. Technique thus places the subject at the centre of activity, whereas technology affirms the independence of production from human subjectivity. Drawing out the contrast, Mitcham notes that 'tools or hand instruments tend to engender techniques, machines technologies. ... Technique is more involved with the training of the human body and mind ..., whereas technology is concerned with exterior things and their rational manipulation. ... Techniques rely a lot on intuition, not so much on discursive thought. Technologies, on the other hand, are more tightly associated with the conscious articulation of rules and principles ... At the core of technology there seems to be a desire to transform the heuristics of technique into algorithms of practice' (1978:252).

Now it is commonly supposed that where there are techniques there must be technology, for if skill lies in the effective application of knowledge, there must be knowledge to apply (Layton 1974). I believe this view to be mistaken. For acting in the world is the skilled practitioner's way of knowing it. It is in the direct contact with materials – whether or not mediated by tools, in the attentive touching, feeling, handling, looking and listening that is entailed in the very process of creative work, that technical knowledge is gained as well as applied. No separate cognitive schemata are required to organize perceptual data or to formulate instructions for action. Thus, skill is at once a form of knowledge and a form of practice, or – if you will – it is both practical knowledge and knowledgeable practice. Moreover, as a form of knowledge, skill (or technique) is different in kind from technology. The former is tacit, subjective, context-dependent, practical 'knowledge how', typically acquired through observation and imitation rather than formal verbal instruction. It does not therefore have to be articulated in systems of rules and symbols. Technological knowledge, by contrast, is explicit rather than tacit, objective rather than subjective, context-independent rather than context-dependent, discursive rather than practical, 'knowledge that' rather than 'knowledge how'. It is, besides, encoded in words or artificial symbols, and can be transmitted by teaching in contexts *outside* those of its practical application.

Historically, as the skilled manipulation of tools has given way to the operation of mechanically determined systems, knowledge of the first kind has gradually been rendered redundant, whilst knowledge of the second kind has become increasingly essential. In other words, far from complementing technique by providing it with a foundation in knowledge, technology forces a *division* between knowledge and practice, elevating the former from the practical to the discursive, and reducing the latter from creative doing or making to mere execution. To see this, one has only to compare the classical, Aristotelian notion of *tekhnê*, with its connotation of skilled craftsmanship, with the modern idiom in which to say of practice that it is 'purely technical' is to intimate that it is merely mechanical. In the dichotomy between discursive knowledge and executive practice, no space remains for the practical knowledge (or knowledgeable practice) of the craftsman. Technology, in short, appears to *do away* with technique, rather than to back it up.

Moreover the transition from technique to technology, on the level of knowledge, has its precise counterpart, on the level of material instruments or devices, in the transition from the tool to the machine. Originally connoting an 'instrument for lifting heavy weights', using the principles of wheel and axle, lever and inclined plane, but empowered by the human body through the hand, in its modern sense the machine is often *distinguished* from the tool on the grounds that it draws on a source of power outside the body, and is not manually operated (Mitcham 1978:235–6, 271–2 fn. 16). For Marx, for example, it was the human 'handling' of tools that set them apart from machines, whose movements are predetermined rather than under skilled constraint (1930:395). Thus the terms 'tool' and 'machine' have come to stand for those aspects of a device that are respectively dependent on, and independent of, human agency (Mitcham 1978:236, see also Mumford 1946:10). Recalling that the original connotation of the classical term *tekhnê* was the skilled making of the craftsman, whilst *mêkhanê* referred to the manually operated devices that assisted its application, it is evident that, overall, the evolution from the classical dualism of *tekhnê/mêkhanê* to the modern dualism of technology/machine has been one in which the human subject – both as an agent and as a repository of experience – has been drawn from the centre to the periphery of the labour process (Ingold 1988a). In other words, it has been a movement from the personal to the impersonal.

In the next section, I mean to show that this movement is tantamount to a *disembedding* of technical relations from their matrix in human sociality, leading to the modern opposition between technology and society. I should like to conclude the present section, however, by returning to the question of speech. I want to suggest that the division forced by technology between discursive knowledge and executive practice extends to the consequences of the 'technologizing' of speech in the form of systems of writing (Ong 1982). In this case, the division appears as one between language and speaking, classically enshrined in the linguistics of Ferdinand de Saussure (1959[1916]), and replicated more recently in the distinction, introduced by Chomsky (1965), between 'competence' and 'performance'. Following Harris (1980: 6–18), I believe that the codification of speech in writing, along with its attendant formal rules for the production of flawless, well-formed and complete sentences – which of couse have no precise counterpart in the ordinary talk of people even in literate societies – has had a profound impact on our own (including linguists') view of what language *is*. Like technology, it appears as an objectively given structure which, though instilled in people's minds, is indifferent to who they are and what they say. Speaking, then, appears as nothing more than the behavioural execution of a language system, and the particular skills and aptitudes that distinguish one speaker from another are relegated to the category of superficial idiosyncrasies of pronunciation, or even of 'mistakes' in their performance. The speaker, in short, becomes the 'operator' of language just as the workman becomes the operator of technology. And in a society imbued with the technology of the word, it seems as hard for us to recover the sense of speech as consisting in skilled verbal artistry, honed by practice and experience, as it is for us to regain an appreciation of the knowledgeable practice of craftsmanship.

The technical and the social

It is commonplace in anthropology to draw an absolute distinction between the domains of technical and social phenomena. This doubtless owes much to the influence of Durkheim. The earliest anthropological reference to the distinction that I know is to be found in a tantalizing footnote to the conclusion of Durkheim and Mauss's essay of 1903 on *Primitive Classification*, where they write of what they call 'technological classifications' as vague and unsystematic constellations of ideas, quite unlike the systematically interconnected categories of scientific classification which are grounded in the structure of social groups. Scientific classifications, Durkheim and Mauss write,

> are very clearly distinguished from what might be called technological classifications. It is probable that man has always classified, more or less clearly, the things on which he lived, according to the means he used to get them: for example animals living in the water, or in the air or on the ground. But at first such groups were not connected with each other or systematized. They were divisions, distinctions of ideas, not schemes of classification. Moreover, it is evident that these distinctions are closely linked to practical concerns, of which they merely express certain aspects (1963:81–2 fn. 1).

This idea of technical classification seems, in many ways, to anticipate the notion of the 'constellation of knowledge' discussed by Wynn in this volume (Chapter 17).

What is important for my present argument is the way technological classification is linked here to the experience of individuals in practical activity, as opposed to the structuring force of society. From the start, technology was placed firmly on the individual side of a pervasive dichotomy between individual and society, whilst science was set apart on the social side. In the subsequent elaboration of the Durkheimian paradigm, the distinction between technology and science was referred back to that between magic and religion, the former issuing from the individual and pragmatic in intent, the latter issuing from society and fundamentally expressive. The same distinction was later taken up by Edmund Leach, in a series of attempts to force a division between *technical* and *ritual* types or aspects of behaviour. Technical behaviour is defined in purely pragmatic, means–ends terms, and 'produces observable results in a strictly mechanical way'. Ritual behaviour, by contrast, is essentially communicative, and serves to convey information, in a symbolic code, about group membership or social identity (Leach 1966:403, cf. 1954:12, 1976:9). The division, then, is between a mechanics of technical systems and a semiotics of social systems. All practical action is 'fully mechanical' in the sense that its effects are entirely predictable from its initial conditions (1976:23), whereas all social action, since it is designed to communicate a state of affairs but not to change it, is inherently non-practical.

To illustrate the effects of applying this conceptual framework across the board of human societies, I shall consider how it has influenced the anthropological understanding of societies of hunters and gatherers. The activities of hunting and gathering, we are frequently told, are 'purely technical', a characterization that carries the implications that they are not only 'fully mechanical' but also residually *non*-social. Thus the work of subsistence production is effectively removed from the sphere of social action, becoming merely a 'need-satisfying process of individual behaviour' (Sahlins 1972: 186

fn. 1). When human beings hunt and gather, even when they do so in co-operation, they can act only in their 'natural' capacity as individuals, rather than as social persons. Social action seemingly begins not with production but with distribution, when hunters and gatherers turn to sharing out the produce brought back to their home base, a movement that expresses and affirms social relations of band membership. If, as Durkheim maintained, there are two parts to a man, the individual and the social being, it is apparently the individual who hunts and gathers, and the social being who shares (Ingold 1988b:275, cf. Durkheim 1976: 16; see Palsson 1991: 8–9, for a critique). And in Leach's terms, every act of hunting and gathering would be a mechanical event, and every act of sharing a communicative or semiotic event.

This view of the separation of production and distribution has been reinforced by a peculiarly Durkheimian reading of the distinction, taken from Marx, between social relations and technical forces of production, according to which these constitute mutually exclusive domains. Representing a widely held position in Marxist anthropology, Friedman writes that 'the social relations of production are not, nor can they be, technical relations' (1974:447). Included in the latter are the forces mechanically exerted by human bodies, when set to work, whether singly or in conjunction. Relations of co-operation in the work of hunting and gathering are thus built into the operation of the technical system – they are *technical* relations, as distinct from the *social* relations activated in the distributive practices of sharing. Yet as Marx surely recognized, the externalization of the forces of production was a historical consequence of the development of the machine. Where, as in hunting and gathering, food production depends on the skilled handling of tools, and indeed of one's own person, the productive forces appear as the embodied qualities of human subjects – as their technical skills. Such qualities cannot be generalized: whereas a technology is indifferent to the personhood of its operators, techniques are active ingredients of personal and social identity. Thus the very practice of a technique is itself a statement about identity (see also Wynn, this volume); there can be no separation of communicative from technical behaviour.

Our conclusion must be that in hunting and gathering societies, the forces of production are deeply embedded in the matrix of social relations. That is to say, the 'correspondence' between technical forces and social relations is not external but *internal*, or in other words, the technical is one *aspect* of the social. The modern semantic shift from technique to technology, associated with the ascendance of the machine, is itself symptomatic of the disembedding of the forces of production from their social matrix, transforming the correspondence between forces and relations of production from the internal to the external, and setting up the now familiar opposition between technology and society.

For as I have already shown, the concept of technology signifies the withdrawal of the person from production, which is consequently reduced to the operation of a quasi-mechanical system comprising human bodies, instruments and raw materials. If persons, human subjects, are external to production, then the sphere of social relations (between persons) must be external to the sphere of technical relations which, if they involve human beings at all, involve them as the bearers of natural and not personal powers (on this distinction, see Shotter 1974:225).

The danger is that we are inclined to read back into history the modern separation of technology and society, identifying the forces of production with all that is *external* to the human subject. Hence we imagine the primitive precursors of the machine to have been such items of material culture as the hand-axe, spear and digging-stick. And this, in turn, leads us to view technical evolution as a process of *complexification*, accompanied perhaps by a simplification in the social spheres of kinship and ritual. However the machine is not simply a more advanced substitute for a tool, nor were hand-tools the original forces of production. For the development of the productive forces has transformed the entire system of relations between worker, tool and raw material, replacing subject-centred knowledge and skills with objective principles of mechanical functioning. In other words, technical evolution describes a process not of complexification but of *objectification* of the productive forces (just as writing represents an objectification rather than a complexification of speech).

This conclusion suggests a radical recasting of the relation between technology and kinship. Instead of seeing an evolution in parallel, in which the former becomes ever more dominant and elaborate as the latter declines in significance, the view I have proposed suggests that the technical forces of production were originally *consubstantial* with the social relations of kinship. Only subsequently, as kinship was disengaged from the organization of production, did the forces 'split off' and acquire separate institutional identity as a 'technology'. At the same time the objectives of production were themselves transformed from the constitution of persons to the manufacture of things. In short, to find the antecedents of technology, we should look to the sphere of artifice, contained in social relations, rather than to the artefacts of material culture (Ridington 1982: 470).

What tools are for

The next step in my argument is to show how this view of the embeddedness of technical relations in social relations affects our understanding of the nature and use of the tool. In itself, of course, the tool is nothing. 'Being a tool' is not at all the same as, say, 'being a stone' or 'being a piece of wood'. For whereas the latter refer to intrinsic properties of the object itself, the former refers to what it affords for a user. An object – it could be a stone or a piece of wood – *becomes* a tool through becoming conjoined to a technique, and techniques, as we have seen, are the properties of skilled subjects. The presence of such a subject is already presupposed in our description of the object as a tool of a certain kind. Thus the tool is not a mere mechanical adjunct to the body, serving to deliver a set of commands issued to it by the mind; rather it extends the whole person.

Indeed there is a certain parallel between the use of tools in production and the giving and receiving of gifts in exchange. The tool has an impact on raw material, as the gift has an impact on its recipient, only so long as it is animated by an *intention* that issues from the person of the user or donor. Divorced from the context of production, the tool reverts to its original condition as an inert object; likewise the gift is inert outside the social context of exchange (Mauss 1954[1925]: 10). The parallel, moreover, extends to the use of words in dialogue. In themselves, words have no meaning, they merely afford the possibility of meaning. Only by their being harnessed to the intention of a speaker, in the context of a verbal exchange, do they come to have a meaningful impact upon the listener. To seek meaning in the isolated word is – to adopt a vivid metaphor suggested by Vološinov (1986: 103) – to expect light from a bulb after having switched off the current. Just as the meaning of words arises from the situational context of utterance, which provides the current that lights the bulb, so also the use-value of tool-objects arises from the situation of the agent in a context of practical action. Thus both the tool, the gift and the word mediate an active, purposive and meaningful engagement between persons and their environments.

Returning to hunters and gatherers, we can ask how this mediation is effected in the context of their relations with their environments. As Ridington (1982:471) has pointed out, hunter–gatherers 'typically view their world as imbued with human qualities of will and purpose'. From their perspective, tools are indeed like words: they mediate relations between human subjects and the equally purposive non-human agencies with which they perceive them-selves to be surrounded. The tool is thus a link in a chain of personal rather than mechanical causation, which serves to deliver intentional action and not merely physical or bodily force. Moreover, unlike herdsmen and farmers, whose tools are used to establish some degree of domination over their environments, hunters and gatherers do not regard their tools as instruments of control. Thus in hunting, it is commonly supposed that the animal gives itself to be killed by the hunter who, as a recipient, occupies the subordinate position in the transaction (see, for example, Tanner 1979:136; Feit 1973:116). The spear, arrow or trap serves here as a vehicle for opening or consummating a relationship. If the arrow misses its mark, or if the trap remains empty, it is inferred that the animal does not as yet intend to enter into a relationship with the hunter by allowing itself to be taken. In that way, the instruments of hunting serve a similar purpose to the tools of divination, revealing the otherwise hidden intentions of non-human agents in a world where causality, as Feit remarks, 'is personal, not mechanical or biological' and where it is 'always appropriate to ask "who did it?" and "why?" rather than "how does that work?"' (1973:116).

In short, whereas for farmers and herdsmen, the tool is an instrument of control, for hunters and gatherers it would better be regarded as an instrument of revelation. The contrast is neatly encapsulated in Ridington's observation that hunter–gatherers, 'instead of attempting to control nature, . . . concentrate on controlling their relationship with it' (1982:471). Evidently, hunters and gatherers choose to conduct their dealings with the environment on the basis not of *domination* but of *trust*. Both terms imply dependency, but 'trust implies the acceptance rather than the denial of the autonomy of the other on whom one depends' (Ingold 1990, see Gambetta 1988). While both herdsmen and hunters depend upon animals for their subsistence, the herdsman attempts to impose his own will on his animal charges by the systematic repression of their own powers of autonomous action. The hunter, by contrast, allows that animals, like human persons, should be free to act of their own volition. And this means that although the hunter's actions are predicated upon the expectation that the animals will respond favourably to his initiative, there is always the risk that they may not. Moreover any attempt to curb the autonomy of the animals, along the lines of pastoral domestication, is regarded as showing lack of proper respect towards them, which would significantly increase the likelihood of their unfavourable response. The very means by which the herdsman aims to ensure access to animals would, for the hunter, entail a betrayal of trust which would have the opposite effect of causing them to desert.

This understanding that hunters and gatherers have of their relations with non-human components of their environments is fundamentally at odds with one of the most basic premises of Western thought, namely that the mission of mankind is to achieve mastery over nature. The world of nature, in this view, is pitted against the world of humanity – the former a material world of things, the latter a social world of persons. The concept of technology is firmly fixed within this polarity, serving to establish the epistemological conditions for society's control over nature by maximizing the distance between them. But through their tools and techniques, hunter–gatherers strive to *minimize* this distance, drawing nature into the nexus of social relations, or 'humanising' it (Pfaffenberger 1988). This 'drawing in' has as its object the establishment of conditions not of control but of a kind of mutualism, and the tools by which it is achieved do not mediate a struggle but a dialogue. Far from having failed, on account of the alleged simplicity of their equipment, to assert their will over 'wild' nature, hunters and gatherers have succeeded rather well in establishing a working basis for co-existence.

Nature and society

At this point, we can return to the issue from which I began, that is, whether the pressures behind the evolution of intelligence lay in the sphere of technical or social relations. There is a striking paradox at the heart of the Machiavellian intelligence hypothesis, for in speaking of the management of social interaction it uses a dominant metaphor – that of 'manipulation' – which is actually drawn from the domain of technical operations. The force of the metaphor is to import into the social domain the kind of antagonism that is presumed to exist in people's relations with non-human components of the environment. In the Machiavellian perspective, other individuals, just like natural resources, exist to be 'exploited', to be turned to one's own account, by a mixture of coercion and deception. In the worldview of hunters and gatherers, a parallel is also drawn between dealings with other humans and with the non-human environment, but in the opposite direction. Its force is to import into the domain of environmental relations the qualities of mutualism, trust and sociability that are supposed to exist in relations among people (see Bird-David 1990). Where Machiavellians claim that we should manipulate other people as we manipulate things in the environment, hunter–gatherers claim that we should display the same sympathy in our dealings with the environment that we display towards other people.

There do not seem to me to be any *prima facie* grounds for privileging the logic of Machiavelli over that of the hunter–gatherers. It is true that Humphrey, in his original argument for the social function of intellect, accords a role to 'sympathy' in social behaviour, suggesting vaguely that it may be biologically adaptive. But he goes on to argue that when people apply an intelligence, fashioned by natural selection for dealing with social partners, to other kinds of entities in the environment, they are bound to make mistakes. Such applications are examples of the 'fallacious reasoning' in which 'primitive and not so primitive peoples' indulge. 'In doing so, they are explicitly adopting a social model, expecting nature to participate in a transaction, but nature will not transact with men; she goes her own way regardless' (Humphrey 1976: 313).

I believe that Humphrey's reasoning, and not that of the primitive, is fallacious. The fallacy lies in the assumption that what is 'given', *ab initio*, is an absolute separation between 'nature' and 'man'. In reality, this separation is the *consequence* of disengagement, of transactional failure, not its cause. For all animals, ourselves included, a condition of life is an active engagement with the world. And in that engagement, the world is no longer 'nature' but an environment. A human environment is constituted in relation to the people whose environment it is, in and through their activities; it is – in that sense – essentially *historical*. Only when people *cease* to participate, adopting a stance of contemplative detachment, does the environment revert to 'nature', a physical world of objects confronting the isolated human subject (see Gibson 1979: 8). Although humans are capable of adopting such a stance from time to time, no-one can permanently live like that.

Now it is through the deployment of technical skills that people act in the world, and in so doing constitute their environments. Thus environments never come ready-made, but are always in the process of creation. But to repeat a point I made earlier, acting in the world is also the skilled practitioner's way of knowing it. The perceptual knowledge so gained is, as we have seen, an integral part of personal identity. Hence, in the constitution of their environments, agents reciprocally constitute themselves as persons. Moreover, what applies to non-human components of the environment applies equally to the environment of other humans. Our knowledge of others is obtained through an active, practical engagement with them, through exploratory touching, looking, listening and – of course – speaking. There are techniques of speech just as there are techniques of hunting and gathering; through speaking and listening (using words) one gains a perceptual knowledge of the human environment, as through hunting and gathering (using tools) one gains a perceptual knowledge of the environment of animals and plants – although, as noted earlier, one may also speak to animals and plants and use tools on other humans. And again, knowledge of both human and non-human worlds is integral to personal identity. It makes us who we are.

I contend that the division between social and technical skills is an artefact of the modern Western distinction between nature and society. Moreover neither, strictly speaking, are particularly exclusive to human beings. Where humans undoubtedly do excel over other species is in their capacity to construct *imagined* worlds, and to formulate ideal standards against which their own conduct and that of others can be monitored, described and interpreted. However such imagining or discursive representation, far from being a necessary prelude to intelligent, practical action, is rather an epilogue, and an optional one at that. Though the exercise of skill depends on perceptual knowledge, such knowledge is obtained directly, in the course of active engagement with other persons and things in the environment; it does not presuppose the processing into images, by some device known as 'the intellect', of 'sense-data' delivered by the receptor organs of the body. In most practical contexts, tool-using entails touch and vision (for we work with our hands, and with our eyes open), just as speech entails hearing (since we listen as we speak). Thus both speaking and tool-using, as forms of skilled activity, are ways of perceiving as well as of shaping the environment. Perception, in short, inheres in action. It must surely follow that the emergence of tool-use and speech is a chapter not in the evolution of some specific mental capacity, going by the name of 'intelligence', but in the evolution of human beings as animals endowed with specific capabilities of action.

References

Bird-David, N. (1990). The giving environment: another perspective on the economic system of gatherer–hunters. *Current Anthropology*, 31, 183–96.

Chomsky, N. (1965). *Aspects of the Theory of Syntax*. Cambridge, MA: MIT Press.

Darwin, C. (1874). *The Evolution of Man and Selection in Relation to Sex* (2nd edn). London: John Murray.

Durkheim, E. (1976 [1915]). *The Elementary Forms of the Religious Life* (trans. J. Swain), 2nd edn. London: Allen & Unwin.

Durkheim, E. and Mauss, M. (1963 [1903]). *Primitive Classification*. London: Routledge & Kegan Paul.

Ellen R. F. (1982). *Environment, Subsistence and System: The Ecology of Small-scale Social Formations*. Cambridge: Cambridge University Press.

Feit, H. (1973). The ethnoecology of the Waswanipi Cree: or how hunters can manage their resources. In *Cultural Ecology; Readings on the Canadian Indians and Eskimos*, ed. B. Cox, pp. 115–25. Toronto: McClelland & Stewart.

Friedman, J. (1974). Marxism, structuralism and vulgar materialism. *Man* (N.S.), 9, 444–69.

Gambetta, D. (ed.) (1988). *Trust: Making and Breaking Co-operative Relations*. Oxford: Blackwell.

Gibson, J. J. (1979). *The Ecological Approach to Visual Perception*. Boston: Houghton Mifflin.

Harris, R. (1980). *The Language Makers*. London: Duckworth.

Humphrey, N. (1976). The social function of intellect. In *Growing Points in Ethology*, ed. P. P. G. Bateson and R. A. Hinde, pp. 303–17. Cambridge: Cambridge University Press.

Ingold, T. (1986a). *Evolution and Social Life*. Cambridge: Cambridge University Press.

Ingold, T. (1986b). Tools and *Homo faber*: construction and the authorship of design. In *The Appropriation of Nature: Essays on Human Ecology and Social Relations*. Manchester: Manchester University Press.

Ingold, T. (1988a). Tools, minds and machines: an excursion in the philosophy of technology. *Techniques & Culture*, 12, 151–76.

Ingold, T. (1988b). Notes on the foraging mode of production. In *Hunters and*

Gatherers I: History, Evolution and Social Change, ed. T. Ingold. D. Riches and J. Woodburn, pp. 269–85. Oxford: Berg.

Ingold, T. (1990). Comment on J. S. Solway and R. B. Lee, 'Foragers, genuine or spurious?'. *Current Anthropology*, 31, 130–31.

Layton, E. T. (1974). Technology as knowledge. *Technology and Culture*, 15, 31–41.

Leach, E. R. (1954). *Political Systems of Highland Burma*. London: Athlone.

Leach, E. R. (1966). Ritualization in man. *Philosophical Transactions of the Royal Society of London, Series B*, 251, 403–8.

Leach, E. R. (1976). *Culture and Communication*. Cambridge: Cambridge University Press.

Marx, K. (1930 [1867]). *Capital*, vol. I, trans. E. and C. Paul from 4th German edition of *Das Kapital* (1890). London: Dent.

Mauss, M. (1954 [1925]). *The Gift*, trans. I. Cunnison. London: Routledge & Kegan Paul.

Mauss, M. (1979). *Sociology and Psychology: Essays*. London: Routledge & Kegan Paul.

Mitcham, C. (1978). Types of technology. *Research in Philosophy and Technology*, 1, 229–94.

Mumford, L. (1946). *Technics and Civilization*. London: Routledge.

Ong, W. J. (1982). *Orality and Literacy: the Technologizing of the Word*. London: Methuen.

Palsson, G. (1991). *Coastal Economies, Cultural Accounts: Human Ecology and Icelandic Discourse*. Manchester: Manchester University Press.

Parker, S. T. and Gibson, K. R. (1979). A developmental model for the evolution of language and intelligence in early hominids. *The Behavioral and Brain Sciences*, 2, 36–408.

Pfaffenberger, B. (1988). Fetishised objects and humanised nature: towards an anthropology of technology. *Man* (N.S.), 23, 236–52.

Ridington, R. (1982). Technology, world view and adaptive strategy in a northern hunting society. *Canadian Review of Sociology and Anthropology*, 19, 469–81.

Sahlins, M. D. (1972). *Stone Age Economics*. London: Tavistock.

Saussure, F. de (1959 [1916]). *Course in General Linguistics*. New York: Philosophical Library.

Shotter, J. (1974). The development of personal powers. In *The Integration of a Child into a Social World*, ed. M. P. M. Richards, pp. 215–44. Cambridge: Cambridge University Press.

Tanner, A. (1979). *Bringing Home Animals: Religious Ideology and Mode of Production of the Mistassini Cree Hunters*. London: Hurst.

Vološinov, V. N. (1986). *Marxism and the Philosophy of Language*, trans. L. Matejka and I. R. Titunik. Cambridge, MA: Harvard University Press.

Vygotsky, L. S. (1962). *Thought and Language*. Cambridge, MA: MIT Press.

Wallace, A. R. (1870). *Contributions to the Theory of Natural Selection*. London: Macmillan.

Washburn, S. L. (1960). Tools and human evolution. *Scientific American*, 203(3), 63–75.

Whiten, A. and Byrne, R. (1988). The Machiavellian intelligence hypothesis: editorial. In *Machiavellian Intelligence: Social Expertise and the Evolution of Intellect in Monkeys, Apes, and Humans*, ed. R. Byrne and A. Whiten, pp. 1–9. Oxford: Clarendon Press.

Epilogue

Epilogue

Technology, language, intelligence: A reconsideration of basic concepts

TIM INGOLD

Introduction

The chapters in this volume speak for themselves, and my purpose is not to summarize them, let alone draw out from them some triumphant new synthesis. We are still far from reaching final answers to any of the principal questions we set out to address, if indeed final answers are possible at all – for every advance merely opens up new horizons. I have, rather, another purpose in mind, which is to interrogate our own questions. The main approaches represented in this book, when put together, constitute a very powerful combination – of neo-Darwinian biology, cognitive and developmental psychology, and formal or structural linguistics. As a social anthropologist, perched precariously on a narrow ledge whilst buffeted by contrary winds from the humanities and the natural sciences, I view this combination with increasing unease. I am disturbed by its apparent obliviousness to the intellectual ferment that has accompanied the contemporary critique of modernism, by the commitment to 'normal science' that brooks no challenge to fundamental paradigmatic assumptions, and by its readiness to frame the accounts it yields – of the entire career of humanity from earliest origins to the present day – in terms of concepts which, like the disciplines to which they belong, are recent products of a very specific history in the Western world.

Of the concept of society, Eric Wolf (1988: 757) has observed that to use it is not to denote a thing but to advance a claim. The same, I believe, goes for the three key concepts that have framed the discussions in this book: technology, language and intelligence. We cannot just assume that the meanings of these terms are unproblematic, that they label objectively given properties of the world (or the mind), and proceed from there. We have rather to attend to their implications, and to make clear the claims we are making when they are invoked. In the tradition of thought that, as a kind of shorthand, we call 'Western', these claims have concerned the ultimate supremacy of human reason. Thus intelligence is the faculty of reason, language its vehicle, and technology the means by which a rational understanding of the external world is turned to account for human benefit. The concept of (civil) society, moreover, has a place within this set of terms as denoting the mode of association of rational beings. Now of these properties or modalities the first two – intelligence and language – are nowadays assumed (by us) to be human universals; even to suggest that any ethnographically known population may be deficient in either or both respects would be regarded as tantamount to racism. Interestingly, the same stricture does not apply to technology: indeed in their often contrived attempts to avoid the derogatory connotations of their references to 'primitive' or 'simple' societies, anthropologists have frequently resorted to the expedient that it is *technological* simplicity to which they refer, and that this has no connotations as regards intellectual or linguistic ability, or sophistication in any other aspects of culture or social organization.

For the sake of argument, let me propose a radically alternative claim: namely, that there is no such thing as technology, nor language, nor intelligence, at least in non-Western societies. By that I do not for one moment mean to suggest that people in such societies do not make common use of tools in their everyday activities, that they do not engage with one another in the verbal idioms of speech, or that these and other activities do not represent creative ways of coping in the world. My concern is rather to focus attention on what it means to say that everyday tool-using is a behavioural instantiation of technology, or that spoken dialogue is the instantiation of language, or that creative activity is the instantiation of intelligence. Even in our own society, in which these propositions form a part of received wisdom, they are not immediately or obviously borne out in experience.

For example, I am presently writing with a pen, I am wearing spectacles which help me to see, I carry on my wrist a watch which tells me the time, a chair and table provide supports respectively for my body and my work, and I am surrounded by innumerable other bits and pieces that come in handy for one thing and another. I incorporate these diverse objects into the current of my activity without attending to them *as such*: I concentrate on my writing, not the pen; I see the time, not my watch. Indeed it could be said that these and other instruments become truly available to me, as things I can use without difficulty or interruption, at the point at which they effectively vanish as objects of my attention. And if anything links them together, it is only that they are brought into the same current, that of my work. Drawing an explicit parallel with tool-use, Wittgenstein made much the same point about the use of words in speech (1953, para 11) – different words have different uses, just as do the pen, watch and spectacles; one normally attends not to the words themselves but to what the speaker is telling us with them, and they are bound together solely by virtue of the fact that the various situations of use are all embedded within a total pattern of verbal and non-verbal activity, a form of life.

There are, then, words, and activities that people do with words (i.e. speaking). And there are tools, and activities that people do with tools (i.e. tool-using). But is there language? Or technology? What is entailed in the assumption that for people to speak they must first 'have' language, or for people to use tools they must first 'have' technology – or indeed for people to engage in intelligent activities of any kind they must first 'have' intelligence? If, on the other hand, we drop the assumption, what further need do we have of these concepts? Suppose, to pursue my alternative claim, that we set ourselves the task of examining the relation, in human evolution, not between technology, language and intelligence, but between craftsmanship, song and imagination. The resulting account, I suspect, would be very different. Without prejudging the issue of which is the better conceptual frame, I shall attempt in what follows to indicate where some of the differences might lie. I begin with language and song.

Language, music and song

In the voice, human beings are equipped with a wonderfully expressive and versatile instrument. We use it to speak, and we use it to sing. But how, if at all, can we demarcate speaking from singing? In the modern conception the answer is simple: speaking is essentially linguistic, singing is essentially musical. Of course, speech may be present *in* the song, in the words that accompany the music – thus the song is conceived as it is written on paper, in two registers proceeding in parallel: the musical sequence written as a series of notes, and the linguistic sequence as a concurrent series of words. So what is the difference between these two sequences, between the melodic line and the syntagmatic chain? The question is closely analogous to the one about the relation between 'ordinary speech' and the manual gesture that normally accompanies it, except that in song the voice conveys both the words and the gesture, the latter taking an auditory rather than a visual form. And the answer suggested by this analogy is that the melody, like the manual gesture, delineates its own meaning, a meaning that is *presented* to the audience – in the sense of being 'made present' for them in the surrounding ambience of sound. In the context of the relations between singer and audience, musical sounds do not *have* meaning, they *are* meaning, standing for nothing other than themselves. Whereas word sounds, we might argue, encode meanings that are separable, and hence decipherable from the phonetic rendition: thus they convey meaning but do not embody it in themselves, their symbolism is not presentational but *re*presentational (on this important distinction, see Langer 1942).

If we follow this logic, then music is song minus the linguistic component, speech is song minus the musical component, whilst poetry is ambiguously situated somewhere in between. To produce a song, it seems that we have to *put together* two things that are initially separated, music and language; whereas in poetry we stretch spoken words beyond the limits of normal utterance so that – like musical sounds – they become expressive in themselves. But on what grounds do we assume this initial separation? Could we not, equally well, put the argument in reverse, and suggest that music and language, as distinct symbolic registers, are the products of a movement of analytic *decomposition* of what was once an indivisible expressive totality, namely song? To support such a reverse argument, we would need to be able to demonstrate that the difference between speech and melodic gesture is one of degree rather than kind, that to speak is indeed – in a sense – to sing, and hence that no absolute line can be drawn between them.

The issue here hinges, in part, on the question of how words acquire meaning. The orthodox view is that words refer to concepts. And concepts are the building blocks of comprehensive mental representations. At once there is presupposed a division between a subject, in whose mind these representations are to be found, and an objective world 'out there'. Meaning is in the mind, not in the world – it is *assigned* to the world by the subject. As I move around physically in the world, and advance through time, I carry my concepts with me – rather as I might carry a map in navigating the landscape. In different times and places I experience different sensations, but like the map, the system of concepts which organizes these sensations into meaningful patterns remains the same, regardless of where I stand. But if the world exists for me only as I have thus constructed it from the data of perception, how can it be shared? How can subjects inhabit a common world of meaning? Again, the orthodox account argues that meanings are shared through verbal communication. Thus, my pre-prepared thought or belief has to be 'encoded' in words, which are then 'sent' in the medium of sound, writing or gesture to a recipient who, having performed a reverse operation of decoding, finishes up with the original thought successfully transplanted into his mind. Of course every act of communication takes place in a context, involving a particular speaker and a particular listener (or listeners) in a given environmental setting. But since words refer to abstract concepts rather than real-world objects, the relation of signification (between word and concept) is itself context-independent. The logic of this account therefore entails that signs can achieve the status of words, i.e. become properly 'linguistic', only at the end point of a process of decontextualisation. At this point, the sign severs all connection with the external world, such that the relation between sign and meaning is wholly interior to the subject.

Not only must this relation of signification be context-free, it must also be conventional. Agreement on the conventional meanings of words is clearly a condition for the faithful transcription of ideas from one mind to another, according to the model of communication presented above. Such conventions, moreover, are presumed to be arbitrary – again on the grounds of the severing of iconic links between verbal signs and the properties of the exterior world. Linguists are fond of reminding us, naïve speakers all, that one word is as good as another for signifying the same concept, so long as the pattern of phonemic contrast that serves to set each word off from each and every other in the language is retained. To me it may seem that a quality of hardness is presented in the very utterance of the word 'hard', just as it is presented in a passage of music played staccato. And likewise, the word 'smooth' *sounds* smooth, as does the same passage played legato. But that, says the linguist, is an illusion born of the frequent association, in experience, of words and their 'real-world' deno-tata. To clinch the argument, he points to the sheer diversity of natural languages, to the fact that the different words – say – for 'dog', in these different languages, may bear not the slightest resemblance to one another, nor indeed to the real-world animal of that name.

Perhaps it is time for naïve speakers to put linguists in their place. For what the former can provide, which the latter cannot, is the perspective of a being who, quite unlike the dislocated, closed-in subject confronting an external reality, is wholly immersed, from the start, in the relational context of dwelling in a world. For such a being, this world is already laden with significance: meaning inheres in the *relations* between the dweller and the constituents of the dwelt-in world. And to the extent that people dwell in the same world, and are caught up together in the same currents of activity, they can share in the same meanings. Such communion of experience, the awareness of living in a common world of meaningful relations, establishes a foundational level of sociality which exists – in Bourdieu's (1977: 2) phrase – 'on the hither side of words and concepts', and that constitutes the baseline on which all attempts at verbal communication must subsequently build. For although it is indisputable that verbal conventions are deployed in speech, *such conventions do not come ready made.* They are forever being built up over time, through a cumulative history of past usage: each is a hard-won product of the hazardous efforts of generations of predecessors to make themselves understood. When we speak of the conventional meaning of a word, that history is simply presupposed or, as it were, 'put in brackets', taken as read. And so we are inclined to think of use as founded on convention when, in reality, convention can only be established and held in place through use. Thus to understand how words acquire meaning we have to place them back into that original current of sociality, into the specific contexts of activities and relations in which they are used and to which they contribute. We then realize that, far from deriving their meanings from their attachment to mental concepts which are imposed upon a meaningless world of entities and events 'out there', *words gather their meanings from the relational properties of the world itself*. Every word is a compressed and compacted history.

Armed with this 'dwelling perspective', how should we view the difference between the spoken word and the musical gesture? It is no longer possible to argue that the former carries a conventional meaning that can be detached from the sound whereas the latter embodies its meaning in itself. We should rather argue that in words, the process of sedimentation and compression of past usage which contributes to the determination of their current sense has advanced to an exceedingly high degree, whereas in melody it is still incipient. But this is a difference of degree rather than kind, one that has perhaps been stretched to its maximal extent in the West by virtue of a cultural emphasis on the novelty of music as against the conventionality of language. One cannot expect the difference to be everywhere, and at all times, to be so clear-cut. For all music, viewed in this light, is on its way to becoming speech, and there is no Rubicon beyond which we can say that it is unequivocally one thing rather than the other. Conversely, all speech has its origins in vocal music, that is in song. As Merleau-Ponty put it, once we put speech back into the current of intercourse from which it necessarily springs, 'it would then be found that the words, vowels and phonemes are so many ways of "singing" the world' (1962: 187) – not, it should be stressed, in the naïve sense of producing an onomato-poeic resemblance between particular sounds and particular aspects of the world, but in the sense of entering intentionally and expressively into it, of 'living' it.

Emotion and reason

The decomposition of song into the two 'compartments' of language and music has come about, I believe, through the assimilation of vocal gesture to a particular view of the human constitution, one that has long held a central place in Western thought, and that reached its apotheosis in the rationalism of Descartes. According to this view, every human being is a composite creature made up of body and mind, susceptible, on the one hand, to emotions and feelings (i.e. bodily sensations), but capable, on the other, of rational deliberation (i.e. mental operations). Thus the musical phrase is envisaged as a feeling shaped in sound, the verbal utterance as the representation of a thought. One is visceral, the other cerebral; one is experienced directly, the other presupposes a mental processing of received sound to extract the 'message'. In music (and more obviously still in dance) the body *resonates* with the world, in language one mind *communicates* with another. Music, assumed to be devoid of propositional content, is placed on the 'purely expressive' side of human existence; language is placed on the 'purely rational' side – all expressive aspects of speech being removed from language itself and assigned to contingent aspects of performance. Moreover, the rational is normally ranked above the expressive, as an index of 'higher' cognitive faculties that enable their pos-sessors to step outside the world and – from this decentred vantage point – to take a cool, dispassionate view of it.

Such, of course, is the professed aim of natural science. Since the ascendancy of reason over emotion is implicated in science's claim to deliver an objective account of the natural world, it comes as no surprise to find the same principle of ranking at the basis of scientific accounts of the evolution of language (for it is surely language that enables humans to be scientists). Early formulations of the gestural theory of language origins, for example, rested on claims that the vocalizations of non-human primates (and by imputation, those of early hominids) were purely emotional or affective, and were therefore unlikely candidates as precursors for linguistic communication, whose key property was taken to be the conveyance of purely propositional information. Neuro-physiologists, for their part, claimed to find empirical proof of the existence of a dichotomy between 'volitional' and 'emotional' behaviours and body move-ments, and proceeded to map these onto different regions of the brain (e.g. Myers 1976). Language was unequivocally ascribed to the former category of behaviour: thus Myers could assert that 'the use of words in verbal communica-tions is clearly volitional'. What, then, are we to make of those words that are uttered without deliberate, prior intent? Myers is at least dimly aware of the problem. He continues:

> The existence of a second type of use of the voice, i.e. in emotional expression, remains uncertain, and its neurology poorly defined. Indeed the neurologist, when confronted with the proposition of an emotional use of the voice, inevitably thinks of curse words or interjections (1976: 746).

The implication is that what are rather primly called 'curse words' really do not merit inclusion within the domain of language at all! Language proper comes to be marked out, through the exclusion of all vocal expression of emotion, as a realm of propositional statements delivered completely free from emotional or affective overtones. Hewes suggests an example: 'The message "the house is on fire" can, if need be, be conveyed with no more excitement than the information that Paris is a city in France' (Hewes 1976: 490).

This may be so. Yet in practice, anyone who says 'the house is on fire' does so in a context, and in a tone of voice that may vary from a level monotone to a high-pitched shout. In the context of utterance the former tone is as expressive of indifference as is the latter of urgency or anguish, and each is liable to evoke a quite different response on the part of the audience, from a detached contemplation of the conflagration to a rush to evacuate the building. How, then, can these possibly be regarded as alternative renderings of the same proposition? Only by abstracting the verbal phrase from its context, by treating it as though – like words printed, as they appear here, on paper – it had a separate existence of its own. In reality, regardless of whether I utter the words with excitement or indifference, or of whether or not I have already rehearsed my speech beforehand in thought, my speaking is an intentional act which can only artificially be broken down into propositional and expressive components. And the same, of course, goes for the utterance of a swearword, which may indeed be no more premeditated than my cry, 'the house is on fire', but which nevertheless launches my intention into the world and carries it forward towards its goal.

In short, whether I speak, swear, shout, cry or sing, I do so with feeling, but feeling – as the tactile metaphor implies – is a mode of active and responsive engagement in the world, it is not a passive, interior reaction of the organism to external disturbance. We 'feel' each other's presence in verbal discourse as the craftsman feels, with his tools, the material on which he works; and as with the craftsman's handling of tools, so is our handling of words sensitive to the nuances of our relationships with the felt environment. Thus, far from characterizing mutually exclusive categories of behaviour – namely 'volitional' and 'emotional' – intentionality and feeling are two sides of the same coin, that of our practical involvement in the dwelt-in world. Only by imagining the human organism to be an isolated, preconstituted entity, given in advance of its external relations, do we come to regard feeling as an inner, affective state that is 'triggered' by incoming sensations. And by the same token, we are led to recover the intentional (or 'volitional') character of speech by supposing that what makes it so is that it does *not* arise in reaction to external stimulus but is rather caused by an internal mental representation – by a thought, belief or proposition pressing to make itself heard (cf. Chomsky 1968: 10–11).

What, then, is language? Or more precisely, how do we come to have the idea that such a thing as language exists, and that it therefore has an evolution that we can attempt to describe and explain? One answer might be that the idea is a by-product of the process of 'interiorization' of personhood that has marked the emergence of the modern Western concept of the individual (Mauss 1985, Dumont 1986). It is this concept that leads us to look within the human being, rather than to the sphere of its involvement in a wider field of relations, to discover the ultimate, generative source of purposive action. Thus every individual is supposed to come independently equipped with a 'built-in' language capacity (or at least a device for its acquisition), located somewhere inside the brain, which is the generative source of speech. Another possible answer, related to the first, is that the idea of language is necessarily entailed by a rationalism that is unable to conceive of action except as the mechanical replication, in a physical medium, of assemblies already constructed in thought. To language, then, is accorded the responsibility for constructing those assemblies, namely sentences, which are merely *executed* in speech. Yet a third answer might be that the idea of language is an invention of linguists who have sought to model the activities of speaking as the application of a coherent system of syntactic and semantic rules, derived by abstraction from observed behaviour. To be able to do this, they have to 'stand back' from the current of discourse, focusing on speech *as* speech whilst the rest of us concentrate on what other people are telling us *in* their speech. But they have gone on to transfer, onto the speakers themselves, their own external relationship to the object of study, imagining the abstractions derived from this 'view from the outside' to be implanted within the speakers' minds and to constitute the essence of their competence. Hence, speaking is seen to consist in the implementation of linguistic rules. Inside the head of every speaker there appears a miniature linguist (see Bourdieu 1977: 94 and Ingold 1986: 94 for closely comparable arguments regarding the anthropological derivation of 'culture' from observations of practice).

   Irrespective of which of the three answers presented above we might favour, the idea of language is a relatively recent one in the annals of human history. Yet it has had a profound impact, not only on the way we interpret our own activities of speaking, *but also on those activities themselves.* For the explicit codification of lexical conventions and grammatical rules sets standards against which utterances may be judged more or less correct or linguistically well-formed, standards which – to varying degrees – may be emulated or enforced. That is to say, language has acquired the status of an institution. Children not only learn to speak, as they have always and everywhere done, through immersion in an environment of vocally accomplished caregivers, they also receive formal schooling in the principles of language, as formulated by those appointed by society to act as its guardians – the grammarians and dictionary-makers. Above all, they are taught *to write*. The influence of writing on modern ideas and practices of language cannot be overestimated. Its advent, as Harris justly claims, 'was the cultural development which made the most radical alteration of all time to man's concept of what a language is' (1980: 6). For writing is not simply the equivalent of speech in an alternative medium. It is rather a kind of reconstructed, *as if* speech: as if the verbal utterance were fully amenable to systematic analysis in terms of syntactical rules; as if the tone of voice and pronunciation were entirely dispensable to meaning; as if the utterance had an existence in its own right, independently of the context of its production.

None of these things are actually true of speech, except perhaps for some kinds of 'reading aloud'. Yet modern linguistics has operated largely on the assumption that they are. Thus it turns out that the prototypical instance of the linguistic utterance, a rule-governed, context-independent proposition delivered without expression or affect, is that artefact so familiar to us but unknown to non-literate societies: the sentence of writing. Every theory of language evolution that holds up this prototype as its point of culmination, as the exemplar of a fully evolved language capacity, has an inbuilt 'scriptist' bias, treating speech that emulates or imitates writing as more perfect than speech that does not, and regarding the latter's deviations from the ideal as imperfections or errors. It is no wonder that in modern society, where the practices of speech have come to be modelled on writing and where speakers are taught to observe a rationalized system of rules and conventions (i.e. to apply language), it has fallen to a specialized branch of verbal craft, namely poetry, to attempt to make up for the resulting expressive and aesthetic impoverishment by producing forms which – whilst approaching the rythmic and tonal patterns of music – are lexically and syntactically aberrant. But as Gell has argued, in a brilliant analysis of the vocal artistry of the Umeda, a society of Papua New Guinea, for a non-literate people whose speech has retained its expressive, song-like quality, unexpurgated by the rationalizations of the language-makers, all speaking is inherently poetic. 'What need of poets then?' (Gell 1979: 61).

Technology, art and craftsmanship

I have argued that song, far from being put together from separate linguistic and musical components, is rather an expressive unity that is decomposed into these components through the imposition of a concept of language of recent, Western origin. Exactly the same argument can be made for the kind of skilled, technical artistry that I denote by the term 'craftsmanship'. Consider the concept of technology, a compound formed from the classical Greek roots *tekhnê* (meaning 'art' in the traditional sense of craft and skill) and *logos* (meaning a framework of principles derived from the application of reason).

The compound is of fairly recent derivation, and did not come into regular use with anything like its present meaning until well into the seventeenth century (Mitcham 1979, Ingold 1988). What it does is to recast the technical skills of the craftsman in terms of an objective system of rational principles, a *logos*, in just the same way that the idea of language recasts the verbal art of speaking in terms of the rules of grammar. And as practice comes to be seen as the mechanical application of technological rules, so its expressive, aesthetic aspects are consigned to a separate domain of 'art' – a concept once synonymous with technical skill but whose meaning is now constituted by its *opposition* to technology on precisely the same grounds that music, in the modern conception, is constituted by its opposition to language.

In a technologically literate society, tool-using is assimilated to the operation of artificial systems, much as speaking is assimilated to writing. Hence the prototypical tool appears as the mechanical gadget which embodies in its own construction the principles of its operation. As an antidote to the scriptist bias of formal linguistics, I have suggested (following Merleau-Ponty) that we regard speech as a species of song. To follow up this suggestion into the analogous field of tool-use, I propose that we consider, as a prototypical instance, the kind of tool using that comes closest of all to song – that is, playing a musical instrument. For if to speak is to sing, then surely to use a tool is to play. Since, as every anthropologist knows, it is helpful to be able to draw on first-hand experience, I shall consider the example of playing the 'cello. As a reasonably proficient 'cellist, my experience is that when I sit down to play everything falls naturally into place – the bow in my hand, the body of the instrument between my knees – so that I can launch myself directly, and with the whole of my being, into the music. I dive in, like a swimmer into water, and lose myself in the surrounding ambience of sound.

This is not to say that I cease to be aware, or that my playing becomes simply mechanical or automatic: quite the contrary, I experience a heightened sense of awareness, but that awareness is not *of* my playing, it *is* my playing. Just as with speech or song, the performance embodies both intentionality and feeling. But the intention is carried forward in the activity itself, it does not consist in an internal mental representation formed in advance and lined up for instrumentally assisted, bodily execution. And the feeling, likewise, is not an index of some inner, emotional state, for it inheres in my very gestures, in the pressure of my bow against the strings, in the vibrato of my left hand. In short, to play is itself to feel, so that in playing, I put feeling *into* the music. It makes no more sense, then, to split off a 'rational-technical' component from the (residually) expressive component of playing a musical instrument than it does to split off a propositional component from the expressive component of speech or song.

I do not claim, of course, that all of what I have described above happens spontaneously, without preparation or rehearsal. A great deal of practice is required, and there are puzzles to be solved. To get around awkward passages, complex configurations of fingering and hand position have to be worked out in advance, and bowing movements have to be planned so that at the end of one phrase the bow is in the right place on the strings for the beginning of the next. At such times, as also when something goes wrong in the performance, one becomes painfully aware both of oneself and of the instrument, and of the distance that separates them. The instrument is felt to be obdurate or resistant; it sticks. My point, however, is that this opposition between player and instrument is collapsed in the instant when the former begins actually to *play*. In that instant, the boundaries between the player, the instrument and the acoustic environment appear to dissolve.

Lest my choice of example may seem to force the issue – for in playing a musical instrument one does not achieve any direct, practical effect beyond the rapidly fading tapestry of sound – let me suggest another instance of tool-use, again drawn from my own experience, this time of anthropological fieldwork among reindeer herdsmen in northern Finland. The tool I have in mind is the lasso, and the herdsman uses it to capture selected deer from the throng of animals circulating in the round-up enclosure. In construction, the lasso is extremely simple: no more than a length of rope with a sliding toggle. When not in use it hangs limply in a coil from the hand, or trails loose on the ground. Yet in the moment of being cast, it assumes the lively form of a flying noose, a form which never stands still even for a single instant. Like the musical phrase shaped in sound, the form hangs suspended in the current of action. Thus, working a lasso, like playing a musical instrument, is pure movement or flow, and everything that I have said applies to the latter applies to the former as well. It involves an embodied skill, acquired through much practice. It carries forward an intention, but at the same time is continually responsive to an ever-changing situation. Just as, with the orchestral 'cellist, the processes of his visual attention to the conductor and his manual handling of the instrument are indissociable aspects of one ongoing process of action, so also the herdsman's handling of the lasso is inseparable from his attention to the movements of the herd in the enclosure. The attentive quality of the action is equivalent to what, in relation to musical performance, I have called 'feeling': to play is to feel; to act is to attend. The agent's attention, in other words, is fully absorbed in the action. Yet things can go wrong in the roundup, as they can in performance: the lasso can miss its mark, ropes can become entangled, the efforts of other herdsmen working in the enclosure may be disrupted, animals can even be injured. The frustrated herdsman then becomes an object of embarrassed self-regard, not to mention abuse from his fellows (I speak from experience). The flow is broken, and one has to begin all over again.

Cognition and practice

So much for the view of the naïve, yet reasonably skilled practitioner. Enter now the cognitive scientist, who claims that where tools are used, there must be a technology – a theory of how the tools are to be operated – lodged, albeit unbeknown to its possessors, inside their heads. The claim is, of course, parallel to that of the linguist who assumes that the 'languages' of non-literate peoples exist fully-formed, implanted in the unconscious minds of speakers, and are merely waiting for him to give them explicit formulation. One wonders, then, what such a *logos* of 'cello-playing or lasso-throwing would look like. It would consist, presumably, in a set of formal rules or algorithms capable of combining elementary motor schemata into complex, patterned sequences which, precisely executed, should produce instrumental gestures appropriate to any given context. The task of representing the technique of 'cello-playing or lasso-throwing in such formal terms would likely be an infinite one, but even supposing it were possible, would an imaginary creature, programmed with this knowledge, and provided with the requisite material equipment, be able to function remotely like a skilled practitioner?

The answer, I believe, is that it would not. It would produce, rather, a sort of 'as if' action, as if what in reality is a continuous flow could be reconstructed in the form of countless steps, each the mechanical execution of a pre-established plan or assembly – analogous to the sentence of language (Bourdieu 1977: 73, Ingold 1986: 209–10). It is as though the quality of attention that, as we have seen, inheres in the skilled practitioner's conduct were to be withdrawn from the conduct itself and concentrated in the operation of a mental constructional device (an 'intelligence'), which, on the basis of a processing of sensory inputs, is supposed to generate plans and place them 'on line' for execution. Thus thought becomes active, action passive. In essence, the 'as if' actor and the skilled practitioner employ different kinds of intentionality. The first is the kind entailed in orthodox Cartesian accounts of volitional behaviour, in which to have an intention is to prefix that behaviour with a thought, plan or mental representation which it serves to deliver. The second is a kind of intentionality that is launched and carried forward in the action itself, and corresponds to the attentive quality of that action. It is the intentionality not of an isolated mind, of the cogitating subject confronting an exterior world of things, but rather that of a being wholly immersed in the relational nexus of its instrumental 'coping' in the world.

There is a certain (though as we shall see, inexact) parallel between the 'as if' actor and the inexperienced novice, and they fail for the same reason. Every act has to be thought out in advance, and once embarked upon, it cannot be changed without further deliberation which, in turn, interrupts the action. Attention *precedes* response, introducing a time lag which would make anything like orchestral playing or capturing reindeer with lassos completely impossible. The skilled practitioner, by contrast, is able continually to attune his movements to perturbations in the perceived environment without ever interrupting the flow of action, since that action is itself a process of attention. Skilled practice cannot, therefore, be understood as the application of objective knowledge in the form of an 'expert system', as though it followed the steps of (say) a 'cello-playing or lasso-throwing programme. This is not to deny that complex neurophysiological processes are involved, which operate on sensory inputs and yield appropriate motor responses. But it is to suggest that whatever goes on in the brain of the practitioner cannot be modelled as entailing anything analogous to mental rules and representations (Dreyfus 1991: 219). It is, of course, entirely tautologous to model neurological processes in this way and then, inverting the relation between model and reality, to claim that neurology provides independent confirmation for the existence of mental representations.

The novice becomes skilled not through the acquisition of rules and representations, but at the point where he or she is able to dispense with them. They are like the map of an unfamiliar territory, which can be discarded once you have learned to attend to features of the landscape, and can place yourself in relation to them. The map can be a help in beginning to know the country, but the aim is to learn the country, not the map. Similarly, the 'cello-teacher may place marks on the fingerboard to show the novice where to put his fingers in order to obtain different notes. The novice is thereby enabled to feel for himself the particular muscular tensions in the left hand, and to hear the resulting intervals of pitch. Having learned to attend to these things, his fingers will find their own place (he can now play in tune), and the marks, which serve no further purpose, can be removed. The same applies to any other branch of apprenticeship in which the learner is placed, with the requisite equipment, in a practical situation, and is told to pay attention to how 'this' feels, or how 'that' looks or sounds – to *notice* those subtleties of texture that are all-important to good judgement and the successful practice of a craft. That one learns to touch, to see and to hear is obvious to any craftsman or musician. As Gibson succinctly put it, learning is an 'education of attention' (1979: 254).

This kind of learning exemplifies what Lave (1990: 310) has called 'understanding in practice', to which she counterposes 'the culture of acquisition'. The latter phrase denotes the theory of learning long favoured by cognitive science (and by Western educational institutions), according to which effective action in the world depends on the practitioner's first having acquired a body of knowledge in the form of rules and schemata for *constructing* it. Learning, the process of acquisition, is thus separated from doing, the application of acquired knowledge. It is implied, moreover, that a body of context-free, propositional knowledge – i.e. a technology or, more generally, a culture – actually *exists* as such and is available for transmission by teaching outside the context of use. Learning, in this view, entails an internalization of collective representations or, in a word, *enculturation*. 'Understanding in practice', by contrast, is a process of *enskillment*, in which learning is inseparable from doing, and in which both are embedded in the context of a practical engagement in the world – that is, in dwelling. According to this theory of learning, the kind of know-how thus gained, 'constituted in the settings of practice, based on rich expectations generated over time about its shape, is the site of the most powerful knowledgeability of people in the lived-in world' (Lave 1990: 323).

By and large, discussions of the relationship between tool using and speech have adopted the unequivocally 'logocentric' perspective of cognitive science and structural linguistics, whose ontological baseline postulates a rational subject positioned vis-à-vis an objective world. The aim has then been to demonstrate a parallel, overlap or even identity between cognitive structures involved in generating representations, on the one hand, of object assemblies (for execution as tool-using behaviour), and on the other, of word assemblies (for execution as speech). The former are glossed as 'technology', the latter as 'language'. The argument sketched above, however, suggests the possibility of a diametrically opposed approach, which takes as its ontological starting point the inescapable condition of human beings' engagement in the world, and that foregrounds the performative and poetic aspects of speech and tool-use that have been marginalized by rationalism. From the vantage point of this approach, the relationship between tool using and speech, far from being the surface manifestation of a more fundamental deep-structural connection between technology and language, is really one between the vocal artistry of speech and song, and the technical artistry of craftsmanship. Moreover, I have found no absolute line of demarcation between speech and song, nor between singing with the voice and 'singing' with an instrument (e.g. 'cello-playing), nor between the latter and other forms of tool-assisted, skilled artistry even of a thoroughly practical, subsistence-oriented kind. One thinks, for example, of the harvester at work, swinging his scythe in a constant, rythmic, dancelike movement and singing as he does so: that, to my mind, is the archetypal situation of human tool-use, not the puzzle-solving scenarios beloved of cognitive psychologists.

Intelligence and imagination

Human beings do, of course, solve puzzles: witness the chess-player devising a strategy of future moves, or the 'cellist working out the fingering for a difficult passage. How, from the point of view of a 'dwelling perspective', is this kind of puzzle-solving to be understood? And how would our account differ from the rationalist argument that regards every solution as the output of a cognitive device, an intelligence, located somewhere within the organism? This latter argument, as we have seen, sets out from the postulate of an original detachment of the intelligent subject, who has then to construct (or reconstruct) the world in his or her mind, prior to bodily engagement with it. The direction in which we proceed is precisely the reverse: postulating an original condition of engagement, of being-in-the-world, we suppose that the practitioner has then to *detach* himself from the current of his activity in order to reflect upon it. Only having achieved such a stance of contemplative detachment can he begin to ask such questions as (of an object) 'What can this be for?' or (of a word) 'What might this mean?' In answering them, he may suppose himself to be contributing meaning or value to an external world that, in itself, is devoid of significance, that is merely *there* for people to do with it what they will. There are, after all, many things you can do with a stone, and if, in response to my own or another's query, I say of that stone that it is a 'missile', am I not contributing my own subjective meaning to an otherwise meaningless, occurrent object?

A being who is dwelling in the world, however, does not encounter stones. He encounters missiles, anvils, axes or whatever, depending on the project in which he is currently engaged. They are available for him to use in much the same way as are the mouth, hands and feet. In the game of football, we use the feet for running and kicking; we do not, however, consider feet *as* feet (i.e. as occurrent anatomical structures) and wonder what to do with them. Such may be the view of the cobbler or chiropodist, but he is playing a different game! As I have already shown in discussing the issue of how words acquire significance, meaning already inheres in the relational properties of the dwelt-in world. In order to release or 'free up' the qualities of objects in themselves, this original meaning has to be stripped away, reducing the 'available' to the 'occurrent' (the terms are Heidegger's; for an excellent discussion of how he uses them, see Dreyfus 1991: Ch. 4). This is done by distancing ourselves from, or stepping outside, the activities in which the usefulness of these objects resides. Only by virtue of such dissociation do we come to confront the spectre of a meaningless environment, the kind of objective world 'out there' that, in the discourse of Western science, goes by the name of 'nature'. Taking nature as a datum of existence, we may then see ourselves as dealing with it by appropriating it symbolically, by attaching cultural significance to its occurrent properties. In so doing, we attempt to recover the meaning that is initially lost through our disengagement from the current of practical action.

What, then, are we doing when we step outside this current? Or to rephrase the question: what kind of activity does *not* involve a palpable engagement in the world? The answer is that it is activity of the special kind we call *imagining*. This is what the chess-player is up to when, sitting apparently immobile and without touching the pieces on the board, he nevertheless proceeds to work out a strategy. Now there are three points I wish to make about this kind of activity. The first is that imagining *is* an activity: it is something people *do*. And as an activity it carries forward an intentionality, a quality of attention that is embodied in the activity itself. Were it otherwise, were every instance of planning supposed to be prefixed by a prior intention in the form of a plan, we would at once be led into the absurdity of an infinite regress (Ingold 1986: 312–13). We have already seen that skilled practice cannot be understood as the mechanical execution of prefigured design; it is now clear that the same applies to the design process itself. Where this process of imagination differs from other forms of activity, and what makes it so special, is that attention is turned inwards on the self: in other words, it becomes reflexive. I dwell, in my imagination, in a virtual world populated by the products of my own imagining.

The second point, which follows from the first, is that whatever we call these products – whether plans, strategies or representations – their forms are generated and held in place only within the current of imaginative activity. The same, moreover, is true of material forms which are generated in the practical activity of craftsmanship. It is said colloquially, yet with good reason, that the craftsman *casts* the material into its projected form: the form, that is, arises out of a practical movement depicted metaphorically as a 'throw' (though in the case of the herdsman casting his lasso, this is quite literally true). Thus, as the musician casts sound into the form of a phrase, so likewise the potter casts clay into the form of a vessel. Yet unlike sound, clay congeals, and as it does so the form, generated in movement, is 'frozen' in the shape of a static artefact that endures beyond the context of its production. It is this, perhaps, that inclines us to think that in the making of artefacts, forms pre-existing as images in the mind are simply transcribed onto the material, as though the movement issued directly from the form and served only to disclose it. The reality is more complex, since both the image of the projected form and the material artefact in which it subsequently comes to be embodied are each independently generated and 'caught' within their separate intentional movements, of imagination and practice respectively. The problem, then, is to understand the relationship between these two generative movements, a relationship that might be characterized, provisionally, as *rehearsal*. One may, in imagination, 'go over' the same movement as a preparation or pre-run for its practical enactment. But the enactment no more issues from the image than does the latter from an image for imagining.

The third point is that imagining is the activity of a being who nevertheless dwells in an actual world. However much he may be 'wrapped up' in his own thoughts, the thinker is situated in a time and place and therefore in a relational context. The scientist may indeed think himself to be an isolated, rational subject confronting the world as a spectacle, yet were he in reality so removed from worldly existence he could not think the thoughts he does. 'We do not have to think the world in order to live in it, but we *do* have to live in the world in order to think it' (Ingold 1991: 17). This is why, as I mentioned earlier, the parallel between the novice practitioner, who has to work out his movements in advance, and the 'as if' actor whose behaviour is the output of a mental constructional device, is an inexact one. The 'as if' actor is the (fictitious) pure subject, possessed of a rational intelligence that delivers thoughts for execution. Such a subject can only dwell within a space circumscribed by the intellectual puzzles it sets out to solve (as against the objective world in which its solutions are applied). The novice, by contrast, though one step removed from the uninterrupted engagement of the skilled practitioner, nevertheless carries on his deliberations 'against a background of involved activity' (Dreyfus 1991: 74). He continues to dwell in a world that provides, above all in the presence of other persons, a rich source of support for his deliberations. The same is true of the scientist, who confronts nature in rather the same questioning way that the novice player confronts his instrument, as a domain of occurrent phenomena whose workings one is out to understand. Here, then, we have the final, essential difference between intelligence and imagination. The former is the capacity of a being whose existence is wrapped up within a world of puzzles, the latter is the activity of a being whose puzzle-solving is carried on within the context of involvement in a real world of persons, objects and relations. And of all the historical products of the human imagination, perhaps the most decisive and far-reaching has been the idea that there exists such a thing as an 'intelligence', installed in the heads of each and every one of us, and that is ultimately responsible for our activities.

Evolution and history

Let me conclude by raising what is perhaps the most intractable problem in our efforts to conceptualize the origins and evolution of tool use and speech. We may put the problem in the form of two questions: When does the evolution of hominid tool use become a history of human tool use? When does the evolution of hominid speech become a history of human speech? By asking these 'when' questions, I do not expect a chronological answer. My concern is rather to identify what it means to state of certain processes of change that they are 'now' in the realm of history rather than evolution, or that they are 'still' in the realm of evolution rather than history. The chronological question cannot be sensibly asked, let alone answered, unless or until we know what we mean by such statements. By and large, palaeoanthropology has deftly shelved the issue by inventing a notion that pretends it has been resolved – this is the notion of 'anatomically modern humans'. It is assumed that everything that went on up to the emergence of anatomically modern humans belongs to evolution, and forms the proper domain of palaeoanthropological study. *Biologically*, it is claimed, these humans differed in no significant ways from contemporary human populations. Everything that went on after their appearance is thus assigned to a separate, historical or civilizational process, eventually leading to populations that are not only biologically but also *culturally* modern. Somewhere along the line, therefore, history must have 'started up', initiating what is sometimes regarded as a second track – cultural rather than biological – in the career of humanity.

Biological anthropologists who appear content with the notion of anatomically modern humans are also inclined to declare that the differences between humans and other animals are of degree rather than kind. This, in itself, is a strange claim to make, and can only be understood in the context of an ancient tension in Western thought between the thesis of man's absolute separation from, and domination over, the world of nature (including animals), and the counterthesis that all living forms (including humans) can be ranged in a single continuum or chain of being (see Ingold 1990: 209). Only in terms of this latter notion can one claim, for example, that the orang-utan is of 'higher degree' than the sea urchin, or the human of 'higher degree' than the orang-utan. The major contribution of the Darwinian theory of evolution was of course to refute this claim, and to replace the image of the chain with that of an ever-branching tree. The fact of the matter is that the human is neither more nor less, in degree, than an orang-utan, it simply *is not* an orang-utan; likewise the orang-utan *is not* a sea urchin. Humans, orang-utans and sea urchins are just different sorts of organisms. Nevertheless, in their encounter with representatives of the humanities (including social and cultural anthropologists) biologists of professed Darwinian persuasion still find themselves retreating into a pre-Darwinian, indeed pre-biological rhetoric in their assertion of differences of degree rather than kind between humans and non-human species. What has driven them to this predicament?

The answer, I believe, may be found in the fact that they have more in common with their opponents than meets the eye. What they share is an identical view of nature (including the human organism as an entity constituted within nature) as an objectively given, ahistorical, material world. As we have seen, anyone who claims to witness such a world, as such, must also imagine himself to be well out of it. The rational observer is presupposed when we speak of nature observed, and this, in turn, implies a separation between the subjective domain of the observer – the problem space within which he moves in analyzing the data of observation – and the objective domain 'out there' which furnishes these data, between microcosm and macrocosm, mind and matter. It is because all *qualitative* distinction is shifted onto the boundary between these subjective and objective domains that differences *within* nature come to be seen as ones of degree. The qualitative difference is implicit even in scientific denials that any such difference exists – scientists, at least, place themselves above nature, even if the rest of humankind is supposed to be immersed within it. One might caricature the position thus: humans differ only in degree from other animals, but scientists differ in kind from other humans. The difference between the chimpanzee and the hunter–gatherer is made into one of degree by turning that between the hunter–gatherer and the scientist into one of kind. (This, incidentally, is why chimpanzees that have had long contacts with human hunter–gatherers are still regarded as manifesting their 'natural' behaviour, whereas the behaviour of chimpanzees that have had contacts with scientists is said to be 'unnatural' or 'human-influenced'.)

There are arguments about whom to admit to the spectators' gallery of nature-watchers. Is it open to all and only members of the human species? Or only to Western scientists? Or to non-human animals as well? Social and cultural anthropologists have traditionally inclined to the first alternative, evolutionary biologists to the second or third, depending on the position they take on the contested issue of animal minds. I would only note in passing that biologists seem rather more prepared to admit animals reared in a Western environment than non-Western humans, and that, moreover, the admission of certain non-human animals does not, in itself, dissolve the boundaries of exclusion, it merely shifts them. Thus the assertion of a difference of degree between humans and chimpanzees may rest upon the assumption of a difference in kind between humans plus chimpanzees on the one hand, and all other animals on the other. Whatever position is adopted, the basic Cartesian split between mind and nature is retained, as is the principle of ranking which places the former above the latter.

We can now appreciate why biological anthropologists who take a 'degree rather than kind' view of human–animal differences nevertheless adhere to a notion of 'anatomically modern humans' which implies the existence of a historical process that differs in kind, not degree, from the process of evolution. No-one seems to find the need to speak of 'anatomically modern chimpanzees' or 'anatomically modern elephants'. This is because the vast majority of chimpanzees and elephants have not gone on to become 'modern Westerners', and none (yet) are scientists. The historical process implicit in, but tucked out of sight by, the notion of anatomical modernity is none other than that conventionally known as 'the ascent of reason', leading from our primitive hunter–gatherer past to modern Western science, technology and propositional language. That is why biologists who have explicitly refuted the doctrine of progress in relation to the evolution of species nevertheless resort, quite unabashedly, to terms like 'complexity' and 'sophistication' when referring to the faculty of intelligence and its products. For the measure of progress is taken to lie in the degree of approximation to rational science.

Modern evolutionary biology presents us with an account of how human beings evolved as animals equipped with certain *capacities*: the capacity to speak, to use tools, or more generally, the capacity for culture. The logic of the argument, that these capacities arose through a process of variation under natural selection, requires that they – or more accurately the programmes channeling their development – are received by each and every individual at the point of conception. That is, they are encoded genetically, and are in that sense innate. Their emergence is considered to be the consequence of a series of evolutionary events that occurred *in* ancestral populations, but which were in no sense produced *by* these populations: thus we are told that although the evolution of human capacities may have been affected by what our ancestors did, they did not *evolve* these capacities. Indeed it is normal practice in biological writing to refer to animals as sites *where* evolutionary, cognitive or behavioural events happen, and by means of this device to avoid any possible suggestion that the animals may themselves be the *originators* of these events.

What, then, are we to make of my capacities to speak English, play the 'cello and throw a lasso? Are these the products of an evolutionary process? The biologist replies: 'Certainly not. *Biologically*, English speakers are no different from Japanese speakers, 'cello players no different from sitar players, lasso throwers no different from archers'. The cultural anthropologist is inclined to agree: 'One of the most significant facts about us', writes Geertz, 'may finally be that we all begin with the natural equipment to live a thousand kinds of life but end in the end having lived only one' (1973: 45). What both these statements imply is that the capacities, know-how and equipment that enable me to live one particular kind of life are not natural or biological but *cultural*; they are the outcomes of an historical rather than an evolutionary process; they are acquired rather than innate.

I do not believe we can remain satisfied with this conclusion. No more than organisms of any other species do human beings come into the world biologically pre-equipped with 'natural capacities', ready to be topped up from the environment. These capacities emerge, in the life history of every individual, through a process of development. This process, moreover, is none other than that by which human beings acquire the specific skills appropriate to a particular form of life. Learning to speak *is* learning to speak one's mother tongue; learning to use tools *is* learning to use particular tools in the particular ways current in the environment of the learner; even learning to walk is learning to walk in one way rather than another. In short, the acquisition of culturally specific skills is part and parcel of the overall developmental process of the human organism, and through this process they come to be literally *embodied* in the organism, in its neurology, its musculature, even in features of its anatomy. Biologically, therefore, English speakers *are* different from Japanese speakers, 'cello players *are* different from sitar players, lasso throwers *are* different from archers. And by the same token, what are commonly designated as cultural processes *are* biological, and historical processes *are* evolutionary. Any comprehensive theory of evolution should therefore be able to deal with differences such as those listed above, without siphoning them off into a separate bucket labelled 'culture history'.

An evolutionary theory that would meet this requirement cannot, obviously, be one that traces patterns of change 'ultimately' to changing gene frequencies in ancestral populations. No-one is suggesting that people have different skills because they have different genes; nevertheless these differences are biological and they have evolved. What is required, then, is a much broader conception of evolution than the narrowly Darwninian one embraced by the majority of biologists. Central to this broader conception is the organism-person as an intentional and creative agent, coming into being and undergoing development within a context of environmental relations (including social relationships with conspecifics), and through its actions contributing to the context of development for others to which it relates. In this account, behaviour is generated not by innate, genetically coded programmes, nor by programmes that are culturally acquired, but by the agency of the whole organism in its environment.

This, finally, is the difference between an account of human evolution that attempts to relate language, technology and intelligence, as built-in capacities of individuals, and one that attempts to relate song, craftsmanship and imagination, as aspects of the practical engagement of beings in a world. The change of perspective that this implies requires nothing less than a new theory of evolution.

References

Bourdieu, P. (1977). *Outline of a Theory of Practice*. Cambridge: Cambridge University Press.

Chomsky, N. (1968). *Language and Mind*. New York: Harcourt Brace Jovanovich.

Dreyfus, H. L. (1991). *Being-in-the-world: A Commentary on Heidegger's 'Being and Time, Division I'*. Cambridge, MA: MIT Press.

Dumont, L. (1986). *Essays on Individualism: Modern Ideology in Anthropological Perspective*. Chicago: University of Chicago Press.

Geertz, C. (1973). *The Interpretation of Cultures*. New York: Basic Books.

Gell, A. (1979). The Umeda language-poem. *Canberra Anthropology*, 2(1), 44–62.

Gibson, J. J. (1979). *The Ecological Approach to Visual Perception*. Boston: Houghton Mifflin.

Harris, R. (1980). *The Language-makers*. London: Duckworth.

Hewes, G. W. (1976). The current status of the gestural theory of language origin. In *Origins and Evolution of Language and Speech*, ed. S. R. Harnad, H. D. Steklis and J. Lancaster, pp. 482–504. Annals of the New York Academy of Sciences, Vol. 280.

Ingold, T. (1986). *Evolution and Social Life*. Cambridge: Cambridge University Press.

Ingold, T. (1988). Tools, minds and machines: an excursion in the philosophy of technology. *Techniques & Culture*, 12, 151–76.

Ingold, T. (1990). An anthropologist looks at biology. *Man* (N.S.), 25, 208–29.

Ingold, T. (1991). Against the motion (1). In *Human Worlds are Culturally Constructed*, ed. T. Ingold, pp. 12–17. Manchester: Group for Debates in Anthropological Theory.

Langer, S. K. (1942). *Philosophy in a New Key*. Cambridge, MA: Harvard University Press.

Lave, J. (1990). The culture of acquisition and the practice of understanding. In *Cultural Psychology: Essays on Comparative Human Development*, ed. J. W. Stigler, R. A. Shweder and G. Herdt, pp. 309–27. Cambridge: Cambridge University Press.

Mauss, M. (1985). A category of the human mind: the notion of person; the notion of self (trans. W. D. Halls). In *The Category of the Person: Anthropology, Philosophy, History*, ed. M. Carrithers, S. Collins and S. Lukes, pp. 1–25. Cambridge: Cambridge University Press.

Merleau-Ponty, M. (1962). *The Phenomenology of Perception* (trans. C. Smith). London: Routledge & Kegan Paul.

Mitcham, C. (1979). Philosophy and the history of technology. In *The History and Philosophy of Technology*, ed. G. Bugliarello and D. B. Doner. Urbana: University of Illinois Press.

Myers, R. E. (1976). Comparative neurology of vocalization and speech: proof of a dichotomy. In *The Origins and Evolution of Language and Speech*, ed. S. R. Harnad, H. D. Steklis and J. Lancaster, pp. 745–57. Annals of the New York Academy of Sciences, Vol. 280.

Wittgenstein, L. (1953). *Philosophical Investigations*. Oxford: Blackwell.

Wolf, E. R. (1988). Inventing society. *American Ethnologist*, 15, 752–61.

Index